Soviet and Russian
Testbed
Aircraft

Soviet and Russian
Testbed
Aircraft

**Yefim Gordon and
Dmitriy Komissarov**

HIKOKI
PUBLICATIONS

First published in 2011 by
Hikoki Publications Ltd
1a Ringway Trading Est
Shadowmoss Rd
Manchester
M22 5LH
England

Email: enquiries@crecy.co.uk
www.crecy.co.uk

© Yefim Gordon, Dmitriy Komissarov

Original translation by Dmitriy Komissarov
Layout by Polygon Press Ltd., Moscow

Colour profiles: © Aleksandr Gavrilov, Valentin Vetlitskiy,
Andrey Yurgenson, Aleksandr Rusinov, AVICO Press.
Line drawings: © the late Vladimir Klimov, AVICO Press,
Andrey Sal'nikov, Vyacheslav Zenkin.

ISBN 9 781902 109183

Printing managed in China by World Print

Contents

One of the Tu-2LL engine testbeds. Note the bomb bay doors modified to fit around the test engine pod.

Acknowledgements

Special thanks are due to Aleksandr Gavrilov, who prepared numerous colour side views of Soviet/Russian test and research aircraft for this book. The authors also wish to thank Mikhail Gribovskiy, Viktor Drushlyakov, Vladimir Rigmant and Nigel Eastaway, one of the key figures in the Russian Aviation Research Trust (RART), all of whom supplied a lot of unique photos.

This book is illustrated with photos by Yefim Gordon, Viktor Drushlyakov, Mikhail Gri-bovskiy, Dmitriy Komissarov, Sergey Komissarov, Sergey Sergeyev, the late Sergey Skrynnikov, Martin Novak, as well as from the archives of the Antonov ASTC, TANTK Beriyev, RSK MiG, AVPK Sukhoi, the Tupolev JSC, the Yak Aircraft Co., the M. M. Gromov Flight Research Institute (LII), AVICO Press, RART and the personal archives of Yefim Gordon, Sergey and Dmitriy Komissarov, Sergey Popsuyevich, and from Internet sources (www.aviaforum.ru)

The IL-18D Tsiklon weather research aircraft resting between missions.

Introduction

Test and research aircraft of every nature are always a fascinating subject – simply because they are something out of the ordinary and often differ a lot from the standard airliners or combat aircraft that can be seen at various airports and military bases. Also, consider the 'forbidden fruit syndrome': such aircraft are often used in more or less 'black' military programmes and are therefore kept well away from people who are not supposed to see them – including plane spotters.

First of all, it has to be said that Russian aeronautical vernacular includes the term *letayushchaya laboratoriya* – literally 'flying laboratory'. This Russian term, which is often abbreviated as LL, is used indiscriminately and can denote any kind of flying testbed (for engines, avionics, equipment, weapons and the like), an aerodynamics research aircraft or control-configured vehicle, a weather research aircraft, a geophysical survey aircraft, an ecological monitoring aircraft and so on. In a broad sense, this term may be applied not only to manned fixed-wing and rotary-wing aircraft but also to unmanned aerial vehicles, including research rockets and gliding models. For reasons of space, however, the authors have chosen not to include UAVs in this book.

Speaking of which, the LL acronym was/is commonly used in the USSR/Russia as a designation suffix to denote aircraft modified for test and research tasks; for instance, a Tupolev Tu-16 bomber converted into an engine testbed was designated Tu-16LL. A direct analogy is the US system with the J and N designation prefixes denoting respectively temporary and permanent testbed status (for instance, Boeing JKC-135A and NKC-135A).

Some flying testbeds/research aircraft are merely transports carting a load of special equipment around; an example is the mobile accident investigation laboratories which travel to the airfield nearest to the scene of an accident so that the crashed aircraft's 'black boxes' (flight data recorder and cockpit voice recorder) can be analysed promptly. Others are modified more extensively – sometimes to the point that reverting to standard configuration is no longer possible.

Flying testbeds are invaluable when it comes to testing new systems and equipment items which find use on new aircraft types or new versions of existing types. Using them shortens the development and test cycle. Sometimes they are downright indispensible if the design bureau (OKB – *opytno-konstrooktorskoye byuro*) is breaking new ground. This was the case when OKB-23 headed by Chief Designer Vladimir Mikhaïlovich Myasishchev was developing the M-4 strategic bomber (NATO reporting name *Bison-A*) – the first such aircraft in the Soviet Union to feature a bicycle landing gear – at the height of the Cold War. In order to test the M-4's innovative 'jump strut' facilitating take-off two other bombers – a Tupolev Tu-4 *Bull* and a Tu-16 *Badger-A* – were suitably modified. Had these testbeds not been used, the risk of the M-4 being lost during initial flight tests would have been much higher. The same is true for new engines – especially if an all-new aircraft is powered by all-new engines (as is the case with Russia's latest offering on the commercial aviation market, the Sukhoi SuperJet 100). Similarly, research aircraft carrying a crew of specialists performing an experiment in real time have proved invaluable for studying various natural phenomena.

The Soviet and Russian flying testbeds and research aircraft described here are grouped by application, in as much detail as possible. Some of the aircraft listed here are sometimes described as experimental aircraft, rather than testbeds; the Mikoyan/Gurevich MiG-21I is a case in point. This dual classification is acceptable and even proper in some cases, since aircraft used for verifying new design features may be both conversions of production types and purpose-built one-off airframes. Yet, by no means all experimental aircraft qualify as testbeds. The MiG-21I does, since it was built with ogival wings as a subscale technology demonstrator for a different aircraft – the

'01 Blue' (c/n 6401401), one of the Tu-16LL engine testbeds, in a demo flight.

Tupolev Tu-144 *Charger* supersonic airliner – and is therefore included here. Conversely, an experimental rocket-powered aircraft, such as the Florov '4302' (not included here), does not rate as a testbed.

Not all flying testbeds and research/survey aircraft operated in the interests of Soviet/Russian aircraft design bureaux, avionics houses, equipment manufacturer and scientific institutions have been included here. Lack of space is one of the reasons; also, very little information is available on some testbeds and research aircraft – or, in some cases, the programmes in which they have been used are still classified.

As the reader will see, some testbeds and research aircraft have been used at various points of their career (or even simultaneously!) in programmes of a very different nature and thus fall into several categories at once – for example, a combined avionics/weapons testbed fitted with a new fire control radar and carrying new missiles. In each chapter and sub-chapter the aircraft are generally listed in alphabetical order by manufacturer/designation. This may appear a bit surprising, as it might seem more logical to arrange them chronologically. However, this creates a problem; in many cases the exact dates when a particular testbed was in use are unknown or not disclosed yet. Also, some aircraft have been repeatedly modified over the years, and describing them all over again would be illogical.

Since the book is concerned with Soviet and Russian testbeds, it includes local conversions of imported aircraft for test purposes. Conversely, Soviet/Russian aircraft converted into testbeds abroad are not included.

IL-18E RA-75598 seen on short finals to Moscow-Sheremet'yevo as a geophysical survey aircraft with a SLAR on the port side.

1 Aerodynamics testbeds

The first major class of test and research aircraft encompasses aircraft modified for exploring aerodynamics issues. These are aircraft with either non-standard airfoils or specially reshaped wings and tail surfaces, and sometimes with additional control surfaces. They are mostly used for verifying the aerodynamic features of future aircraft, though occasionally they serve to refine the basic type from which they have been converted.

aircraft was coded '35' and the c/n was quoted in LII's papers as '301'. Judging by the early-style overall silver finish with red nosecone, wingtips and fin cap, this was an early-production example (later examples were delivered in two-tone camouflage finish with yellow nosecone, wingtips and fin cap) and the full c/n was 430301 or 931301.

'35', an Aero L-39C fitted experimentally with winglets at LII in Zhukovskiy.

Aero L-39C winglet testbed

In 1981-85 the Flight Research Institute (LII – *Lyotno-issledovatel'skiy instituut*) named after Mikhail M. Gromov used a Czechoslovak-built Aero L-39C Albatros – the Soviet Air Force's then-standard advanced jet trainer – for verifying the winglets envisaged for Soviet fourth-generation medium-haul and long-haul airliners (the Tupolev Tu-204 and Ilyushin IL-96 respectively). The trapezoidal winglets, which did not replicate the winglet shape of either of these types, were grafted onto the rear portions of the standard wingtip fuel tanks. The

This DB-3 coded '13' carries a laminar-flow test article with a narrow chord. The Cyrillic letters IYe on the tail denote an aircraft operated by TsAGI. Note the pitot boom.

Ilyushin DB-3 aerodynamics testbeds

An example of the Ilyushin DB-3 long-range bomber sporting the tactical number 13 and the letters ИE ('IYe' in Cyrillic script) on the vertical tail, was used by the Central Aero- & Hydrodynamics Institute named after Nikolay Ye. Zhukovskiy (TsAGI – *Tsentrahl'nyy aeroghidrodinamicheskiy instittoot*) in 1940-41 for studying new laminar-flow airfoil sections in real flight. Mounted vertically above the centre fuselage were rectangular wing sections featuring different airfoils and different aspect ratio. An air data boom was installed on the navigator's station glazing. The design work proceeded under the direction of Matus R. Bisnovat. Some sources cite LII as the organisation that conducted the experiments (presumably in co-operation with TsAGI).

Another DB-3 was used by LII for studying the behaviour of heavy aircraft in sideslip flight mode. The aircraft performed flights with a small drogue parachute deployed from one of the outer wing panels.

Ilyushin IL-18 aerodynamics research aircraft

Between 1976 and 1978 LII reportedly used an unidentified IL-18 *Coot* four-turboprop long-haul airliner for flutter research. This aircraft was fitted out with electrically operated vibration inducers which were supposed to provoke flutter.

Kamov Ka-26 aerodynamics research aircraft

In 1972-76 the OKB-938 design bureau led by Chief Designer Nikolay I. Kamov and LII used an early-production Ka-26 *Hoodlum* utility helicopter, CCCP-24084 (c/n 7000903), for exploring dangerous flight modes (including the vortex ring mode) and determining the required engine power in various flight modes. This allowed the technique of piloting helicopters with the Kamov OKB's distinctive coaxial layout to be refined.

Mikoyan/Gurevich MiG-8 (Ootka) experimental aircraft

Shortly before the termination of hostilities in the Second World War the leaders of the Mikoyan design bureau (OKB-155), anticipating the Soviet Union's entry into the jet age, began verifying some of the design features envisioned for future jet fighters. To this end it was deemed necessary to build an experimental aircraft (in today's terms, a technology demonstrator) featuring an unusual layout – the tail-first or canard layout – and moderate-

The same aircraft in a different configuration with a broad-chord test article and no pitot boom.

The logo of TsAGI, the main Soviet aerodynamics research establishment which operated the testbed shown on this page.

ly swept wings. In their explanatory note accompanying the demonstrator's preliminary design (PD) project, Chief Designer Artyom I. Mikoyan and Deputy Chief Designer Mikhail I. Gurevich wrote:

'We have designed and built the tail-first aircraft as a proof-of-concept vehicle for the purpose of checking the stability and handling of aircraft utilising this layout and checking out how highly swept wings work. The aircraft has a pusher propeller, which makes it possible to assess low-speed controllability unaffected by the prop wash. This is of particular interest when developing pure jet aircraft. This aircraft will enable us to study the issues of controllability, taxiing, take-off and landing (including go-around) on an aircraft whose control surfaces are not affected by the prop wash.'

As they developed the propeller-driven canard, the Mikoyan OKB designers planned to follow up with a jet-powered aircraft utilising the same layout. In this aircraft the engine would be located in such a way that the jet efflux would not come into contact with the aircraft's structure. The technology demonstrator was developed in co-operation with TsAGI's scientific staff.

Officially designated MiG-8 (the even number indicated a non-fighter type), the demonstrator was known in house simply as *** Oot****ka* (duck – or, in the aeronautical sense, canard layout). The aircraft was of all-wooden construction. The high-set strut-braced wings were swept back 20° and had 2° anhedral. They were of two-spar construction, with a fabric-covered wooden framework, and the thickness/chord ratio was 12% throughout.

The MiG-8 Ootka experimental aircraft in ultimate configuration with the vertical tails at mid-span and equal-size landing gear wheels. Note the canted wingtips.

Here the MiG-8 is shown in penultimate guise (with relocated fins but with ordinary wingtips).

The MiG-8 experimental aircraft in early configuration with wingtip-mounted vertical tails

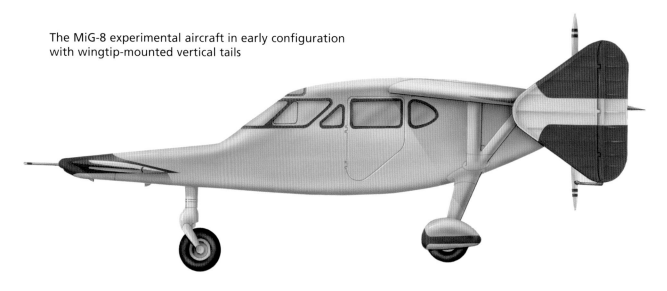

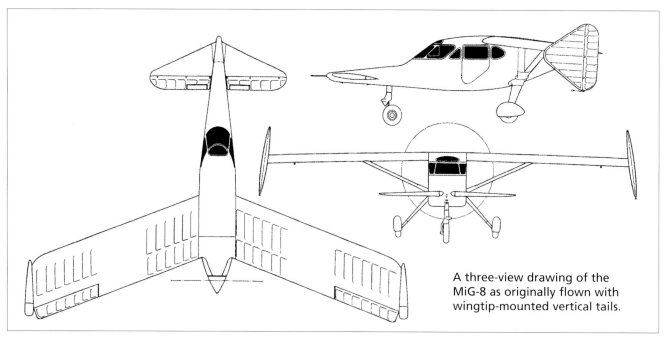

A three-view drawing of the MiG-8 as originally flown with wingtip-mounted vertical tails.

The MiG-8 in revised form with the vertical tails at mid-span

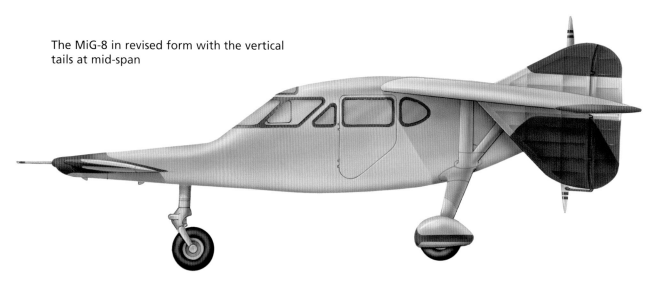

Here the MiG-8 is shown with wool tufts all over for airflow visualisation. The airflow pattern was filmed from a chase aircraft. Note the absence of the main-wheel spats.

Small fixed leading-edge slats were installed on the outer portions. The low-set canard foreplanes had a sweptback leading edge and no trailing-edge sweep. The twin triangular vertical tails were initially mounted at the wingtips. The elevators were operated via rigid linkages (push-pull rods), while the ailerons and rudders were actuated by cables.

The fuselage nose was long and pointed, carrying the canard foreplanes; the rear end of the fuselage accommodated the engine bay. The aircraft was a three-seater, with two passengers sitting side by side behind the pilot. The cabin was extensively glazed, giving the pilot a good forward and lateral field of view, and accessed via a car-type door to port.

The MiG-8 had a non-retractable tricycle landing gear – another 'first' for the Mikoyan OKB. All three units were of welded steel construction and provided with pneumatic shock absorbers. The nose unit, initially fitted with a 300 x 150 mm (11.81 x 5.9 in) wheel, featured an oil damper. The main gear units were cantilever and carried 500 x 150 mm (19.68 x 5.9 in) wheels fitted with spats. Later a 500 x 150 mm wheel was fitted to the nose unit as well.

The powerplant was a 110-hp Shvetsov M-11FM five-cylinder radial driving a two-blade fixed-pitch wooden pusher propeller of 2.36 m (7 ft 8$^{59}/_{64}$ in) diameter; it was completely cowled except for the cylinder heads. The wing centre section accommodated two duralumin fuel tanks with a total capacity of 195 litres (42.9 Imp gal); the oil tank holding 18 litres (3.96 Imp gal) was located aft of the cabin.

The MiG-8 entered flight test in 1945, Aleksey N. Grinchik making the maiden flight. Various versions of the aerodynamic layout were explored; the vertical tails were soon moved from the wingtips to a position at 55%

span, the LE slats were deleted and the rudders were fitted with servo tabs. Downward-canted wingtips were also tested for a while.

The MiG-8 proved to be extremely stable and absolutely spin-proof, refusing to enter a spin even at extremely high angles of attack (AOAs). Mikoyan test pilots Aleksandr I. Zhookov and Arkadiy N. Chernoboorov and project engineer Ye. F. Nashchokin (who was a qualified pilot) also flew the MiG-8 a lot. The Mikoyan OKB offered the aircraft to Aeroflot, the Soviet airline, citing the MiG-8's good flight performance, safe and easy handling and low manufacturing costs as its strong points; yet Aeroflot showed a total lack of interest in the machine. Still, the MiG-8 demonstrated its simplicity and high reliability during its lengthy operation as a company-owned liaison aircraft.

Mikoyan/Gurevich MiG-15*bis* (*izdeliye* SYe, LL) aerodynamics research aircraft

In order to eliminate the MiG-15 *Fagot* fighter's reverse roll reaction to rudder inputs at high speeds LII engineers I. M. Pashkovskiy and D. I. Mazoorskiy proposed offloading the ailerons and increasing rudder area. Following TsAGI recommendations, a Mikoyan OKB team under Vladimir P. Yatsenko designed new wingtips with a modified airfoil and an angular trailing edge. The ailerons were also modified so that their span and area was increased but overall wing span remained unchanged. Also, the height and area of the vertical tail were increased to improve controllability. The redesign was initiated by Ministry of Aircraft Industry (MAP – *Ministerstvo aviatsionnoy promyshlennosti*) order No. 939 issued on 2nd December 1950.

'510 Blue' (c/n 125010), one of the two MiG-15*bis* (*izdeliye* SYe) aerodynamics research aircraft with redesigned wingtips and fin.

This view shows the shape of the *izdeliye* SYe's vertical tail. Note the faired ciné cameras installed aft of the seat headrest to capture the airflow pattern on the wings.

Blueprints for the changes were issued in December 1950. Since the MiG-15's mid-set horizontal tail required the fin and rudder to be built in two sections to simplify manufacturing, the engineers chose not to design an all-new vertical tail but simply fitted a new upper fin and rudder section to the existing lower fin. The new assembly was taller and the upper fin section had increased and constant chord (unlike the tapered standard fin). This resulted in a kinked leading edge and made the aircraft look rather incongruous with its outsize tail.

The aircraft received the manufacturer's designation '*izdeliye* SYe'. *Izdeliye* (product) such-and-such was a common way of coding military hardware items; the letter Ye probably stood for *yedinitsa* (lit. 'single unit', or rather 'one-off'). The aircraft was also known as LL and has also been referred to in some sources as MiG-15LL, though the latter designation is doubtful. Three MiG-15*bis Fagot-Bs* – two flying prototypes, including 510 Blue (c/n 125010), and a static test airframe – were built in this form by the Kuibyshev aircraft factory No. 1 in March 1951; the static

MiG-15*bis* (*izdeliye* SYe) aerodynamics research aircraft (c/n 125010)

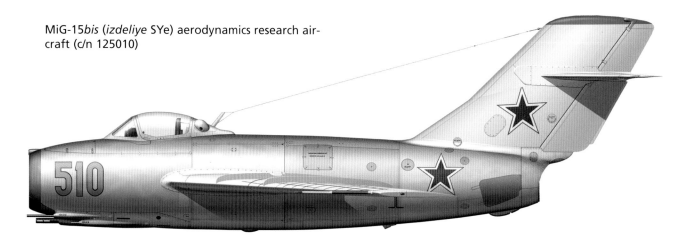

Upper view of the *izdeliye* SYe showing the reshaped wings and the grid pattern on the upper surface used in conjunction with wool tufts for airflow visualisation.

test airframe was delivered to TsAGI on 23rd March.

The aircraft were tested at Zhukovskiy in June and July 1951 by LII test pilot Anatoliy M. Tyuterev. Tests revealed that the structural changes did not resolve the reverse roll reaction problem and the programme was abandoned.

Mikoyan/Gurevich MiG-15*bis* aerodynamics research aircraft/ flight control system testbeds

In 1952 two other *Fagot-Bs* were modified to test flight spoilers assisting the ailerons for roll control. The spoilers were 1 m (3 ft 3⅜ in) long on one aircraft and 0.2 m (7⅞ in) long on the other one.

Mikoyan/Gurevich MiG-15*bis* aerodynamics research aircraft

An early Kuibyshev-built MiG-15*bis* serialled '172 Blue' (c/n 121072) was converted for aerodynamics research by LII. The aircraft had a non-standard rounded fin tip and redesigned upward-canted wingtips with a modified airfoil. Unfortunately, nothing is known about the time frame and the results of these tests.

Mikoyan/Gurevich MiG-17 (*izdeliye* SI-10) aerodynamics research aircraft

Benefiting from Korean War experience and studies of a North American F-86A-5-NA Sabre (USAF serial 49-1319) captured in Korea, the Mikoyan OKB attempted to improve the MiG-17 fighter's manoeuvrability, stability at high AOAs and field performance by means of effective high-lift devices. To this end an early-production Gor'kiy-built MiG-17 *sans suffixe* (*Fresco-A*) with 0.522-m² (5.612-sq ft) airbrakes serialled '214 Red' (c/n N54210214) was converted into an aerodynamics research aircraft designated *izdeliye* SI-10.

Here, the *izdeliye* SYe is seen with a retractable 'surfboard' of unknown purpose and an instrumented nose probe.

This MiG-15 has had the starboard wing painted black with a wave pattern for an aerodynamic research programme.

sections had black zebra-stripe markings applied for icing visualisation during tests.

The standard flaps were replaced by Fowler flaps occupying the entire trailing edge between the ailerons and the fuselage. The flaps moved on four tracks each and had two settings, 16° for take-off and 25° for landing. For roll control the ailerons were assisted by spoilers; these were located on the wing undersurface ahead of the flaps(!) and opened 55 mm (2¹¹⁄₆₄ in) when aileron deflection exceeded 6°. Finally, the MiG-17's customary wing fences were deleted. In all other respects the SI-10 was a standard MiG-17 powered by a 2,700-kgp (5,950 lbst) Klimov VK-1A non-afterburning turbojet and armed with a single 37-mm (1.45 calibre) Nudel'man N-37D cannon and two 23-mm (.90 calibre) Nudel'man/Rikhter NR-23 cannons.

The SI-10's wings were its main new feature. While retaining the standard basic structure and size, they had constant leading-edge sweep, lacking the MiG-17's characteristic leading-edge kink at half-span. The wings were equipped with four-section automatic leading-edge slats occupying the outboard 67% of each half-span. The slats deflected 12°; each section moved on two guide rails made of Type 30KhGSA steel and the outer

The new wings were designed in December 1952 according to TsAGI recommendations and manufactured at the Mikoyan OKB's prototype construction facility, MMZ No. 155 (*Moskovskiy mashinostroitel'nyy zavod* – Moscow Machinery Plant); the aircraft was completed in late 1954. The redesign incurred a sizeable weight penalty. The slats and flaps alone added 120 kg (264.5 lb) to the aircraft's empty weight, the spoilers another 14 kg (30.86 lb), and 70 kg

The MiG-17 (*izdeliye* SI-10) aerodynamics research aircraft

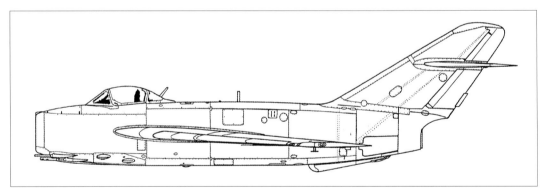

'214 Red' (c/n N54210214), the SI-10 aerodynamics research aircraft which served to investigate ways of improving the MiG-17's agility. The deployed leading-edge slat and lack of wing fences are clearly visible.

'638 Blue' (c/n 1406038), a MiG-17 used for aerodynamics research by LII. Note the wingtip fairings, the frame with eight pitot heads on the port stabiliser and the ciné camera faired into the canopy.

(154 lb) of ballast had to be installed in the forward fuselage to maintain the CG position.

The SI-10 was completed on 10th November 1952; the maiden flight, however, was delayed until January 1953 because TsAGI took so long to prepare the report clearing the aircraft for flight tests. The manufacturer's tests were performed by LII test pilot Sultan Amet-Khan and Mikoyan OKB test pilots Gheorgiy K. Mosolov, Gheorgiy A. Sedov and Arkadiy N. Chernoboorov between 17th February and 25th April 1953; Yuriy I. Korolyov was the engineer in charge. The tests were part of a plan drawn up by MAP in order to eliminate the deficiencies of the SI-02 and SI-01 (the MiG-17 *sans suffixe* prototypes) noted during state acceptance trials.

In the fourth quarter of 1953 the standard horizontal tail was replaced by variable-incidence stabilisers which could be adjusted from −5° to +3° for longitudinal trim. A BU-14 irreversible actuator was added in the pitch con-

Another LII research aircraft, MiG-17 '813 Blue' (c/n N54210813). The starboard wing is covered in wool tufts for airflow visualisation, suggesting a modified airfoil; again, a ciné camera is built into the canopy's rear end.

The MiG-19SVK's port wing with wool tufts for airflow visualisation. The extended outer wing leading edge and taller wing fence are obvious.

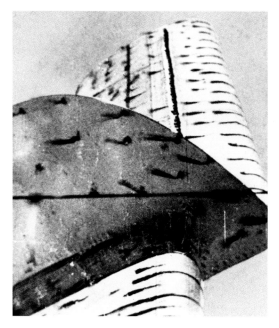

trol circuit. This modification added another 28 kg (61.72 lb) to the aircraft's empty weight. At this stage the aircraft was flown by Gheorgiy A. Sedov.

In June-July 1955 the SI-10 passed its state acceptance trials which included a special spin trials programme. The aircraft was flown by test pilots Stepan A. Mikoyan, Vladimir N. Makhalin, Anatoliy P. Molotkov and Nikolay A. Korovin from the Red Banner Air Force Research Institute (GK NII VVS – *Gosudarstvennyy Krasnoznamyonnyy naoochno-issledovatel'skiy institoot Voyenno-vozdooshnykh seel*; subsequently GNIKI VVS) who logged a total of 32 hrs 10 min in 47 flights. The trials showed that the spoilers and the variable-incidence stabilisers improved manoeuvrability and handling considerably, especially at high speed and high altitude. Conversely, the automatic LE slats, for all their extra weight, gave no great improvement in manoeuvrability. Performance was very similar to that of the SI-02 pre-production aircraft. Hence the SI-10 remained a one-off.

Mikoyan/Gurevich MiG-17 aerodynamics research aircraft

In 1952 another early-production MiG-17 was fitted with a redesigned fin swept back 55° at quarter-chord instead of 45° in an attempt to improve longitudinal stability. Hence this particular *Fresco-A* can be regarded as an aerodynamics research aircraft. Flight tests showed that the idea was a good one; the new fin worked as it should, improving controllability at high Mach numbers.

Two other MiG-17s *sans suffixe* were used for aerodynamics research. The first, a Kuibyshev-built MiG-17 serialled '638 Blue' (c/n 1406038), had bulbous fairings on the wingtips, a massive vertical frame with eight pitot heads (the thing looked every bit like a garden rake!) carried on a strut ahead of the port stabiliser, a long pitot on the intake upper lip and a ciné camera faired into the aft portion of the cockpit canopy. Black stripes were painted on the port wing upper surface, possibly for icing visualisation. The other aircraft built by plant No. 21 in Gor'kiy ('813 Blue', c/n N54210813), had the entire starboard wing covered in wool tufts and an identical ciné camera installation, suggesting a modified airfoil.

Mikoyan/Gurevich MiG-19SVK experimental high-altitude interceptor (*izdeliye* SM-9/3-VK)

At first the Soviet Air Force was quite happy with the climb rate and service ceiling of the MiG-19 *sans suffixe* (*Farmer-A*) and MiG-19S *Farmer-C*. However, no sooner had the type entered service than Western nations began sending reconnaissance balloons en masse across the Soviet borders, followed by Boeing RB-47 Stratojet high-altitude spyplanes. Much the worse for wear, Soviet intelligence sources reported that Lockheed was working on a reconnaissance aircraft capable of flying at 25,000 m (82,020 ft) – which subsequently emerged as the famous (or notorious, if you choose) U-2.

No Soviet aircraft of the time could cope with intruders flying that high. As a first step to remedy the situation, the Mikoyan OKB brought out the lightened MiG-19SV (*izdeliye* SM-9V, for *vysotnyy* – high-altitude) powered by uprated Mikulin RD-9BF afterburning turbojets (F = *forseerovannyy* – uprated). This entered flight test in 1956, but experiments aimed at increasing the MiG-19SV's service ceiling continued at LII in 1957. Major structural modifications were made to the second prototype SM-9V (converted Gor'kiy-built MiG-19 *sans suffixe* '644 Red', c/n N59210644) and the first SM-9/3-V (c/n N61210101) as recommended by TsAGI. Wing span and wing area were enlarged by fitting an extended cambered leading edge outboard of the boundary layer fences, thus creating a prominent dogtooth, and new extended wingtips with a modified airfoil.

Wing span was 10.3 m (33 ft 9³³⁄₆₄ in) instead of 9.0 m (29 ft 6²¹⁄₆₄ in) on the standard aircraft. Wing area increased from 25.16 to 27.7 m² (from 270.5 to 297.8 sq ft), aspect ratio from 3.24 to 3.8 and wing taper from 3.04 to 3.4. The root chord remained unchanged at 4.19 m (13 ft 8³¹⁄₃₂ in) while tip chord decreased from 1.378 m (4 ft 6¼ in) to 1.239 m (4 ft 0⁴⁹⁄₆₄ in). Mean aerodynamic chord (MAC) decreased accordingly from 3.023 m (9 ft 11 in) to 2.963 m (9 ft 8²¹⁄₃₂ in).

(Note: According to some documents, the aircraft was initially fitted only with the new wingtips increasing wing span by 2.013 m (6 ft 7¼ in), wing area by 2 m² (21.5 sq ft) and aspect ratio to 4.5. Later the span was slightly reduced concurrently with the installation of the extended leading edge; thus, wing area was increased by a further 0.54 m² (5.8 sq ft) despite the reduction in span.)

The converted aircraft was redesignated MiG-19SVK (the K stood for [*nov*oye] *kry**lo** –* new wings). The modified wings gave greater lift and had a better lift/drag ratio at high AOAs. To reduce weight the wing cannons were removed; however, additional equipment had to be installed, and the bottom line was that the aircraft's empty weight increased by 65 kg (143 lb).

The MiG-19SVK was flown by LII test pilots Valentin P. Vasin (project test pilot) and Vladimir S. Ilyushin. Originally the aircraft was powered by RD-9BF engines (c/ns G631013 and G631033), replaced in early May 1957 by RD-9BF-1s (c/ns 6506 and 6511) even before the wing modification. With the latter powerplant, empty weight was 7,250 kg (15,990 lb). With standard wings the SM-9V's centre of gravity had been at 43.7% MAC; after the conversion, however, it moved forward to 35% MAC. Prior to the first flight a speed limit of Mach 1.3 or 800 km/h (444 kts) IAS and a G limit of 5 were imposed on the aircraft. Later, the original engines were reinstalled for comparison purposes.

Tests showed that with RD-9BF engines and new wings the MiG-19SVK had a service ceiling of 19,100 m (62,660 ft) as compared to 18,800 m (61,680 ft) with standard wings. With RD-9BF-1s, performance was marginally worse: the service ceiling was 18,900 m (62,010 ft) as compared to 18,550 m (60,860 ft) with standard wings. On 29th April 1957 Vasin attained a top speed of 1,420 km/h (767.56 kts) at 16,250 m (53,310 ft); the aircraft was powered by RD-9BF engines and had a 6,130-kg (13,514-lb) TOW. In a subsequent test flight on 17th May with RD-9BF-1 engines and a 6,430-kg (14,175-lb) TOW, top speed at the same altitude dropped to 1,275 km/h (689.18 kts).

Vasin reported that the MiG-19SVK was almost identical to the standard MiG-19S as regards piloting techniques and was *'easier to fly at 320-380 km/h [177-211 kts] IAS, lacking*

Front view of the Mikoyan Ye-6T/3 (MiG-21F-13 prototype) after conversion as a CCV with canard foreplanes. The canards were mounted below the air intake axis.

the [standard aircraft's] tendency to pitch up and roll'. Nor had field performance changed.

Thus, the increase in service ceiling was only 300 m (980 ft) rather than the anticipated 1,000 m (3,280 ft). The basic aircraft had reached its limit. However, several liquid-fuel rocket boosters were by then approaching the test stage and could be fitted to the MiG-19 in future to increase total thrust at high altitude. This would not only increase the zoom altitude but enable the aircraft to put on a burst of speed when necessary, and the Mikoyan OKB decided to concentrate on this new and promising design area.

Mikoyan/Gurevich Ye-6T/3 control-configured vehicle

Once the KAP-2 autopilot (***kren****ovyy avtopi**lot*** – bank-only autopilot giving autostabilisation in the roll channel) developed by the Mikoyan OKB had been put through its paces on the Ye-6T/3 (one of the MiG-21F-13 *Fishbed-C* prototypes), the aircraft was refitted with the new KAP-3 autopilot; the latter passed its tests with good results in September 1963 and was cleared for production as the AP-155. Even as the KAP-2 was being tested, the aircraft was

fitted with movable canard foreplanes (called 'destabilisers' in Russian terminology of the time) on the sides of the extreme nose; thus the Ye-6T/3 can be regarded as a control-configured vehicle (CCV). The canards were locked neutral before the aircraft went supersonic.

Other changes included removal of the port cannon and ammunition box in order to provide room for an extra 140 litres (30.8 Imp gal) of fuel. The RSIU-4V *Min**dal'*** (Almond) radio was replaced with a more modern RSIU-5 *Doob* (Oak) radio.

The CCV was tested in 1961-62 as part of the efforts to create two very different aircraft – the Ye-8 interceptor derived from the MiG-21 and the Mikoyan Ye-152P heavy interceptor. A total of 58 flights had been made when the programme was completed in July 1962.

Mikoyan MiG-21SM aerodynamics testbed

In 1975 a single production MiG-21SM *Fishbed-J* was converted into an aerodynamics testbed; the wings featured a modified airfoil, the standard rounded leading edge being

The third prototype MiG-21F-13 (Ye-6T/3) after conversion for aerodynamic research with all-movable canards on the extreme nose.

The Ye-6T/3 retained the missile pylons and is seen here carrying K-13 missiles.

These views of the modified Ye-6T/3 show the cropped-delta shape of the canards, their anti-flutter booms and the photo calibration markings near the trailing edges.

Side view of the Ye-6T/3 CCV. The vertical tail was modified, with a dielectric insert at the top not found on the standard MiG-21F and MiG-21F-13.

replaced by a sharp one. The tests revealed a considerable improvement of flight performance; yet for various reasons the new wing design was not introduced on the production lines.

Mikoyan MiG-21I supersonic transport technology demonstrator (*izdeliye* 21-11, Analogue)

In the mid-1960s the Soviet aircraft industry was busy with a high-priority and highly prestigious programme – the Tupolev Tu-144 *Charger* supersonic transport (SST). The Tu-144 utilised a tailless-delta layout, and the original version known at Andrey N. Tupolev's OKB-156 as *izdeliye* 044 had unconventional wings of ogival planform. Before the Tu-144

prototype registered CCCP-68001 could take to the air, a lot of the features embodied in its design had to be verified. In particular, it was necessary to check the stability and handing of the ogival-wing, tailless-delta aircraft in all flight modes. Hence MAP deemed it necessary to build a subscale proof-of-concept vehicle based on the production MiG-21 fighter. This aircraft would be used for checking out the new wing design in flight and for developing the optimum piloting techniques.

The Mikoyan OKB's head office in Moscow was not in a position to do the job, having higher-priority programmes for the Air Force and the Air Defence Force (PVO – **Pro**tivovoz-**doosh**naya obo**rona**) to take care of. Therefore, development of the Tu-144 technology demonstrator was entrusted to the

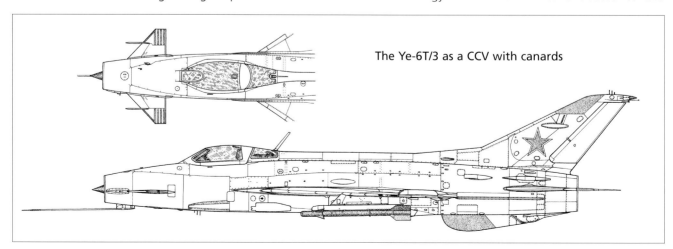

The Ye-6T/3 as a CCV with canards

special design office of the Gor'kiy aircraft factory No. 21. An MAP order to this effect was issued on 1st August 1964. At OKB-155 the programme was supervised by I. V. Froomkin, while Yevgeniy I. Mindrov (head of the special design office) headed the design effort in Gor'kiy. The scientists Maks A. Taïtz, Arseniy D. Mironov and V. Grachov, as well as project engineer V. Startsev (representing LII), participated actively in drawing up the specifications to which the technology demonstrator would be designed.

The demonstrator was officially designated MiG-21I (for *imitahtor* – 'simulator', or rather emulator) and bore the in-house product code '*izdeliye* 21-11'. Unofficially the machine was known as the *Analog* (Analogue); this was a reference to its similarity to the Tu-144 in layout and wing planform. Some sources erroneously referred to the aircraft as the 'A-144', which supposedly stood for 'analogue of the Tu-144'.

The machine was based on the production MiG-21S. The fighter's fuselage and landing gear were mated with new thin wings of ogival planform and changes were made to the control system. The armament, weapons control system and part of the equipment were deleted.

Thus, much of the airframe had to be designed from scratch; also, appropriate modifications had to be made to the fuselage and the aircraft systems. The Gor'kiy factory's design office prepared the blueprints for the new airframe components and equipment items in 1965-66; N. A. Limanov was the aircraft's project chief.

Since the MiG-21I's wings represented a scaled-down version of the *izdeliye* 044's wings, the design team made use of the Tu-144 blueprints supplied by OKB-156. A wind tunnel model of the MiG-21I was tested at TsAGI; this made it possible to determine the machine's principal aerodynamic parameters in all flight modes and issue recommendations for the test pilots long before the demonstrator actually flew.

The new wings had a gross area of 41.1 m² (442.4 sq ft) and an aspect ratio of 1.62. The leading-edge sweep was 78° on the root portions or leading-edge root extensions (LERXes) and 55° on the outer portions; the LERXes ran almost all the way to the air intake lip, the port LERX blending with the AoA sensor fairing on the side of the nose. The wings utilised a thin symmetrical airfoil with a sharp leading edge; the thickness/chord ratio varied from 2.3% to 3.5%.

Upper view of the MiG-21I/1's forward fuselage.

There were no high-lift devices. The entire trailing edge was occupied by two-section elevons; each section had its own actuator enclosed by an elongated fairing on the wing undersurface. All four sections moved in concert when the control stick was moved back and forth; when it was moved from side to side the elevons deflected differentially. The elevons' travel limits were +7°/–20° in elevator mode and +15°/–28° in aileron mode. The rudder had a maximum deflection of ±25°; all

The MiG-21I/1 shows off its tailless-delta layout and the ogival wings. Cloth 'walkways' are placed on top of the wing roots to protect the skin during maintenance.

Front views of the MiG-21I/1, showing the straight LERXes and the elevon actuator fairings under the wings.

Opposite page: The MiG-21I/1 (CCCP A-144) was finished in two shades of grey, with blue trim and a Soviet flag in typical 1960s 'flying' style. The registration was repeated on the underside of the wings. Note how the port LERX merges with the AoA sensor fairing. Note also the data link aerial aft of the nose gear unit.

control surfaces were operated by irreversible hydraulic actuators.

The pitch control circuit incorporated a device allowing the pilot to select the stick-to-elevons gearing ratio between 0.735 and 1.59. The roll control circuit featured a non-linear ratio between stick travel and elevon travel, just like on the MiG-21S. Spring-loaded artificial-feel devices were provided to emulate the aileron/elevator forces on the stick. The control system included a three-channel (pitch/roll/yaw) damper which could be selected on or off in flight. With the damper in operation, the elevon and rudder deflection limits were ±0°30' in the pitch channel, ±1°12' in the roll channel and ±4° in the yaw channel.

MiG-21I CCCP A-144 is prepared for a flight, with Su-11 and Su-15 interceptors in the background.

Two views of CCCP A-144 as it taxies out for take-off.

Front and upper views of the MiG-21I/2 with the definitive LERXes. The lower, rear and side views depict the MiG-21I/1.

As compared with the standard MiG-21S, the fuselage of the MiG-21I was 0.75 m (2 ft 5$\frac{17}{32}$ in) shorter; in contrast, the wing span was 60% greater. The Analogue was powered by a Tumanskiy R13F-300 turbojet delivering 6,490 kgp (14,310 lbst) in full afterburner. The six bladder tanks in the fuselage and the integral wing tanks held a total of 3,270 litres (719.4 Imp gal); the aircraft was also able to carry the MiG-21's standard 490-litre (107.8 Imp gal) drop tank on a centreline pylon.

Three examples of the MiG-21I (two prototypes and a static test airframe) were built; interestingly, the prototypes were completed

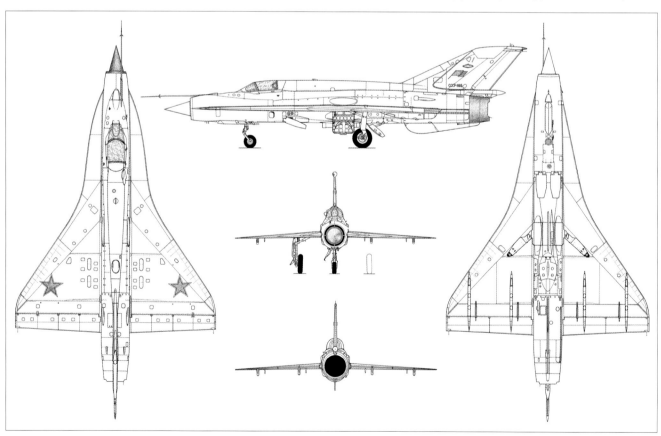

The MiG-21I/1 on final approach

with almost two years between them. The airframes were largely manufactured in Gor'kiy, with the exception of the wings, which were 'subcontracted out' to aircraft factory No. 64 in Voronezh. The reason for this was that plant No. 64 was due to commence series production of the Tu-144 shortly and had to master the structurally complicated ogival wings, which were quite difficult to manufacture; building the MiG-21I wing panels would give the plant its 'first taste' of the design.

Final assembly of the first prototype (which, somewhat surprisingly, bore the c/n N010104 – that is, Batch 1, 4th aircraft in the batch) took place in Moscow at MMZ No. 155. The second prototype (c/n N010103) and the static test article were assembled in Gor'kiy.

(Oddly, regarding the last digits of the construction numbers, there appears to be 'one c/n too many'; yet the c/ns of the two prototypes are confirmed both by official documents and by photo proof. It remains unclear whether the static test article was c/n N010101 or N010102.)

The first prototype was finished in two shades of grey (the fuselage spine, fin leading edge and brake parachute housing were painted a lighter colour). Unusually, the machine sported the non-standard civil registration CCCP A-144 and a Soviet flag on the tail in lieu of the normal star insignia. On 18th April 1968 the aircraft made its maiden flight at the hands of LII test pilot Oleg V. Goodkov, who had been appointed the MiG-21I's project

An air-to-air of the first MiG-21I.

CCCP A-144 flies chase with the Tu-144 prototype, using the airbrakes to keep formation. The wing shapes make an interesting comparison.

test pilot. The first flight was preceded by the usual series of ground checks and taxi tests, including a short hop to an altitude of 1-2 m (3-6 ft). It was during this hop that Goodkov discovered that the Analogue was reluctant to land; the aircraft bounced several times as the pilot tried to bring it down to terra firma. The pilot and the specialists involved in the tests

The second flying MiG-21I had an overall silver finish and wore Air Force insignia.

were worried at first, but the reason turned out to be simple – the large wings created an air cushion effect that increased wing lift appreciably as the aircraft approached the runway surface. At a session of MAP's Methodical Council that was to give the go-ahead for the maiden flight, LII specialists demanded that a second short hop should be made before the first real flight could be attempted. Still, Goodkov got his way and obtained permission for the flight – and the flight went successfully, with nothing untoward. By the end of 1968, when Tu-144 CCCP-68001 was due to enter flight test, MiG-21I CCCP A-144 had made 64 flights; 57 of them were performed by Goodkov.

The second prototype MiG-21I was not completed until the end of 1969. This machine was painted silver overall and wore Air Force insignia but no tactical code. Instead, the aircraft proudly wore the Komsomol (Young Communist League) badge on both sides of the nose, along with the legend 'KME' (*Komsomol'sko-molodyozhnyy ekipazh* – Komsomol youth crew).

It was decided that the aircraft would make its first flight from Gor'kiy-Sormovo (the factory airfield of plant No. 21). To this end a support team headed by engineer I. Zhidenko was dispatched from Moscow. The team also included V. Startsev (the engineer in charge of the second prototype's tests) and Igor' P. Volk, a young LII test pilot who was appointed this machine's project test pilot. Upon arrival they were joined by I. V. Froomkin, the MiG-21I project chief at OKB-155.

The aircraft differed from the first prototype in several respects. Preparations for the first flight were completed at the end of

Another view of the MiG-21I/1, showing the wing underside.

January 1970, and soon Volk flew the MiG-21I without any trouble.

The first prototype, the bulk of whose test flights took place in 1968-69, served for evolving the techniques of piloting a tailless-delta aircraft controlled by means of elevons. The second prototype, whose test programme at LII began in 1970, served for aerodynamic experiments and was subjected to modifications. In particular, the original straight LERXes with 78° leading-edge sweep were replaced by curved ones. A cigar-shaped pod housing a ciné camera was installed at the tip of the fin and a small fairing housing a second ciné camera looking aft at the starboard wing was added to the fuselage spine. The cameras captured the airflow pattern on the wings in various flight modes; to this end the starboard wing was covered with wool tufts. Additionally, the upper and lower surfaces of the wings were rigged with air pressure sensors (pitot heads and static ports) and a laser system visualising the airflow over the wings was fitted.

This view of the MiG-21I/2 shows the new curved LERXes, the cigar-shaped fairing on the fin and the camera housing on the spine.

The MiG-21I/2 had wool tufts for airflow visualisation. Note the new shape of the LERXes, the KME badge and the photo calibration 'target' on the nose.

The MiG-21I reached a maximum altitude of 12,000 m (39,370 ft) and was flown at indicated airspeeds from 212 km/h (131 mph) at 3,000 m (9,840 ft) to 1,205 km/h (748 mph) at 11,500 m (37,730 ft); it attained a maximum speed of Mach 2.06. The CG range in these flights was 40.2% to 47.7% MAC; the CG position was selected by installing special weights of 277-290 kg (610-640 lb) apiece in the nose or rear fuselage. The first prototype had an all-up weight of 6,550-7,930 kg (14,440–17,480 lb) and had a wing loading of 160-193 kg/m² (32.8-39.5 lb/sq ft). The appropriate figures for the second aircraft were 6,260-7,640 kg (13,800-16,840 lb) and 152-186 kg/m² (31.16-38.13 lb/sq ft).

Large-scale flight tests with the MiG-21I were undertaken by LII jointly with the Tupolev and Mikoyan bureaux and TsAGI in 1968-72. The research work was directed by Igor' M. Pashkovskiy, who was assisted by engineers in charge I. V. Froomkin (OKB-155) and V. Ya. Molochayev (LII); Oleg V. Goodkov was project test pilot. The results of these tests provided a virtually complete picture of a tailless-delta aircraft's flight dynamics and control characteristics within a wide range of AoAs and speeds (up to Mach 2). The machine's sta-

bility and control characteristics were determined, its handling in subsonic and supersonic flight at varying speeds and altitudes was assessed, and the aircraft's stall and spinning characteristics were explored. Additionally, a lot of aerophysical research was undertaken with the MiG-21I.

A total of 16 pilots made test and familiarisation flights in the Analogue. In addition to the ones mentioned above, they included LII test pilots Aleksandr A. Shcherbakov, Arkadiy P. Bogorodskiy, Eduard P. Knyaginichev, Vladimir V. Podkhalyuzin and Viktor S. Konstantinov, Mikoyan OKB test pilots Aleksandr V. Fedotov, Pyotr M. Ostapenko and Boris A. Orlov, MAP test pilots Eduard V. Yelian, Mikhail V. Kozlov and Sergey T. Agapov and others. By July 1969 these pilots had made 142 flights in the first prototype MiG-21I alone; eight of these were performed by Yelian and Kozlov, the very pilots who performed the Tu-144's maiden flight on 31st December 1968. Thus, the Analogue also served as a trainer, allowing the aircrews of various MAP subdivisions (including the Tupolev OKB, the State Civil Aviation Research Institute (GosNII GA – *Gosudarstvennyy naoochno-issledovatel'skiy institoot grazhdahnskoy avi-*

Two views of the second MiG-21I in front of LII's hangar during modifications. The tufted starboard wing retains the original straight LERX, while the port LERX has been removed and is about to be replaced with a new curved section, which can be seen lying on the ground alongside.

ahtsii), LII and the Voronezh aircraft factory) and GNIKI VVS to get ready for their first flights in the supersonic airliner. Speaking of which, when Tu-144 CCCP-68001 entered flight test, the MiG-21I/1 was sometimes used as a chase aircraft. Eventually the ogival-wing *izdeliye* 044 turned out to be unsatisfactory and the Tu-144 underwent a major redesign, entering production as *izdeliye* 004 with double-delta wings, retractable canard foreplanes, repositioned engines and other changes – but that's another story.

The Analogue was fairly easy to fly and exhibited pleasant handling, which is why it was popular with the pilots. The larger wings gave the MiG-21I high manoeuvrability; however, acceleration was slower compared with the production fighter. Eduard V. Yelian commented thus on the Analogue: *'I had flown straight-wing aircraft before, and I have to say that in take-off and landing mode swept wings have provided the pilot largely with a lot of inconvenience, such as all manner of stalls and oscillations. [...] Well, at subsonic speeds*

the tailless MiG-21 derivative proved to be quite similar in handling to straight-wing aircraft; it possessed absolutely brilliant controllability. Its landing speed was around 220 km/h [136 mph] versus the MiG-21's 320 km/h [198 mph]. I had a really good time flying the

A beautiful shot of the Tu-144 prototype, CCCP-68001, in an early test flight, with the MiG-21I in echelon port formation.

Analogue. Oh, there were a few minor complications in supersonic flight – the machine was prone to Dutch roll up to Mach 1.7, but that was easily corrected.' Boris A. Orlov noted that *'…the aircraft was easy to fly on take-off and landing; it handled well during manoeuvres. During deceleration, however, as soon as the angle of attack reached a certain limit the aircraft would pitch up spontaneously, becoming unstable. Pushing the stick all the way forward gave no effect; the only option was to stall the machine by applying a bootful of rudder and then recover from the stall.'*

Determining the relationship between the MiG-21I's lift coefficient and its angle of attack (especially at high alpha) turned out to be a tough task, as no calculation technique for this existed in the Soviet Union at the time. Such a technique was developed by a team of physicists at LII under the direction of lab chief V. Grachov and section chief Yu. Zavershnev. The results of the research flights depended to a large extent on the pilot who had to perform the deceleration (right down to stalling speed) strictly in accordance with the mission plan. He had to remember to switch on the test equipment at the right time and maintain radio contact with the ciné theodolite measurement stations – all the while taking care not to exceed the AoA so as not to stall the aircraft prematurely. For safety's sake the spin was initiated at around 10,000 m (32,810 ft); this was important because, to complicate the matter, the MiG-21I was not fitted with spin recovery rockets which were incompatible with the tailless-delta layout. The results of the research were promptly forwarded to the Tupolev OKB.

After exploring the Analogue's entire speed and AoA range and meticulously performing

Overall view of the MiG-21I/2 at the Central Russian Air Force Museum in Monino.

all flight modes envisaged by the MiG-21I's test programme, Igor' Volk finally took the second prototype to stall mode and the aircraft flicked into a spin. After five turns of the spiral, Volk made a clean recovery and landed safely. The tests revealed that a tailless-delta aircraft with ogival wings retains high wing lift up to extreme angles of attack; moreover, the wing lift remains virtually constant.

Tragically, CCCP A-144 crashed fatally at Zhukovskiy when the main part of the test programme was already completed. On 28th July 1970 LII test pilot Viktor S. Konstantinov was flying the first prototype. At that time he was practising a display programme for the 1971 Paris Air Show where he was to fly a production MiG-21. That was to be the first appearance of a Soviet combat aircraft at a major international airshow; until then, Soviet participation in such events had been strictly civilian, reflecting the Soviet Union's ostensibly peaceful foreign policy. Konstantinov had polished his display programme to perfection in a standard MiG-21 (well in advance of the event, it should be noted). Now he decided to repeat it partly on the MiG-21I/1 as he came in to land after a test mission and rashly started a series of aerobatic manoeuvres right above the runway – forgetting that the MiG-21I's handling was rather different from that of the standard tailed-delta fighter. One error on the part of the pilot was enough – the aircraft lost speed and started pancaking. CCCP A-144 impacted right in the middle of the runway and exploded; the pilot ejected at the last moment but the altitude was too low for his parachute to open.

The tests continued on the surviving second prototype. Once the test and crew training programmes had been completed, the aircraft was donated to the Soviet Air Force Museum (now Central Russian Air Force Museum) in Monino. Curiously (and quite appropriately), it is parked next to a production Tu-144 (*izdeliye* 004), CCCP-77106.

Mikoyan MiG-23 *sans suffixe* aerodynamics testbed

In 1982-83 a MiG-23 *sans suffixe* ('1971-model MiG-23') fighter coded '04 Blue' (the c/n ended …604) was used by LII for studying active airflow control methods. To this end the aircraft was fitted with a specially made pneumatic system comprising 22 spherical air bottles charged to 320 kg/cm² (4,570 psi) housed in an oval-section pod resembling a drop tank

which was attached to the centreline pylon. Pipelines of 30 mm (1³⁄₁₆ in) diameter connected this pod with special manifolds in the outer wings featuring numerous ejector nozzles; the air was forced through these at a pressure of 150 kg/cm² (2,140 psi). Special pipeline joints had to be developed to allow the wings to swing, and the pipes were partly visible on the outer faces of the wing glove pylons, which were unusually large and deep. The wings were provided with data recording system sensors and airflow visualisation tufts, and a ciné camera was installed in a small pod at the top of the fin.

Mikoyan MiG-23M (modified) testbed

A MiG-23M *Flogger-B* fighter coded '01 Blue' (c/n 0390206596, f/n 7901) served as a testbed for the aerodynamic features of the MiG-23MLD *Flogger-K* – the ultimate fighter version. The wing glove fairings were cut back at the root to create the MiG-23MLD's characteristic additional wing dogtooth (the so-called lambda extensions), and the pitot was fitted with vortex generator plates 36.5 cm (1 ft 2³⁄₈ in) long improving high-alpha behaviour, but the *Flogger-B*'s long dorsal fin was retained. This aircraft is now preserved.

Mikoyan MiG-23B aerodynamics testbed

A unique example of the MiG-23 survives in the aviation museum at Riga-Skulte airport. Coded '60' (the c/n was reported as 3910601), it is a converted MiG-23B fighter-bomber combining the Lyul'ka AL-21F-3 afterburning turbojet (with its characteristic long nozzle) with fixed-area intakes as used on the MiG-23BM (MiG-27) *Flogger-H* powered by the Tumanskiy R29-300 afterburning turbojet! Apparently this aircraft served as an aerodynamics testbed for the new intake design.

Mil' Mi-4 aerodynamics testbeds

In 1965-70 LII used a Mil' Mi-4 *Hound* utility helicopter (identity unknown, c/n 0581) for exploring flight modes involving main rotor blade stall. The tests gave valuable information which was used in developing the Mi-6, Mi-8 and Mi-24 helicopters' main rotors.

Concurrently another Mi-4 (c/n 04122) was used by LII in 1965-69 for testing various

MiG-23 *sans suffixe* ('1971-model MiG-23') '04 Blue' was used by LII for studying active airflow control methods. The ventral pod is not a drop tank – it houses air bottles for the active airflow control system.

A Mi-4GF used by LII as an aerodynamics testbed (the registration is unfortunately retouched). Note the air data boom and the fairing on the main rotor head.

versions of the main rotor blades and studying the rotor's vortex system. One of these two helicopters was a civil-registered Mi-4GF (that is, a demilitarised ex-Air Force example retaining the ventral 'bathtub' for the gunner and the olive drab military colour scheme). The helicopter sported an instrumented test boom on the nose and a conical fairing on the main rotor head.

Mil' Mi-6 rotor system testbed

In 1975 the Mil' design bureau (OKB-329) converted a Mi-6 *Hook* heavy-lift helicopter coded '88 Red' (c/n 8683904V) into a testbed for the rotor system of the Mi-6's successor, the Mi-26 *Halo*. The *Hook*'s five-blade main rotor and four-blade tail rotor were replaced with the *Halo*'s eight-blade main rotor and five-blade tail rotor. Only the new main rotor was fitted originally, the standard tail rotor being retained. This helicopter also served for testing a number of other systems and units of the Mi-26.

Mil' Mi-8 aerodynamics testbeds

A very early-production Kazan'-built Mil' Mi-8 *Hip* utility helicopter (identity unknown, c/n 0102) was used by LII in 1966-77 for exploring flight modes involving main rotor blade stall. The tests gave valuable information which was used in the development of the Mi-6, Mi-8 and Mi-24 helicopters' main rotors.

In 1972-85 another Mi-8 was used for exploring speed measuring systems optimised for low speeds and studying the helicopter's low-speed stability and handling. The c/n was quoted in abbreviated form as '127', which

This V-24 with a short cockpit section was converted into a testbed with a fenestron. The stub wings have been removed.

could mean anything – 0127 (first aircraft in Batch 27 under the first Kazan' c/n system), 3127 (27th aircraft in Batch 31 under the second Kazan' c/n system) or even an Ulan-Ude-built Mi-8T c/n 97...127.

One more Mi-8T delivered to LII in Soviet Air Force markings (no tactical code, c/n unknown) was modified in an attempt to improve the helicopter's performance and reduce fuel consumption. The aircraft incorporated a number of low-drag modifications, the most obvious of these being a new clamshell cargo door design; the doors were lengthened and reshaped to give the rear fuselage a profile similar to that of the Bell 206 JetRanger/LongRanger light helicopter. Two-piece teardrop fairings were fitted around the engine jetpipes, the main rotor head was faired, and the standard external fuel tanks, KO-50 cabin heater and DISS-2 Doppler speed/drift sensor fairing under the tailboom were removed, the fuel being carried in the cabin.

Mil' Mi-24 rotor system testbeds

A number of Mi-24 attack helicopters served as systems and avionics testbeds for the new-generation Mi-28 *Havoc* attack helicopter. One of them, possibly an uncoded early-production Mi-24A *Hind-A* with starboard-side tail rotor, was used to test the new main rotor. The helicopter in question featured a large lattice-like boom with air data sensors on the fuselage nose (the machine gun had to be removed, of course). There was also a similarly equipped Mi-24A *sans* stub wings and missile director antenna; quite possibly it was the same aircraft at a later stage. Another Hind served as a testbed for the Mi-28's Model 286 squashed-X tail rotor.

Close-up of the fenestron assembly; the number of blades is surprisingly small.

Mil' Mi-24 fenestron testbed

In 1975 an uncoded Mi-24 *sans suffixe* (*Hind-B*) attack helicopter – one of the original V-24 prototypes or possibly one of the ten pre-production aircraft – was fitted experimentally with a large-diameter eight-bladed fenestron (fan-in-fin) replacing the standard three-blade tail rotor and its pylon. The Mi-24's distinctive stub wings were removed, probably to save weight. However, development was discontinued because the fenestron was ill-suited for helicopters in the Mi-24's weight class.

Sukhoi Su-7B aerodynamics research aircraft (S-25)

Between 1965 and 1973 LII used a modified Su-7B *Fitter-A* fighter-bomber coded '02 Red' (c/n 2502) for laminar flow research at supersonic speeds. The aircraft (called S-25) had a test article looking like a reverse-tapered wing section fitted to the starboard wing leading edge. In another configuration a canted constant-chord test article was fitted instead of the port fuselage pylon, the other pylon carrying a test equipment pod. This research helped to find ways of improving the lift/drag ratio of supersonic aircraft.

A production Su-7B coded '02 Red' (c/n 2502) was converted to the S-25 boundary layer control system (BLCS) testbed. This view shows the aerodynamic surface replacing the port fuselage pylon and the test equipment pod on the starboard pylon.

Two more views of the S-25, showing the highly modified starboard wing with a reduced-sweep leading-edge extension featuring a BLCS.

Sukhoi '100LDU' control-configured vehicle

In 1968 LII converted a production Su-7U *Moujik* fighter-bomber trainer coded '08' (c/n 0408) into a CCV as part of the effort to create a fly-by-wire (FBW) control system for the Sukhoi T-4 strategic supersonic missile strike aircraft. Since the latter had the product code *izdeliye* 100, the research aircraft was designated 100LDU (*laboratoriya [dlya izoocheniya elektro]distantsionnoy [sistemy] oopravleniya* – FBW control system exploration testbed). Canard foreplanes were fitted to the forward fuselage to emulate the T-4's canard-delta layout; setting the canards at a fixed angle in subsonic flight made the machine statically unstable in the pitch control channel. This facilitated designing the architecture of an automatic longitudinal stability augmentation system with an integral control algorithm. The tests began in 1968 and were conducted by LII test pilot Igor' P. Volk, who went on to become an astronaut.

In 1973-74 work was carried out using the 100LDU on verifying and defining the working laws for a combined FBW control system/stability augmentation system developed for the Sukhoi T-10 *Flanker-A* experimental fighter, the prototype/forerunner of the Su-27.

Sukhoi '100L' aerodynamics testbed (100L1 through 100L8)

In 1966-67 a Moscow-built Su-9 *Fishpot-A* interceptor coded '61 Blue' (c/n 100000610) was extensively modified for aerodynamics research as part of the Sukhoi T-4's development programme. Designated 100L (for *[letayushchaya] laboratoriya*), this testbed was intended for exploring the aerodynamics of low aspect ratio wings with a thin, sharply swept leading edge over a wide range of angles of attack and at speeds ranging from Mach 0.3 to Mach 2.0. It was, in effect, a sub-scale demonstrator helping the engineers to choose the optimum wing planform for the heavy T-4's wings. The thin wings with a sharp leading edge selected for one of the T-4's project versions were to combine a high lift/drag ratio in supersonic cruise with acceptable field performance.

The conversion involved rewinging a production Su-9 – well, not exactly: the greater part of the wing structure remained unchanged. The leading edge portion of each wing ahead of the main spar was replaced with a new one featuring greater chord and a more pointed profile; as a result, leading-edge sweep was increased from 60° to 65° and wing area increased accordingly. Several modified leading edge sections with different airfoils were fitted consecutively in the course of the tests.

This was not the only change effected on the 100L. About 1,000 kg (2,200 lb) of ballast was installed in the forward fuselage to shift the CG forward; hence the nose landing gear unit had to be reinforced by fitting a larger nosewheel. The wing tanks were rendered inoperative and converted into test equipment bays, reducing the fuel supply by more than 600 litres (132 Imp gal); more test equipment was housed in the avionics bays fore and aft of the cockpit. Interestingly, to visualise the airflow over the modified wings the aircraft was equipped with a smoke generator, the smoke exiting via special perforations in the wing leading edge. The modifications increased the aircraft's empty weight by 1,300 kg (2,865 lb) and the maximum take-off weight by 750 kg (1,650 lb) as compared to a stock Su-9.

The 100L research aircraft entered flight test in July 1966, the test programme continu-

ing until 1972; LII test pilot Eduard I. Knyaginichev performed the greater part of the programme. The aircraft was flown in eight different configurations, each one being allocated a sequence number as a suffix to the designation (from 100L1 to 100L8). Three wing versions with a sharp leading edge and one with a rounded leading edge were tested, as were different wing sweep angles and modified tailplanes with a sharp leading edge. Tests revealed that top speed remained unchanged but the landing speed was reduced by a sizeable 40 km/h (nearly 25 mph).

Another Su-9 (c/n 1301; the full c/n was probably 1315301) was also converted into a similar research aircraft likewise designated 100L. The test results gave a wealth of valuable data which allowed the OKB to select the most efficient wing design (a cranked delta shape) for the T-4.

The S-25 in flight, showing the ventral aerodynamic surface. The bullet-shaped fairing on the fin trailing edge is non-standard, too.

The 100LDU control-configured vehicle created as part of the effort to develop the T-4 (*izdeliye* 100) strategic missile strike aircraft was based on a Su-7U trainer ('08 Red') and utilised for stability augmentation systems research. The large articulated canards with anti-flutter booms are clearly visible. The badge aft of the tactical code indicates the aircraft was operated by a Young Communist League crew.

Su-9 '61 Blue' (c/n 100000610) was converted into the 100L aerodynamics research aircraft for testing possible wing designs for the Sukhoi T-4. This is the initial configuration designated 100L-1.

The same aircraft in later configuration as the 100L-2M with mildly ogival wings. The starboard wing is covered in wool tufts for airflow visualisation.

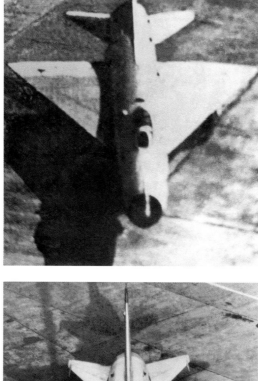

Far right: The same aircraft in slightly modified form with a ciné camera on the maintenance hatch cover for filming the airflow pattern.

Another aspect of the 100L-1. The rudder is not black – it's just a trick of the light.

Sukhoi L.02-10 control-configured vehicle

In 1968-75 LII undertook a massive test and development effort with a Novosibirsk-built Su-9 coded '93 Blue' (c/n 1215393) which had been converted into a CCV. Canard foreplanes are a familiar thing, but have you ever heard of a canard *foretail*? Well, Su-9 c/n 1215393 featured two all-movable vertical control surfaces of trapezoidal planform installed above and below the fuselage ahead of the cockpit. Actually the upper half of the 'foretail' was found to impair cockpit visibility unacceptably and was removed, all flights being made with only the lower section in place.

Originally '93 Blue' was used to investigate the flight dynamics and stability of a directionally unstable aircraft and test an automatic directional stability augmentation system. In 1975-78 the aircraft underwent more modifications performed jointly by LII and the Sukhoi design bureau (OKB-51), receiving a direct

side-force control (DSFC) system and the designation L.02-10 (again, the L stood for [*letayushchaya*] *laboratoriya*). In this guise the aircraft was tested at Zhukovskiy in 1979-84. One source states the test dates as 1966-69 and says the results were used in the development of the Su-27 (T-10).

Sukhoi L.07-10 aerodynamics research aircraft

In 1975-76 the Sukhoi OKB extensively modified a Su-9U *Maiden* trainer (c/n 112001301) as part of the effort to develop a fourth-generation fighter known in-house as the T-10 (NATO reporting name *Flanker-A*) – the precursor of the famous Su-27 *Flanker-B*. Designated L.07-10 (that is, 'flying laboratory' No. 7 under the T-10 programme), this aircraft was intended for exploring the T-10's wing

Above: Su-9 '93 Blue' in standard configuration prior to being modified as the L.02-10 CCV.

The logo of the Sukhoi OKB, which had several aircraft modified under its R&D programmes.

The L.02-10 as originally modified with two all-movable 'fore-tails' of unequal size ahead of the cockpit featuring anti-flutter-booms. The aircraft was never flown in this configuration (with both 'foretails').

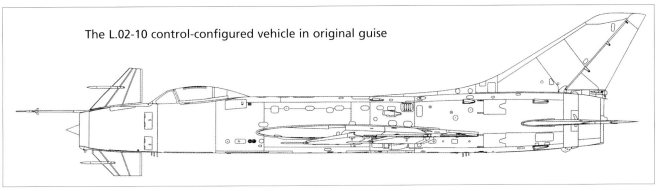

The L.02-10 control-configured vehicle in original guise

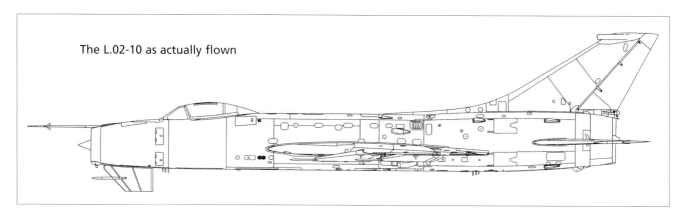

The L.02-10 as actually flown

The L.02-10 hangared for modifications which involved removal of the upper 'foretail'. The letters on the badge on the nose read KME (*Komsomol'sko-molodyozhnyy ekipazh* – Young Communist League crew).

The L.02-10 as actually flown; the axle of the removed upper 'foretail' protrudes ahead of the cockpit like a sight. Note the forward-looking camera in the fairing atop the fin and the non-standard aerial for the communications radio aft of the cockpit replacing the usual fin top antenna.

aerodynamics; it featured ogival wings with a sharp leading edge having compound curvature.

The L.07-10 entered flight test at LII in 1977. Tragically, it crashed after suffering a birdstrike on take-off in 1982, killing the pilot.

Sukhoi T47-6 experimental interceptor

When the Su-11 *Fishpot-B* interceptor (*izdeliye* T-47) was developed, eight prototypes designated T47-1 through T47-8 were built.

However, one of them (the T47-6) did not take part in the main trials – it was set aside for conversion under a different programme. Like most of the prototypes, it was converted from a standard low-rate initial production T-47.

Initially the T47-6 acted as an aerodynamics research aircraft, being fitted with canard foreplanes and a modified air intake featuring a fixed shock cone/radome instead of the standard movable shock cone. In this guise the aircraft flew both with a Lyul'ka AL-7F turbojet rated at 6,240 kgp (13,760 lbst) at full military power and 9,200 kgp (20,280 lbst) in full afterburner and with an AL-7F-1 uprated to 6,800 kgp (14,990 lbst) dry and 9,600 kgp (21,160 lbst) reheat. Later the T47-6 was used in the development of the T-3A-9 aerial intercept weapons system and the ill-starred Sukhoi T-37 heavy interceptor, which was built but never flew.

Sukhoi Su-15 research aircraft for spinning trials and uncontrollable roll research

Two specially-modified Su-15 interceptors were used for extensive aerodynamics research to investigate the aircraft's stability and control characteristics in certain flight modes. Thus the Flight Research Institute twice held spinning tests of the Su-15 – in 1968, using an unidentified early-production *Flagon-A* with pure delta wings, and in 1973, using the second production example of the double-delta Su-15T *Flagon-E* ('37 Red', c/n 1115337). Oleg V. Goodkov was project test pilot in both cases, but Igor' P. Volk performed part of the test programme on '37 Red'. A while earlier, in 1970, this aircraft was used to check the flight performance in minimum afterburner mode; in 1971 it was used to verify landing techniques developed for a fully-forward CG configuration. It was not until 1972 that Su-15T '37 Red' was equipped with PPR-90 spin recovery rockets (***pro**tivo**shtop**ornaya porokho**va**ya ra**ket**a*), enabling high-alpha and spinning tests to be held. A similar spinning test programme was performed by GK NII VVS, with Norayr V. Kazarian as project test pilot.

The tests showed that considerable vibration set in when the aircraft approached critical AoAs, warning the pilot that he was 'pushing it too far'. Actually the Su-15 could enter a spin only due to a grave piloting error or if the spin was initiated intentionally. The spinning characteristics of the two versions were similar, but the double-delta version was more stable during the spin.

Another test programme was held in response to a series of accidents in which the Su-15 suddenly started rolling uncontrollably during vigorous manoeuvres at supersonic speeds, the pilots perceiving this as a critical control system failure. The second production Su-15 (c/n 0115302), which still had a test equipment suite from an earlier trials programme, was set aside to investigate the problem. In 1970-71 the Sukhoi OKB and LII held a joint test programme to determine the conditions in which this phenomenon occurred – first at subsonic speeds and then beyond Mach 1; Yevgeniy S. Solov'yov and Igor' P. Volk were the OKB's and the institute's project test pilots respectively. It was established that there were no control system failures at all – the cause was traced to a specific relationship of inertia forces along different axes; the pilots provoked the uncommanded roll by pulling negative G. The correct course of action in this situation was to reduce speed and set the controls neutral.

Sukhoi T-58K (Su-15K?) aerodynamics research aircraft with modified wings

In 1968 the Sukhoi OKB started work on modified wings for the Su-15 (T-58) featuring a sharp leading edge. By April 1973 this work reached the practice stage – the fourth production Su-15 (c/n 0115304) was converted into the T-58K research vehicle; the K stood for [*modifi**tsee**rovannoy*e] *kry**lo*** – [modified] wings. The standard boundary layer fences and missile pylons were deleted, the pylon attachment points being enclosed by special fairings, and a new extended leading edge section with a more pointed airfoil was installed. The BU-220 hydraulic actuators in the tailplane control circuit were replaced by BU-250 units, and part of the standard avionics was replaced by test equipment and ballast. The T-58K underwent trials at LII in 1972-74. One publication called it, rather confusingly, Su-15K and stated that the purpose of the work was to explore the effect of wing-generated vortices on the aircraft's flight performance.

Sukhoi T10-24 CCV

The idea of fitting canard foreplanes to the Su-27 dated back to 1977 when the original T-10 *Flanker-A* was still in the running. The reason was that the fighter's Phazotron N001 Mech (Sword; NATO codename *Slot Back*) fire control radar turned out to be nearly 200 kg (440 lb) overweight, causing a forward shift of the aircraft's CG and causing the T-10 to become statically stable. This negated the advantages conferred by the Su-27's static instability concept, impairing manoeuvrability and performance in general. The canards could alleviate this problem by shifting the lift focus forward. The decision to scrap the *Flanker-A* and start from scratch with the

T-10S *Flanker-B* (the production form of the fighter) caused the plans concerning canards to be shelved.

The idea was revived in 1982 when the Sukhoi OKB started work on the Su-27M (T-10M, aka Su-35 – the first aircraft to bear the designation). This aircraft was to feature an even heavier NIIP N011 Bars (Leopard) radar, with an attendant shift in CG position, and the ghost of static stability reappeared. Now the canards became an absolute necessity; as a bonus, they would enhance control efficiency at high AoAs when the stabilators could be blanketed by the wings and stabilator authority would become insufficient. Sukhoi OKB General Designer Mikhail P. Simonov was the motive power behind the idea.

The T10-24 at the Sukhoi OKB's experimental plant in the process of conversion to the canard-equipped control-configured vehicle.

As can be seen here, the T10-24 ('24 Blue') had photo calibration markings on the radome.

Lower view of the T10-24 CCV.

In early 1985 an early-production Su-27, the T10-24 ('24 Blue', c/n 36911007...01, fuselage number 07-01), was modified at the OKB's experimental shop and became the first *Flanker* to be equipped with canards. The canards were positioned on the LERXes just aft of the cockpit; they had a span of 6.4 m (21 ft), an area of some 3 m² (32.25 sq ft) and a leading-edge sweep of 53°30'. The aircraft made its first post-conversion flight in May 1985 with Sukhoi OKB test pilot Viktor G. Pugachov at the controls. It was used for aerodynamics research and can thus be classed as a CCV.

Tests showed that the canards did their job, improving field performance and high-alpha handling. The canards were programmed to

The T10-24 control-configured vehicle

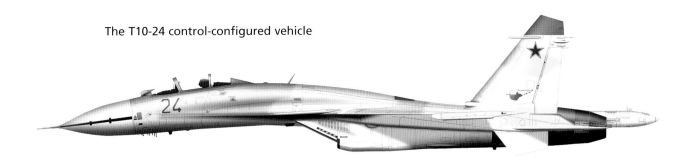

deflect automatically as the AoA grew; they increased the aircraft's static instability, allowing it to pull higher Gs without necessitating reinforcement of the wings and horizontal tail. Moreover, they gave a sizeable increase in lift; this was obtained by carefully choosing the canards' location so as to create favourable airflow interference between the canards and wings. In a nutshell, the canards clearly improved the Su-27's aerodynamics.

The results of the T10-24's trials impressed the engineers so much that the canards were incorporated into the Su-27K/Su-33 (T-10K) *Flanker-D* shipboard fighter and other advanced versions of the *Flanker* family. Similarly, the T10-24 was used for evolving the ski jump take-off technique at the initial stage of the Su-27K programme, operating from Novofyodorovka AB near the town of Saki on the Crimea Peninsula, which hosted the Soviet Naval Aviation's Flight Test Centre. The NIUTK installation (*Na**ooch**no-issle-*

*d*ovatel'skiy i oo**cheb**no-treni**rov**ochnyy **kom-***pleks* – R&D and training complex) featuring a ski jump and an arrestor wire system had been commissioned there in 1982 to help Soviet pilots master carrier operations techniques. The cumbersome acronym did not catch on, but the unofficial name 'Nitka' (Thread) did, and was formalised by 1990. Unfortunately the T10-24 crashed on 20th January 1987; GNIKI VVS test pilot Aleksandr S. Puchkov ejected safely.

Tsybin Ts-1 experimental aircraft (LL-1, LL-2 and LL-3 flying testbeds)

In the second half of 1945 a team headed by Pavel Vladimirovich Tsybin started developing a special aerial vehicle intended for conducting research and practical experiments in the air. These experiments were meant to solve the problem of selecting the most advantageous wing shape for aircraft flying at high subsonic

Here the T10-24 takes off from the T-2 ski jump at the Nitka 'unsinkable carrier' test and training complex at Saki. Note the single nosewheel of the land-based fighter is retained.

The Ts-1 in its original straight-wing form known as LL-1 with the take-off dolly attached.

The same aircraft sitting on its landing skid.

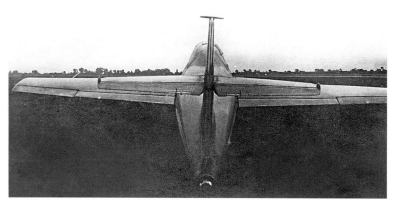

Rear view of the LL-1, showing the cruciform tail unit. The stabilisers are likewise unswept.

and transonic speeds. The projecting and construction of the machine proceeded with due regard to the programme of the future experiments and research which had been developed by TsAGI scientists jointly with prominent specialists of the aircraft industry. In early 1947 the aircraft, which was designated Ts-1, was submitted for testing. It was, in effect a flying testbed, or 'flying laboratory', in Soviet parlance, hence its other designation – LL-1.

The Ts-1 was of all-wooden construction with a plywood skin. It had unswept flat wings with an area of 10.2 m² (109.8 sq ft), with two spars made of *del'ta-drevesina* ('delta wood') – a moulded composite of wood impregnated with phenol formaldehyde resin. The tail unit had a cruciform layout, the horizontal tail being set half-way up the fin; the rudder and elevators featured mass balances. Prior to a

flight in glider mode, special ballast tanks in the fuselage were filled with up to 1,000 kg (2,200 lb) of water.

The Ts-1 was powered by a PRD-1500 solid-fuel rocket motor (*porokhovoy raketnyy dvigatel'* – 'gunpowder rocket motor') developed by the design team of engineer I. I. Kartukov. This powerplant was best suited for the short-duration flight modes of the research and experimental flights as regards size, weight and, primarily, because of its high thrust yield and high acceleration ability. Besides, in 1945 there were no well-developed turbojets that were suitable in terms of their weight and thrust. The rocket motor was installed in the rear fuselage; it yielded a thrust of up to 1,500 kgp (3,310 lbst) and the burn time was eight to ten seconds. This ensured level flight with a speed of up to 900 km/h (559 mph).

The landing gear of the Ts-1 had an unusual design. Instead of a conventional undercarriage the aircraft featured a twin-wheel dolly, which was little more than an axle linking the two wheels, and a landing skid mounted on the fuselage centreline. The Ts-1 took off in glider mode, being towed to altitude by a Tupolev Tu-2 bomber. The take-off dolly was jettisoned immediately after lift-off. At an altitude of 5,000-7,000 m (16,400-22,960 ft) the aircraft cast off and the test pilot put the Ts-1 into a dive, the diving angle being between 45° and 60°. At a straight-line section of the diving trajectory the pilot ignited the rocket motor. In this mode the Ts-1 attained a speed of 1,000-1,050 km/h (621-653 mph).

During the few seconds of the steep dive the onboard equipment registered the airflow parameters, photos were taken of the airflow pattern (to this end the wings were covered with wool tufts), and other tasks envisaged by the test mission were accomplished. Then the 1-tonne water ballast that was no longer needed was drained into the atmosphere. The Ts-1, having thus lost half of its weight, performed flight manoeuvres like a normal glider and touched down on the skid. At a take-off weight of 2,039 kg (4,495 lb), the Ts-1's lift-off speed was 150-160 km/h (93-99 mph); with the water jettisoned and the weight reduced to some 1,100 kg (2,425 lb), the landing speed did not exceed 120 km/h (75 mph).

Initial testing of the Ts-1 was conducted by test pilot M. Ivanov. Subsequently Sultan Amet-Khan, Sergey N. Anokhin, Nikolay S. Rybko and other test pilots flew the experimental machine. Nearly 30 flights were performed in the aircraft's straight-wing configuration (LL-1). That was the first stage of the experiments. Further experiments in the programme presupposed alterations in the aircraft's lifting surfaces.

As early as 1946, a design group led by A. V. Beresnev projected two sets of metal (duralumin) wings to be mated with the second Ts-1 fuselage/vertical tail assembly. The new wings had the same area and aspect ratio as the original wooden wings. One set was swept-back, while the wings of the other set were swept forward, with the same leading-edge sweep angle (+30° and –30° respectively). The horizontal tail was also designed anew; it was swept-back, with 40° leading-edge sweep. The Ts-1 configured with swept-back wings having 4° anhedral was designated LL-2, while the version with forward swept-wings and 12° dihedral was known as the LL-3.

The wing panels of the new wing sets had standardised attachment points – that is to say, they were attached to the fuselage by the same fittings as the wooden straight wings of the LL-1. Therefore the changes in the CG position associated with the new position of the MAC and aerodynamic centre after the installation of new wings were compensated by adjusting the amount of water ballast in the forward and rear fuselage tanks. Calculations of the optimum filling of the tanks were made, providing an acceptable stability margin for both wing versions.

The forward-swept wings were installed on the second Ts-1 airframe. During the flights of the LL-3, which numbered nearly a hundred, somewhat greater diving speeds were attained as compared to the straight-wing LL-1; the aircraft reached 1,150-1,200 km/h (715-746 mph) or Mach 0.95-0.97. As a result, the researchers succeeded in studying

the properties of the little-known forward-swept wings and the properties of FSW aircraft in general.

The LL-2, on the other hand, never flew; there was no point in testing it because two turbojet-powered fighters featuring wings with 35° sweepback at quarter-chord – the

Mikoyan/Gurevich I-310 (MiG-15) and the Lavochkin La-15 – had undergone comprehensive tests in 1948. Furthermore, the wooden structure of the first Ts-1 fuselage already showed signs of wear and no longer could guarantee flight safety; as for the second Ts-1 example, it was decided not to fit the swept-back wings to it. To obtain as much data as possible, a multitude of strain gauges and dynamographs was placed in the fuselage and

The LL-1 creates a local dust storm as a test firing of the rocket motor is performed.

Another view of the LL-1 going through a ground run of the rocket motor.

The LL-1 takes off on a towed flight; the towing line attachment is in line with the canopy's front end.

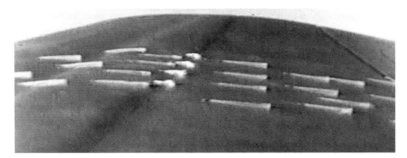

The other config-
uration of the
Ts-1 tested in
flight was the
LL-3 with for-
ward-swept
wings. This is a
close-up of the
starboard wing's
outer section
with airflow visu-
alisation tufts.

In addition to
wool tufts, the
LL-3's wings had
holes along the
trailing edge
through which
smoke was emit-
ted to visualise
the airflow aft of
the wings. Note
how the airflow
curves near the
trailing edge.

The Ts-1 after
after conversion
to the LL-2 con-
figuration with
sweptback wings
and stabilisers.
This version was
not tested in
flight. The air-
craft beyond is a
Ts-25 cargo glider

wings of the Ts-1. Photo cameras were mounted near the wing roots; in flight they registered the oscillations of the wool tufts caused
by local turbulence of the airflow, as well as
the position of shock waves at different
moments of the aircraft's acceleration.

The flights of the Ts-1 in the LL-1 and LL-3
versions and the research conducted on them
provided scientists with unique materials on
the aerodynamic characteristics of aircraft with
different wing types, on the spanwise and
chordwise distribution of the airflow pressure,
on the emergence and travel of shock waves
(compression shocks) and flow separation
areas behind them at critical Mach numbers,
on the relationship between the increase of
the hinge moments of controls and the flight
speed, on the special features and changes of
parameters of boundary layer, and so on.

Apart from studying the properties of transonic flight modes, the Ts-1's interchangeable
sets of wings ensured a uniquely clean experi-

ment providing comparison between different
configurations of the lifting surface. There was
every reason to believe that differences in the
behaviour of the aircraft were caused by the
interchangeable wings.

In the course of the Ts-1's test programme,
in August 1947, Sergey N. Anokhin was
awarded the grade of Test Pilot First Class.

The dimensions of the Ts-1 airframe versions are given on the opposite page.

Tupolev Tu-16 laminar flow research aircraft

A Kazan'-built Tu-16 *Badger-A* bomber coded
'44 Red' (c/n 4200404) was used in several
test and research programmes by LII. In particular, in 1963-70 it served for laminar flow
research with aerofoil-shaped test articles at
subsonic speeds. The test results were used in
refining the Tu-134 *Crusty* short-haul airliner,
which first flew in 1963.

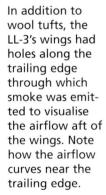

Basic dimensions of the Tsybin Ts-1 in various configurations			
	LL-1 (unswept wings)	LL-2 (swept-back wings)	LL-3 (forward-swept wings)
Length, tail up (less pitot)	8.98 m (29 ft 5½ in)	8.98 m (29 ft 5½ in)	8.98 m (29 ft 5½ in)
Wing span	7.1 m (23 ft 3½ in)	7.1 m (23 ft 3½ in)	7.22 m (23 ft 8¼ in)
Horizontal tail span	2.55 m (8 ft 4½ in)	2.55 m (8 ft 4½ in)	2.55 m (8 ft 4½ in)
Landing gear wheelbase	3.95 m (13 ft 0 in)	3.95 m (13 ft 0 in)	3.95 m (13 ft 0 in)
Wing leading-edge sweep	n.a.	+30°	–30°

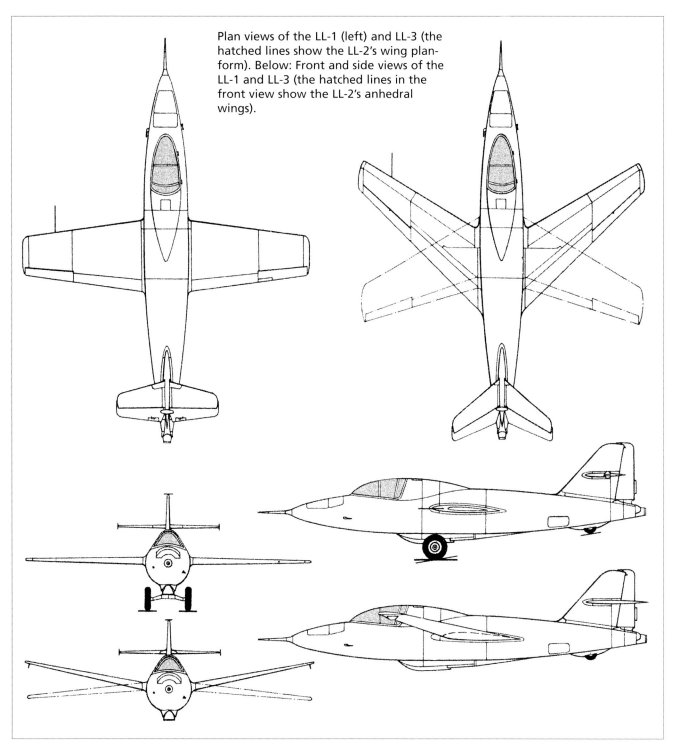

Plan views of the LL-1 (left) and LL-3 (the hatched lines show the LL-2's wing planform). Below: Front and side views of the LL-1 and LL-3 (the hatched lines in the front view show the LL-2's anhedral wings).

47

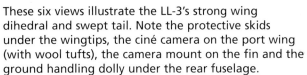

These six views illustrate the LL-3's strong wing dihedral and swept tail. Note the protective skids under the wingtips, the ciné camera on the port wing (with wool tufts), the camera mount on the fin and the ground handling dolly under the rear fuselage.

Two views of Tu-16 '44 Red' (c/n 4200404) with a laminar flow airfoil test article on the centre fuselage. Note the ciné camera on the fin leading edge.

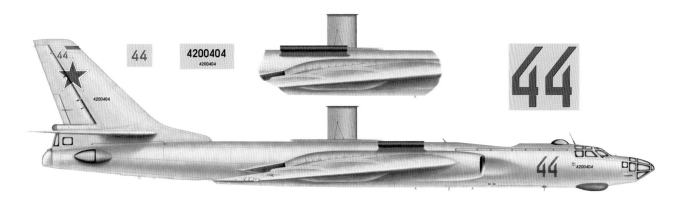

Tupolev Tu-22 aerodynamics research aircraft

In 1972-78 LII used a Tu-22 *Blinder* bomber ('20 Red', c/n ...41605...) for measuring the aerodynamic noise generated by aircraft and working out ways of reducing this noise. Later, in 1984-87, the same aircraft served for laminar flow tests, flying with alternative versions of the wing leading edge aerofoil. The test results proved useful when LII developed another research aircraft, the Tu-154M-ULO (see below). Apparently the same aircraft had the standard radome replaced with a metal nosecone to which test articles were attached from below. It was later recoded '05 Red', serving on until 1991.

The Tu-16 aerodynamics test-bed, '44 Red', with a scrap view of a different configuration.

'05 Red', the Tu-22 aerodynamics testbed, retired at Zhukovskiy.

The same aircraft in earlier guise as '20 Red' with a test article under the nose

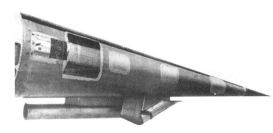

Tupolev Tu-22M3-LL laminar flow research aircraft

Despite being a mean fighting machine, the Tu-22M3 *Backfire-C* long-range bomber found a civil use under the *kon**ver**siya* (adaptation of military technologies for civil use) programme. The bomber's high performance rendered it

49

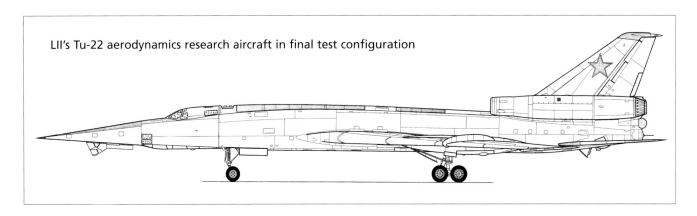

LII's Tu-22 aerodynamics research aircraft in final test configuration

A wind tunnel model of the Tu-22M3-LL carrying a test article piggy-back. On the drawing below, the test article is a model of the Tu-204 airliner.

uniquely suitable for a supersonic airflow research aircraft. Firstly, the Tu-22M3 could reach a speed of 2,300 km/h (1,277 kts) and withstand dynamic pressures up to 5,000 kg/m². Secondly, its variable-geometry wings allowed airflow examination at different sweep angles. Finally, the capacious bomb bay could house a lot of test instrumentation.

In the early 1990s the Tupolev OKB and its branches (the MMZ 'Opyt' experimental plant, the flight test facility in Zhukovskiy and the like), TsAGI and LII began a laminar flow research programme for swept wings. To this end the first prototype Tu-22M3 ('32 Red', c/n 4830156) was fitted with two bolt-on sections on the outer wings. The port one had a standard airfoil section while the starboard one

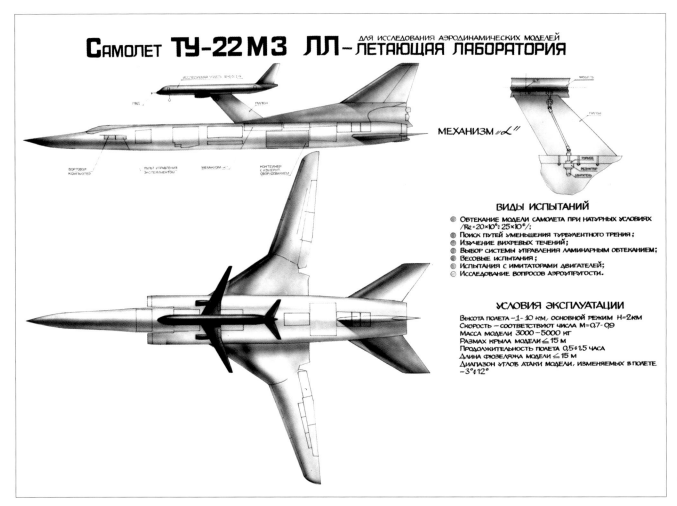

Самолет **ТУ-22М3 ЛЛ**—ЛЕТАЮЩАЯ ЛАБОРАТОРИЯ
ДЛЯ ИССЛЕДОВАНИЯ АЭРОДИНАМИЧЕСКИХ МОДЕЛЕЙ

МЕХАНИЗМ "ᵃ"

ВИДЫ ИСПЫТАНИЙ
- Обтекание модели самолета при натурных условиях /Re=20×10⁶÷25×10⁶/;
- Поиск путей уменьшения турбулентного трения;
- Изучение вихревых течений;
- Выбор системы управления ламинарным обтеканием;
- Весовые испытания;
- Испытания с имитаторами двигателей;
- Исследование вопросов аэроупругости.

УСЛОВИЯ ЭКСПЛУАТАЦИИ
Высота полета –1–10 км, основной режим H=2км
Скорость – соответствуют числа M=0,7–0,9
Масса модели 3000–5000 кг
Размах крыла модели ⩽ 15 м
Продолжительность полета 0,5÷1,5 часа
Длина фюзеляжа модели ⩽ 15 м
Диапазон углов атаки модели, изменяемых в полете –3°÷12°

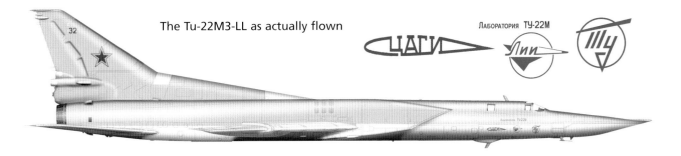

The Tu-22M3-LL as actually flown

Лаборатория ТУ-22М

was a special laminar-flow airfoil developed by TsAGI using the latest techniques. The original fixed refuelling probe was replaced by an elongated fairing to cut drag. This aerodynamics testbed was designated Tu-22M3-LL.

Calculations on boundary layer stability showed that the experimental airfoil could give laminar flow all the way to mid-chord, reducing drag by nearly 50%. The experimental sections were provided with pressure sensors to register pressure distribution and boundary layer departure over a wide range of wing settings, Mach and Reynolds numbers.

On 11-16th August 1992 the Tu-22M3-LL was in the static display at MosAeroShow '92, the first Russian airshow at Zhukovskiy, bedecked with Tupolev, LII and TsAGI logos and showing the gloss black experimental sections on the outer wings. Test flights began in the last days of December 1992; later the aircraft was used in other TsAGI and LII research programmes. Tragically, on 9th September 1994 the Tu-22M3-LL collided with LII's Tu-134AK RA-65760, which was being used as a camera ship during a test flight. The airliner crashed, killing all on board; the bomber landed safely at Zhukovskiy but has been withdrawn from use since then and gradually fell into disrepair.

TsAGI and LII planned to convert a Tu-22M3 for real-flight testing of large scale models carried above the fuselage on a pylon (!). At high Mach numbers, a model with a 10-m (32 ft 9^{45}⁄$_{64}$ in) wing span could attain Reynolds numbers around 20 million, that is, about ten times higher than in an ordinary transonic wind tunnel. This testing technique would be far cheaper than building large supersonic wind tunnels, for which funding is currently unavailable. Yet this plan was not implemented – most probably because of the accident described above.

Tupolev Tu-134LLShP aerodynamics testbed

After replacement of the original 5,800-kgp (12,790-lbst) Solov'yov D-20P-125 turbofans with the definitive 6,800-kgp (14,990-lbst) Solov'yov D-30 turbofans and fitment of an enlarged horizontal tail with a span increased from 9.2 m (30 ft 2¼ in) to 11.8 m (38 ft 8½ in), the first production Tu-134 *sans suffixe* short-haul airliner, CCCP-65600 (c/n 5350002), was used for high-alpha/low-speed tests. To this end it was fitted with a spin recovery parachute, so it can be regarded as an aerodynam-

The Tu-22M3-LL in the static park at MosAeroShow '92.

RA-77114, the Tu-144LL research aircraft, in the paint shop at Zhukovskiy. Note the logos of the programme participants aft of the flight deck.

The logo of the Tupolev JSC which created the Tu-144LL testbed.

The Tu-144LL caught by the camera a second before becoming airborne.

ics testbed if you like. In fact, some Western sources have listed this aircraft under such improbable designations as 'Tu-134LLShP' (which supposedly means *letayushchaya laboratoriya s [protivo]**shtop**ornym para**shoo**tom* – testbed with an anti-spin 'chute) and even 'Tu-134LLChR'. (ChR supposedly stands for *chrezvy**chain**yy re**zhim***, but this Russian term means 'contingency rating' and refers to engines!)

The tailcone was replaced by a long thick fairing with external stiffening ribs which housed the spin recovery parachute canister with a downward-hinging hemispherical cover; electric cables ran along the port side of the fuselage and fairing to the cover/parachute release mechanism. The forward baggage door was modified to serve as an escape hatch for bailing out, should spin recovery become impossible.

Tupolev Tu-144LL research aircraft

The beginning of co-operation between Russia and the Western world in the field of supersonic transport technology dates back all the way to 1965, when neither the Tu-144 *Charger* nor the BAC/Aérospatiale Concorde had flown yet. A long-standing and effective partnership between the Tupolev OKB (ANTK Tupolev) and Aérospatiale began at the 27th Paris Air Show in 1965 where both companies unveiled the projects of their future SSTs – the Tu-144 and the Concorde. Displayed at Le Bourget in model form, the two aircraft had far more in common with each other than they had with American SST projects (the Boeing 2707 and others). At the show, Aérospatiale President Henri Ziegler and the Concorde's chief designer Pierre Sartre had talks with the Soviet Union's Minister of Aircraft Industry Pyotr V. Dement'yev and General Designer Andrey N.

Tupolev; the parties agreed on Soviet-French co-operation in the service introduction of the Tu-144 and the Concorde. A few years later, Aérospatiale and the Tupolev OKB were assigned responsibility for the practical steps aimed at such co-operation.

The basis for practical co-operation was established at the 28th Paris Air Show in 1967 when Pierre Sartre arranged a meeting between Aérospatiale's and Tupolev's chief project aerodynamicists, Fage and Cheryomukhin. Little by little the scope of information exchange widened; both sides concluded that there was no point in playing hide-and-seek and started sharing information freely, aiding each other immensely. Eventually the cessation of Tu-144 operations and the fact that the Concorde was operated on just two trans-Atlantic routes killed the prospects for any further development of the first-generation SSTs.

Still, the interest in a second-generation successor lived on. Eventually, however, the insurmountable hurdle turned out to be not of a technological nature – it was the environmental issues (primarily the sonic boom and its effect on living creatures) which environmental protection crusaders hyped up to such a degree that flights of SSTs over land were banned altogether. An international association known as the Group of Eight – formed by the Boeing Commercial Airplane Group, McDonnell Douglas, British Aerospace, Aérospatiale, Deutsche Aerospace (as part of the Airbus Industrie consortium), Alenia, an association of Japanese aerospace companies (Fuji Heavy Industries, Mitsubishi Heavy Industries, Kawasaki and others) and ANTK Tupolev – came to the conclusion that creating an affordable second-generation SST that

would meet the new tough sonic boom limits was technically impossible. If the SST were to fly at subsonic speeds on the overland legs of the journey to avoid booming someone, it would have to have a non-stop range of some 12,000 km (7,450 miles) to be economically viable, which was likewise impossible for the time being. Thus all further attempts to create a successor to the Tu-144 and the Concorde were shelved for many years.

Since a supersonic airliner is an extremely expensive project, the design methods and actual design features that go into such an aircraft should be verified on flying testbeds if at all possible to reduce the technical risk. The larger the testbed aircraft is and the closer its performance comes to the projected SST's design performance, the more valuable it becomes in its research capacity. Such complex issues as calculating the skin temperature of an integral tank filled with cold fuel in kinetic heating conditions, or calculating the heating intensity of the fuel being fed to the engines, can only be checked out in actual supersonic flight which will yield as authentic a result as

The Tu-144LL makes a slow flypast with the retractable canards deployed.

RA-77114 on short finals to runway 30 at Zhukovskiy.

The Tu-144LL sits on the taxiway at Zhukovskiy after being towed into position by a KrAZ-255B 6x6 lorry, with an SPT-114 self-propelled gangway positioned near the forward entry door. The nose is raised and the canards retracted.

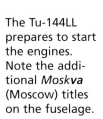

The Tu-144LL prepares to start the engines. Note the additional *Moskva* (Moscow) titles on the fuselage.

you can possibly get, allowing errors in calculation methods to be corrected. This is why in the early 1990s the Tupolev OKB started work on a research aircraft designated Tu-144LL and intended to provide data for the future SST-2.

Working hard and persistently, by 1993 the ANTK Tupolev engineers had finalised the Tu-144LL's outlook and completed a set of drawings for the modification job. At the 40th Paris Air Show in June of that year ANTK Tupolev and the US company Rockwell International struck the first international co-operation agreement concerning development of the Tu-144LL research aircraft.

At that point ANTK Tupolev had three Tu-144Ds in flyable condition – CCCP-77112, CCCP-77114 and CCCP-77115. CCCP-77114 (c/n 10082) was selected for conversion; CCCP-77112 (c/n 10072) was to serve as a ground test rig in support of the programme, while CCCP-77115 (c/n 10091) was in 'hot reserve'.

The conversion involved a change of powerplant: the Tu-144D's 20,000-kgp (44,090-lbst) Kolesov RD36-51A non-afterburning turbojets gave place to Kuznetsov NK-321 afterburning turbofans, a version of the Tu-160's NK-32 engine rated at 13,000 kgp (28,660 lbst) dry and 25,000 kgp (55,115

lbst) reheat. This was a forced measure, as the RD36-51A engine custom-made for the Tu-144D was long since out of production and the surviving examples had a remaining service life of just a few dozen hours. The new engines necessitated manufacturing new rear portions of the engine nacelles and the inlet ducts, installation of additional equipment (the ESUD-32-1 full authority digital engine control system (FADEC) and the SKSU-32-1 engine monitoring system), local reinforcement of the wings to take the more powerful engines and some changes to the aircraft's systems.

It has to be said that even when a Russian-US inter-government commission co-chaired by the then Prime Minister Viktor S. Chernomyrdin and Vice-President Albert L. Gore had passed documents giving the Tu-144LL official status, the aircraft remained the subject of a controversy. 'Why the Tu-144? – some people clamoured. – Why not the Concorde?' In the USA there was a powerful anti-Russian lobby which maintained that the funds allocated by the US Congress for the development of a supersonic research aircraft should not be given to Russia and that the Concorde should be used as the basis for this programme instead. In 1995, responding to

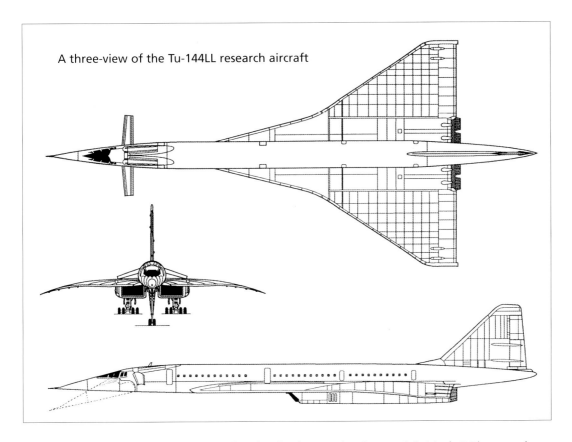

A three-view of the Tu-144LL research aircraft

these statements, Louis Williams, head of NASA's High Speed Research Program, gave several interviews to the world's leading aviation magazines, explaining why the Tu-144 had been selected over the Concorde.

The supersonic research aircraft should come as close as possible to the future SST-2 in size, performance and aerodynamic efficiency, said Williams and gave figures to prove his point. The Tu-144's wing area was 507 m² (5,450 sq ft) versus the Concorde's 425 m² (4,570 sq ft); its cruising speed of Mach 2.35 (versus the Concorde's Mach 2.2) came closer to the SST-2's envisaged speed of Mach 2.4, and its lift/drag ratio at Mach 2.0 was 8.1 versus 7.3 for the Concorde.

The Tu-144 had a record-high share of titanium in its structure accounting for nearly 20% of the airframe weight, retractable canards and a unique wing/fuselage structural design. All this made it a better candidate than the Concorde.

The success of the Tu-144LL programme was in no small part due to good interaction

A fine shot of the Tu-144LL travelling at supersonic speed taken from a fighter used as a chase plane.

The Yakovlev AIR-4MK experimental aircraft registered CCCP–Э–31 (that is, SSSR-E-31 in Cyrillic characters).

This view shows the AIR-4MK's two-section full-span flaps at maximum deflection.

Another view of the AIR-4MK, showing the so-called floating ailerons with anti-flutter booms at the wingtips.

between the partners. The organisational aspects were handled by IBP Corporation headed by Judith DePaul.

The conversion work took place in 1995-96. Reregistered RA-77114 and wearing a white colour scheme with a blue/red stripe on the fin and Russian and US flags, the Tu-144LL made its first flight on 29th November 1996; the last flight took place on 28th February 1998. In the course of 27 test flights the aircraft participated in eight flight experiments. These were:

• defining the aircraft's temperature balance;

• exploring the basic aerodynamic parameters;

• exploring the temperature of the powerplant in different flight modes;

• exploring the influence of ground effect on low aspect ratio wings;

• assessing the aircraft's stability and handling;

• measuring the structure's acoustic loads and cabin noise levels;

• exploring the temperature fields in the fuel system;

• exploring the deformation of the wings in flight.

The Yakovlev AIR-4MK experimental aircraft

Additionally, two unique ground experiments associated with the powerplant were performed. These were:

• defining the optimum configuration of an advanced supersonic air intake as regards pressure ratio, airflow irregularity and pulsation;

• studying the integration of short supersonic inlets and the engines.

The test results made it possible to evolve methods of calculating the thermal load, aerodynamics, acoustic load and so on for the next generation of SSTs and supersonic business jets.

The Tu-144LL was displayed statically at the MAKS-97 and MAKS-99 airshows in Zhukovskiy. On the latter occasion the aircraft wore the legend '*Mos**kva*' (Moscow) on the fuselage.

Tupolev Tu-154 laminar flow testbeds

A Tu-154B-1 *Careless* medium-haul airliner (CCCP-85119, c/n 75A119) was converted for laminar flow research by LII in the 1980s. Confusingly referred to by some sources as a Tu-154LL (although it had nothing to do with the Tu-154s involved in the Buran space shuttle programme), this aircraft was retired by 1992.

Plans were in hand to convert the second prototype Tu-154M (by then reregistered RA-85606, c/n 84A701) into a similar laminar flow research aircraft designated Tu-154M-LL (ULO). ULO stood for *oopravl**en**iye laminarnym obtek**a**niyem* – laminar flow control; the latter was to be achieved by boundary layer suction. Apparently this was never done and RA-85606 languished at Zhukovskiy with an expired certificate of airworthiness.

Yakovlev AIR-4MK (E-31) experimental aircraft

In the early 1930s the Civil Air Fleet Research Institute (NII GVF – *Na**ooch**no-is**sle**dovatel'skiy insti**toot** Grazh**dahn**skovo voz**doosh**novo **flot**a*) began experiments aimed at reducing the landing speed of aircraft by means of various high-lift devices. Among the means envisaged was the use of full-span landing flaps. To make as much of the wing trailing edge as possible available for these flaps, the normal ailerons were to be replaced by so-called 'floating' ailerons which were, in fact, all-movable wingtips rotating on axles. To test this idea, NII GVF researchers modified a Yakovlev AIR-4 light aircraft which is sometimes referred to as AIR-4MK (*mekhani**zee**rovannoye kry**lo** – 'mechanised wings', the Russian term of that period for wings with high-lift devices). The new wings were developed by B. N. Zalivatskiy and Leon M. Shekhter. The aircraft was also known as E-31, after its registration CCCP-Э-31 (SSSR-E-31 in Cyrillic characters).

(Note: Under the Soviet civil aircraft registration system used between 1932 and 1958, the CCCP- country prefix was followed by a letter designating the aircraft's operator plus up to four figures. In this case the operator designator is an E (for *eksperimen**tahl'**nyy* – experimental) denoting NII GVF. Cf. CCCP-Лxxxx (that is, SSSR-L in Cyrillic characters, derived from *lee**ney**nyy* [*samo**lyot**] – aircraft in airline service) denoting the Main Directorate of the Civil Air Fleet (GU GVF – ***Glahv**noye oopravl**en**iye grazh**dahn**skovo voz**doosh**novo **flot**a*) which operated scheduled passenger/cargo services; CCCP-Hxxx (the Cyrillic N) for the Main Directorate of the Northern Sea Route (GU SMP – ***Glahv**noye

The UT-1E, aka UT-1 (15). The 'floating ailerons' are clearly visible.

oopravlenya **Sev***ernovo morsko*vo pu**tee**), that is, the Polar Aviation branch; CCCP-Cxxxx (the Cyrillic S) for Osoaviakhim, the organisation running the Soviet air clubs; CCCP-Axxxx for the agricultural division; CCCP-Kxxx (derived from **Kras**nyy krest – Red Cross) for the People's Commissariat of Health and so on. The rendering of the registrations as actually applied is used throughout.)

The wing span of the AIR-4MK was increased from 11.0 m (36 ft 1⁵⁄₆₄ in) to 12.5 m (41 ft 0⅛ in). 10.3 m (33 ft 9½ in) were occupied by the flaps. Outboard of these, the wingtips, each of them having a span of 1.1 m (3 ft 7⁵⁄₁₆ in), were transformed into 'floating ailerons' with the axis of rotation close to the leading edge. The ailerons could be deflected differentially within ±15° for roll control; upon completion of the manoeuvre they switched to 'floating' mode, aligning themselves with the airflow. With these ailerons, the wing area increased from 16.5 to 18 m² (from 177.6 to 193.8 sq ft).

The aircraft's empty weight was increased by 40 kg (88 lb) and the AUW by 45 kg (99 lb).

The tests, which began in 1933, showed that the use of 'floating ailerons' was a basically sound idea. Thanks to the large flap area the landing speed was reduced from 66 km/h (41 mph) to an unprecedented 34 km/h (21 mph) at the cost of an insignificant drop in the maximum speed, which was reduced from 150 to 145 km/h (from 93 to 90 mph). One source claims, though, that the cruising speed was increased by 7%.

Yakovlev UT-1E, alias UT-1 (15) experimental aircraft

In 1938 a modified example of the Yakovlev UT-1 single-seat advanced trainer, known as UT-1E (the E presumably stands for *eksperimen***tahl'***nyy* – experimental), or UT-1 (15), was used for experiments at TsAGI. Like the AIR-4MK described above, it was fitted with

Another UT-1 used for aerodynamics research. Note the ciné cameras mounted on struts to capture the airflow over the wings covered with wool tufts.

the so-called floating ailerons. The wings of the UT-1E had a momentless airfoil section developed by F. G. Glass which was the subject of studies and experiments at TsAGI. The aircraft designer and historian Vadim B. Shavrov erroneously refers to this aircraft as the AIR-15 or UT-15 in his reference book on Soviet aircraft, describing it as a racing aircraft with the airfoil developed by F. G. Glass which was *'tested in early 1938 and transferred to TsAGI for research'* (he mentions separately a UT-1 version with floating ailerons, as though this were a different aircraft).

Yakovlev UT-1 development aircraft with VVS wing airfoil

A single UT-1 was experimentally fitted with wings incorporating a VVS (Air Force) airfoil. The modification proved to be disappointing because the landing speed increased.

Yakovlev UT-1 development aircraft with retractable undercarriage and enclosed cockpit

In 1938 an example of the UT-1 powered by an unidentified version of the Shvetsov M-11 five-cylinder radial engine was equipped with a retractable undercarriage (the main units retracted inwards) and an enclosed cockpit. It makes an interesting comparison with the later AIR-18 from which it differed only in being fitted with a radial engine rather than an MV-4 (Renault Bengali 4) four-cylinder inverted in-line engine and in having no wheel well doors on the undercarriage legs.

Yakovlev UT-1 development aircraft with automatic flaps (1941 and 1944)

In 1941 another UT-1 was fitted with TsAGI-type automatic landing flaps. Another source mentions the testing of this aircraft or a similarly modified UT-1 at the Flight Research Institute (LII) of the People's Commissariat of Aircraft Industry (NKAP – *Na**rod**nyy komissari-**aht** aviatsi**on**noy pro**mysh**lennosti*) in 1944.

Yakovlev Yak-7L aerodynamics research aircraft

In the early 1940s, trying to improve the performance of the production Yak-7 fighter, the Yakovlev design bureau (OKB-115) fitted a single example of the aircraft with laminar-flow wings utilising low-drag airfoil sections. Designated Yak-7L (*lami**nar**noye kry**lo** –* laminar-flow wings), the aircraft showed a speed improvement of 15-20 km/h (9.3-12.4 mph) over the standard fighter. Nevertheless, the Yak-7L remained a one-off.

Yakovlev Yak-28 aerodynamics testbeds

An unidentified Yak-28 was used for measuring the pressure distribution and pressure fluctuations on the wings and control surfaces in various flight modes. Various means of airflow visualisation were used during these experiments.

Presumably the same aircraft was used for studying the shock waves on the wings under natural conditions.

The logo of the Yakovlev OKB which used a lot of flying testbeds in the development of its aircraft.

The Yak-7L development aircraft with laminar-flow wings, seen here in front of one of LII's hangars.

Above and right: Yak-28P '20 Blue' used for aerodynamic research by LII.

The instrument panel of Yak-28P '20 Blue'.

Far right: Close-ups of the nose of '20 Blue', showing the nose probe and pylon-mounted test article.

Another Yak-28 was used for measuring the airframe's structural strength in flight. The aircraft was modified by installing aerodynamic force generators at the wingtips, supplemented by a set of measuring and data recording equipment.

A Yak-28L *Brewer-B* tactical bomber referred to as 'No. 720' (c/n 2920720?) served in 1967-76 for studying the airflow around suspended models of aircraft fuselages at subsonic and supersonic speeds. The results of these tests were used in the development work on the Tupolev Tu-128 *Fiddler* heavy interceptor.

A Yak-28P *Firebar* interceptor coded '20 Blue' was converted into a testbed of unknown purpose by LII. The radome was replaced by a conical metal nosecone terminating in a short fat probe with an ogival tip, and a cylindrical object resembling a scale model of a supersonic air intake with a conical centrebody was carried under the nose on a short pylon. A similar arrangement had been used on the Tu-22 testbed mentioned earlier.

Yakovlev Yak-28P development aircraft with additional ailerons

For various reasons the Yak-28P interceptor was subject to strict speed limits. Above 7,000 m (22,965 ft) this was due to flutter and structural strength considerations; below 7,000 m it was due to aileron reversal which, as mentioned earlier, had never been completely eliminated. Yakovlev engineers offered an ingenious way of correcting this latter deficiency – by installing small triangular additional ailerons at the extreme wingtips (that is, outboard of the outrigger gear fairings).

The prototype of the 'long-nosed *Firebar*' (with the late-style conical radome replacing the original short ogival radome), '51 Yellow', was converted for testing the idea; hence the aircraft could be described as a CCV. Tests held in 1966 showed that the arrangement worked as intended, but no decision to update new-build or in-service Yak-28Ps was taken.

2 Aircraft systems testbeds

A number of aircraft were modified for experiments with in-flight refuelling (IFR) systems, innovative landing gear designs and so on. These aircraft are listed here, grouped by application.

1. Refuelling system testbeds

Antonov An-12BP refuelling system testbed

In 1972-74 LII used a Tashkent-built An-12BP *Cub* medium transport coded '85 Red' (c/n 6344204) for exploring the parameters of IFR systems and the conditions in which contact with the tanker was possible. It is not known whether any actual IFR system components were installed; fact is that eventually the aircraft was reconverted to standard configuration and returned to the Air Force, ending up in the Ukraine.

Another (or possibly the same) An-12 was involved in the tests of the UPAZ-1A Sakhalin refuelling pod developed by OKB-918 (this system is described in more detail later).

Ilyushin IL-28 refuelling system testbeds

a) fighter IFR system integration: In late 1957 a Voronezh-built IL-28 *Beagle* tactical bomber coded '01 Red' (c/n 2402101) was converted into a makeshift tanker trainer used for testing the probe-and-drogue IFR system developed by OKB-918 led by Semyon M. Alekseyev. The 'tanker' worked with the receptacle positioning testbed converted from a MiG-19 fighter (see below).

An experimental winch emulating a hose drum unit (HDU) was installed in the IL-28's bomb bay, paying out a 5-mm (0³⁄₁₆ in) steel cable with a drogue of 640 mm (2 ft 1³⁄₁₆ in) diameter to a point 42 m (137 ft) beyond the bomber's tail. Initially a 36-kg (79-lb) unstabilised drogue was used. After the first four flights, however, it was replaced with a drogue incorporating a stabilising device 100 mm (3¹⁵⁄₁₆ in) wide mounted at 60 mm (2²³⁄₆₄ in) from the base. Both models had a lock for engaging the probe.

Test pilot Nikolay O. Goryainov (who has the distinction of being the first Soviet pilot to

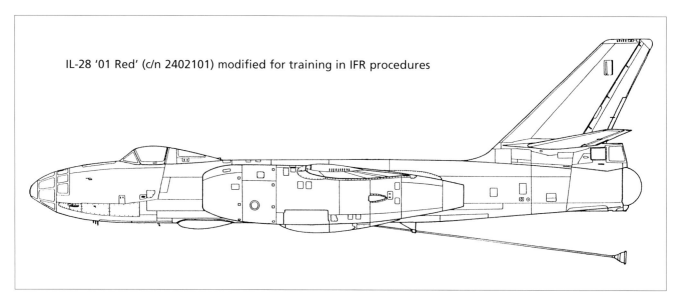

IL-28 '01 Red' (c/n 2402101) modified for training in IFR procedures

successfully refuel a heavy bomber in flight) was appointed project test pilot for the 'tanker trainer'. On 28th August 1957 he made a flight to check the operation of the winch. The drogue was deployed at 7,000 m (22,965 ft) and 400, 450, 500 and 550 km/h (216, 243, 270 and 297 kts) IAS. After that, LII test pilots Sergey F. Mashkovskiy, Pyotr I. Kaz'min and Sergey N. Anokhin made ten 'refuelling' flights on 18th, 20th, 24th and 27th September, 3rd, 16th and 30th October, 1st November, 7th and 27th December (see MiG-19 entry for more details). The tenth flight had to be cut short when the drogue entered the fighter's air intake and collapsed, the debris damaging one of the engines. The trials showed that the chances of making contact depended mainly on the drogue's stability which left much to be desired.

b) bomber IFR system tests: In due course the Soviet military posed more stringent requirements which the IL-28 could no longer meet. One of the *Beagle*'s greatest deficiencies was its inadequate range. However, at that stage it was deemed inadvisable to retire the many IL-28s in Soviet Air Force service, so someone suggested retrofitting the bombers with the probe-and-drogue refuelling system. To this end two more IL-28s were converted for real-life IFR system tests. One of them was a tanker with a dummy HDU in the bomb bay, while the other *Beagle* featured a fixed refuelling probe offset to port above the navigator's station. The two aircraft made successful contacts but the system was not fitted to operational IL-28s because Aleksandr S. Yakovlev's OKB-115 brought out the more promising Yak-129 supersonic tactical bomber which eventually entered production and service as the Yak-28 *Brewer*.

LII materials mention an IL-28 with a c/n ending 418 which was used to explore the possibility of developing a controllable refuelling drogue with a view to facilitating contact with the receiver aircraft. The tests of a non-functional automated IFR system on this aircraft proceeded in 1958-59.

Three stills from a ciné film showing the two IL-28s modified for IFR system trials. The receiver aircraft IFR probe is offset to port.

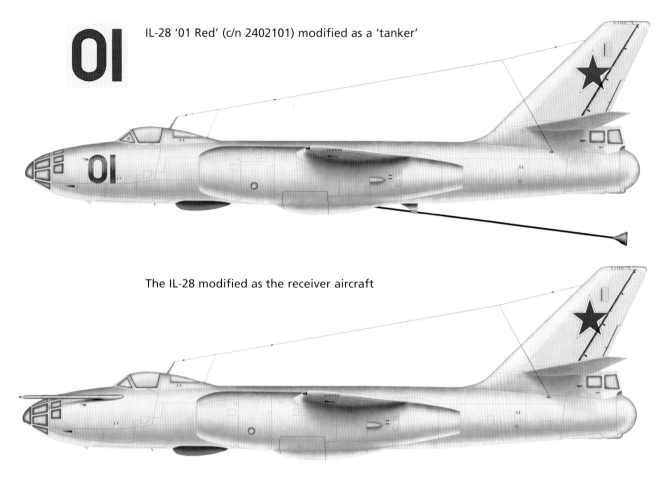

01

IL-28 '01 Red' (c/n 2402101) modified as a 'tanker'

The IL-28 modified as the receiver aircraft

Mikoyan/Gurevich MiG-15*bis* refuelling system testbeds

The first Soviet experiments with in-flight refu-elling (starting in the pre-war years) were con-cerned with bombers. However, the début of jet fighters with thirsty engines evoked an interest in giving tactical aircraft IFR capability as well. This could enhance their combat potential in many ways, such as:

• extending on-station loiter time beyond the aircraft's normal endurance;

• escorting heavy bombers;

• delivering tactical strikes (including nuclear strikes) at ranges exceeding the aircraft's unrefuelled combat radius;

• extending the intercept range of inter-ceptors;

• increasing the chances of coming back safely from a long-range mission by topping up the tanks on the way home.

In-flight refuelling studies in the USSR began in earnest in 1948. First, a team headed by Vladimir S. Vakhmistrov developed the so-called 'system of crossing ropes' based on a sys-

Kuibyshev-built MiG-15*bis* '342 Blue' was one of several converted for testing a probe-and-drogue in-flight refuelling system. The intake lip and fin cap are painted blue. The background has been retouched away here.

Another view of '342 Blue' at GK NII VVS during trials. The fighter carries 250-litre drop tanks.

Close-up of the fighter's nose and the offset IFR probe.

tem developed by the British company Flight Refuelling Ltd., but this proved cumbersome, unreliable and difficult to maintain. The next step was a wingtip-to-wingtip IFR system developed by a team headed by LII test pilots Igor' I. Shelest and Viktor S. Vasyanin in 1948. This was used on several Soviet aircraft types and is described in the appropriate entries below. The then brand-new MiG-15 was considered to have an adequate combat radius and hence initially not regarded as a candidate for IFR. (Note: Some sources mention that the Shelest/ Vasyanin IFR system was tested on the MiG-15 as well, but no proof of this is available.)

However, the wingtip-to-wingtip system had some serious shortcomings, including a

fairly complicated engagement procedure and a low fuel transfer rate. Despite the stabilising parachute, the hose thrashed around in the tanker's wingtip vortex and could get caught in the receiver aircraft's aileron – with disastrous results. Also, the need to provide the Tupolev Tu-4 *Bull* heavy bombers with fighter escort created the need to extend the MiG-15's range; the MiG-15S*bis* (*izdeliye* SD-UPB) escort fighter with enlarged drop tanks could not solve the problem. Hence the Soviet Air Force considered using a probe-and-drogue system based on the *Burlaki* captive escort fighter concept. The latter was co-developed by Aleksandr S. Yakovlev's OKB-115 and OKB-30 (the design office of the MMZ

Seen from the Tu-4 tanker's refuelling system operator's station, '342 Blue' makes contact with the drogue.

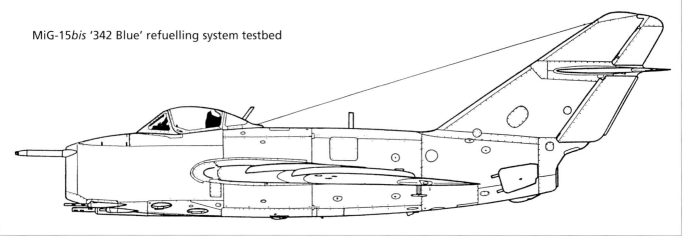

MiG-15*bis* '342 Blue' refuelling system testbed

Tu-4 c/n 2805204 converted into a two-point tanker refuels two *Fagot-Bs* equipped with IFR probes. One is '342 Blue'; the other is '074 Black' (c/n 3810704) sporting a Komsomol Crew badge.

No. 30 aircraft factory at Moscow-Khodynka). It involved fitting the fighter with a telescopic nose probe whose barbed tip was 'fired' into a drogue deployed by the Tu-4; when the probe locked into place the fighter pilot shut down his engine and the fighter was towed by the Tu-4 like a glider. If enemy fighters attacked, the fighter pilot started his engine, broke contact with the bomber and engaged the enemy, subsequently hooking up to the bomber again for the journey home. (In 19th-century Russia, the *burlaki* were teams of strongmen whose job was to haul barges up rivers by means of ropes; the analogy with the towed fighter concept was obvious.)

Initially OKB-115, OKB-30 and Pavel O. Sukhoi's OKB-134 modified the existing Burlaki system by adding new elements. The fighter's 'harpoon' incorporated a valve and plumbing to the fuel system. The bomber was equipped with three kerosene tanks, a pump and an inert gas pressurisation system to reduce the risk of fire and explosion if hit by enemy bullets. Once the fighter had made contact with the tanker's towing drogue, a hose terminating in a smaller drogue was paid out along the towing cable and the fighter accelerated, locking the two drogues together. (The original drogue was modified so as to allow fuel to pass through it into the probe.) 1,210 litres (266.2 Imp gal) of fuel could be transferred in six minutes. When refuelling was completed the smaller drogue was automatically disengaged and the hose rewound.

Two of the aircraft used for service trials of the Burlaki system – Kuibyshev-built Tu-4 '41 Red' (c/n 1840848) and Novosibirsk-built MiG-15*bis* '2204 Red' (c/n 2215304) – were converted for IFR trials which were held at LII between 24th September 1954 and 2nd March 1955. The tanker was piloted by Aleksandr A. Yefimov, with A. I. Vershinin as the refuelling system operator (RSO); the jet was flown by Sergey N. Anokhin and Fyodor I.

Boortsev. V. I. Stepanov was the engineer in charge by the Yakovlev OKB, while V. S. Yolkin was assigned in the same capacity by LII.

The programme involved ten flights on the MiG-15*bis*, including five contacts at 2,000 m (6,560 ft) and 4,000 m (13,120 ft); on three occasions, fuel was actually transferred. An attempt to repeat the performance at 8,500 m (27,890 ft) failed, however, because the system's rubber components froze up and became inflexible. It was established that the modifications had virtually no adverse effect on the performance and handling of both the tanker and the receiver aircraft.

Yet, the 'wet Burlaki' system was considered excessively complex, and as early as in December 1952 another design bureau, OKB-918 led by Semyon M. Alekseyev, took on the flight refuelling problem. This bureau, which absorbed the entire Vakhmistrov team, later became the NPP *Zvezda* ('Star' Scientific & Production Enterprise) company led by Guy Il'yich Severin and best known for the K-36 ejection seat fitted to almost all current Russian combat aircraft.

Another view of the tanker, with '342 Blue' receiving fuel from the starboard drogue. The Tu-4's tail barbette mounts a ciné camera pack instead of cannons to record the procedure.

MiG-15*bis* '074 Black' approaches the tanker\s drogue at high altitude.

Here the fighter has gone forwards a bit too far, bending the hose the wrong way. Note the ciné camera fairing on the fighter's wing.

The contact with the port drogue as recorded by the MiG-15's camera. Too bad that the wing fence obscures the serial, which begins with a 5.

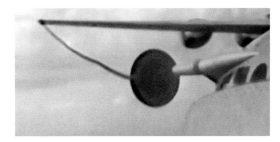

'618 Blue', yet another MiG-15*bis* involved in the probe-and-drogue IFR system trials.

'074 Black' is refuelled at high altitude. Note the stripes on the hose showing the operator how much of it has been deployed.

Again, a late-production Kuibyshev-built Tu-4 (c/n 2805204) – this time not one of the Burlaki testbeds – was converted into a tanker at plant No. 18 in Kuibyshev. The arrangement proposed by OKB-918 differed from the Yakovlev/OKB-30 system in two important respects. Firstly, it was much simpler, with only one drogue and hose (just like the system developed by Flight Refuelling Ltd. which is in worldwide use today). The engagement was much easier to perform and fuel transfer could begin immediately after the probe had locked into the drogue. Secondly, the aircraft was a

two-point tanker. Two HDUs were installed in the forward bomb bay, with the hoses running inside the wings and exiting from specially-modified wingtips. The refuelling system operator sat in the tail gunner's station; the tail guns were replaced by a ciné camera unit to record the refuelling sequence.

Three *Fagot-Bs* serialled '342 Blue' (c/n 123042 or 133042), '5…' and '618 Blue', as well as a modified Moscow-built MiG-15 *sans suffixe* ('074 Black', c/n 3810704), were fitted with fixed telescopic IFR probes offset to port on the intake upper lip; the conversion work was done by the Novosibirsk factory in May 1952. Two aircraft ('5…' and '074 Black') had faired ciné cameras on top of the wings to capture the moment of contact with the drogue. Test flights began in 1953, with a considerable delay because of late equipment deliveries for the tanker conversion; subcontractors, such as OKB-140, OKB-30 and plant No. 279, were responsible for the delay. LII test pilots Sergey N. Anokhin and Vladimir M. Pronyakin flew the fighters. Engineers V. Ya. Molochayev and S. N. Rybakov, as well as LII test pilots Pyotr I. Kaz'min, Stepan F. Mashkovskiy and Leonid V. Chistyakov, participated actively in the refinement of the system.

At first, Mikoyan engineers were apprehensive about having the probe near the intake, fearing the drogue would generate excessive turbulence at the air intake lip and provoke a compressor stall. These fears were possibly caused by knowing that in the USA, a Republic F-84 Thunderjet fitted experimentally with the probe-and-drogue refuelling system had the probe mounted on the starboard wing, well clear of the nose air intake. However, trials showed these fears were unfounded.

Several versions of the hose had to be tried before the system was satisfactory. The original hose incorporating a reinforcing wire spiral proved prone to plastic deformation and not durable enough. A 'soft' hose with no reinforcing wire, on the other hand, flexed excessively and fighter pilots found that just a little turbulence made 'hitting the tanker' very difficult. Another problem was the considerable amount of fuel remaining in the hose after the transfer pumps were shut down; immediately after breaking contact with the tanker the fighter was liberally doused with fuel, some of which even entered the cockpit. Still, the system was simple, reliable, could be automated and offered a high fuel transfer rate.

The combination of Tu-4 tanker and two MiG-15*bis* receivers was presented twice for state acceptance trials but failed both times because of problems with the supporting rollers inside the tanker's wings. These rollers caused hose oscillation and failure of the fighters' IFR probes due to the whiplash effect of the hose. Also, unlike the USAF, the Soviet Air Force had no need to fly its fighters over long distances. However, once again the probe-and-drogue IFR system was used successfully on strategic aircraft – the Myasishchev 3MN/3MS *Bison-B* and 3MD *Bison-C* heavy bombers, most members of the Tupolev Tu-95/Tu-142 *Bear* family, the Tu-126 *Moss* airborne warning and control system (AWACS), the Tu-22KD/RD/PD/UD *Blinder* and Tu-22M2 *Backfire-B* supersonic long-range bombers and so on.

Experiments continued with modified MiG-17 and MiG-19 fighters, using both the wing-to-wing and probe-and-drogue systems (these development aircraft will be described later), but in the 1960s, development work was put on hold due to the higher priority given to missile systems during the Khrushchov era. It was not until the early 1980s that Soviet tactical aircraft received IFR capability at last when a versatile probe-and-

Mikoyan/Gurevich MiG-19 (*izdeliye* SM-9D or SM-10) development aircraft

The range of the standard MiG-19 *sans suffixe* fighter – about 2,000 km (1,240 miles) with two 760-litre (176.2 Imp gal) drop tanks – was insufficient for a number of missions. Hence in May 1954, even before the *Farmer-A* entered production, the Council of Ministers issued a directive ordering the Mikoyan OKB to develop an IFR system and submit a MiG-19 equipped with this system for state acceptance trials. The same document tasked the Tupolev OKB with

'415 Red' (c/n N59210415), the SM-10 IFR system testbed. Note the bullet fairing of the IFR receptacle on the port wing.

The SM-10 makes contact with a Tu-16Z tanker. Note the second bullet fairing on the starboard wingtip; the aircraft was apparently fitted with two receptacles.

drogue system suitable for both tactical and strategic aircraft was developed. Four decades earlier, it seemed that the solution lay just a few years ahead.

Mikoyan/Gurevich MiG-17 refuelling system testbed

A single MiG-17 *sans suffixe* (*Fresco-A*) coded '153 Red' was fitted experimentally with a fixed offset IFR probe on the air intake upper lip in similar manner to the MiG-15s described above. Unfortunately, no information is available on the tests of this aircraft.

Top: The MiG-19 IFR probe testbed ('10 Red', c/n N54210110) as seen by the gunner of the IL-28 'tanker'. This view shows clearly the four dummy probes.

Above: The same aircraft makes contact with the 'tanker', using the nose probe. This location for the probe was clearly the most effective.

developing a tanker version of the Tu-16 twin-jet bomber. This emerged as the Tu-16Z (*zaprahvshchik* – tanker); like the basic bomber version, the tanker was code-named *Badger-A*.

The Tupolev OKB had by then gained some experience with the wingtip-to-wingtip IFR system developed by a team headed by LII test pilots Igor' I. Shelest and Viktor S. Vasyanin. The tanker deployed a hose stabilised by a drogue parachute from one wingtip; the receiver aircraft approached in echelon formation and deployed a special grapple under the opposite wingtip, placing it over the hose. Then the tanker's RSO slowly rewound the hose until a fitting at the end locked into a receptacle under the receiver aircraft's wingtip, activating a switch. The receiver accelerated so that the hose formed a loop and rotated the receptacle through 180°, opening a valve and activating another switch, whereupon fuel transfer under pressure could begin.

Design studies of a similar system for the Tu-16 began as early as the autumn of 1953. On 17th September 1953 MAP issued order No. 44 concerning the design of a flight refuelling system based on the Tu-16. On 26th

May 1954 the Council of Ministers issued directive No. 1013-438 backed up by MAP order No. 354 on 3rd June; these documents required the Tu-16Z tanker to be submitted for state acceptance trials in the fourth quarter of the year. The idea was that the tanker would refuel both ordinary *Badger-A* bombers and the escorting fighters.

The Mikoyan OKB had also experimented with modified MiG-15*bis* and MiG-17; however, these had used the probe and drogue system, so Mikoyan had to start from scratch with the wingtip-to-wingtip system. In mid-1955 two standard *Farmer-As* serialled '316 Red' (c/n N59210316) and '415 Red' (c/n N59210415) were stripped of armament and fitted with a refuelling receptacle under the port wingtip; outwardly the modification could be recognised by a small bullet fairing on the leading edge. Because of the greater endurance the oxygen supply was increased to 18 litres (4 Imp gal). In case of need the aircraft could be easily reconverted to standard configuration in field conditions by removing the refuelling receptacle and extra oxygen bottles and replacing the port wingtip. The tanker was standard, except that the hose diameter was reduced from 88 to 50 mm (from $3^{15}/_{32}$ to $1^{31}/_{32}$ in).

Bearing the in-house designation '*izdeliye* SM-10', the modified MiG-19s entered flight test in the autumn of 1955; '316 Red' and '415 Red' were designated SM-10/1 and SM-10/2 respectively. Some sources refer to the aircraft as *izdeliye* SM-9D (*dozaprahvka* – refuelling). Mikoyan OKB test pilot Vladimir A. Nefyodov flew the SM-10s during Stage I of the manufacturer's tests; A. Komissarov and Igor' I. Shelest were appointed project engineers at the Mikoyan OKB and LII respectively.

The Tu-16Z's powerful fuel transfer pumps delivered about 1,000 litres (220 Imp gal) per minute, topping off the SM-10's tanks in about three minutes. When they were full an indicator light illuminated in the fighter's cockpit; the pilot shut the fuel transfer valve and decelerated to rotate the receptacle back, breaking contact with the tanker.

Refuelling was possible at 450-500 km/h (279-310 mph) and 9,000-10,000 m (29,530-32,810 ft). The tanker's operator could terminate the fuel transfer at any moment if need arose. Multiple refuellings were possible in a single sortie, both day and night (in clear weather conditions only); for night refuelling the fighter's port wingtip was illuminated by a spotlight. In one of the test flights the SM-10 'hit the tanker' twice, staying airborne for six hours.

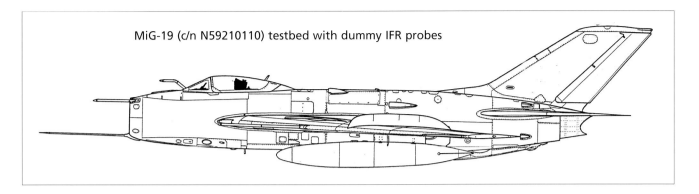

MiG-19 (c/n N59210110) testbed with dummy IFR probes

The chief problem during refuelling was the tanker's wingtip vortex. Therefore in 1956 the SM-10 was transferred to LII for Stage II of the trials. The tanker's hose was modified in order to get a more secure connection with the receiver aircraft, avoid hose folding, reduce the effect of the Tu-16's wake turbulence and reduce the required hose length; no changes were made to the fighter. During Stage II the SM-10 was flown by LII test pilot Vladimir M. Pronyakin. At this stage the refuelling process was fully automated; the SM-10 made the rendezvous with the tanker, using the **Oo**zel (Knot) identification friend-or-foe (IFF) transponder which doubled as a short-range radio navigation (SHORAN) system. Endurance with one top-up was six hours.

After the manufacturer's tests, which involved a total of 49 flights, between 17th December 1956 and 14th March 1957 the SM-10 passed its state acceptance trials with flying colours; the aircraft were flown by Nefyodov and Pronyakin. Still, the IFR system was not fitted to production MiG-19s because the Powers That Be decided it was unnecessary for tactical aircraft; also, the wingtip-to-wingtip system did not allow simultaneous refuelling of two fighters. Then the Soviet leader Nikita S. Khrushchov got his famous 'missile itch', and in the 1960s Soviet defence

spending priorities shifted to intercontinental ballistic missile development and production.

Mikoyan/Gurevich MiG-19 refuelling receptacle positioning testbed

In the late 1950s OKB-918 adapted the probe-and-drogue IFR system for the Myasishchev M-4 bomber. Once again, the tanker would be a converted bomber – this time with a removable HDU in the bomb bay, known as KAZ (**kom**pleksnyy agre**gaht** za**prahv**ki = all-in-one refuelling unit). This would be lowered when the bomb bay doors were opened, deploying the drogue.

Apart from checking the operation of the HDU proper, the best location of the receiver aircraft's IFR probe had to be determined. To this end the tenth production Gor'kiy-built MiG-19 sans suffixe coded '10 Red' (c/n N59210110) was converted at LII in late 1957. The aircraft had no fewer than four dummy IFR probes. Probe No. 1 was a dog-leg tube bent at right angles and attached to the port wing gun mount so that the tip was level with the top of the canopy. It was located 1,400 mm (4 ft 7⁷⁄₆₄ in) from the air intake lip and 550 mm (1 ft 9²¹⁄₃₂ in) from the fuselage side, protruding 750 mm (2 ft 5³³⁄₆₄ in) beyond the wing leading edge.

Test flights with MiG-19 c/n N59210110 and IL-28 c/n 2402101				
Flight	Date	Pilot	Probes used	Successful attempts
No. 1	18-9-57	S. F. Mashkovskiy	all	2 of 4 (nose probe)
No. 2	20-9-57	S. F. Mashkovskiy	Nos. 1, 2, 3	2 of 7 (nose probe)
No. 3	24-9-57	S. F. Mashkovskiy	Nos. 1, 2, 3	2 of 5 (nose probe)
No. 4	27-9-57	S. F. Mashkovskiy	No. 3	No contact
No. 5	3-10-57	S. F. Mashkovskiy	No. 4	1 of 33
No. 6	16-10-57	S. N. Anokhin	No. 3	2 of 41
No. 7	30-10-57	S. F. Mashkovskiy	No. 3	1 of 12
No. 8	1-11-57	S. F. Mashkovskiy	No. 1	2 of 34
No. 9	7-12-57	P. I. Kaz'min	No. 1	9 of 30
No. 10	27-12-57	P. I. Kaz'min	Nos. 1, 2	No contact (incident)

The straight No. 2 probe was located in the fuselage nose, offset 250 mm (9^{27}⁄$_{32}$ in) to port and protruding 800 mm (2 ft 7½ in) beyond the intake lip. The other two probes were attached to the port wing 1,800 mm (5 ft 10^{55}⁄$_{64}$ in) and 2,615 mm (8 ft 6^{61}⁄$_{64}$ in) from the centreline – that is, 1,050 mm (3 ft 5^{21}⁄$_{64}$ in) and 1,850 mm (6 ft 0^{55}⁄$_{64}$ in) from the fuselage side respectively. Probe No. 3 protruded 2,000 mm (6 ft 6^{47}⁄$_{64}$ in) beyond the wing leading edge and was located 1,800 mm from the air intake; for probe No. 4 these distances were 1,000 mm (3 ft 3⅜ in) and 4,000 mm (13 ft 1^{31}⁄$_{64}$ in) respectively.

The MiG-19 testbed was flown by Test Pilots (1st Class) Stepan F. Mashkovskiy, Pyotr I. Kaz'min and Sergey N. Anokhin. These pilots were familiar with the hose-and-drogue IFR system, having taken on fuel from a Tu-4 tanker in the modified MiG-15*bis* fighters. The test programme was supervised by project engineer N. N. Koormyatskiy.

By then the Tu-16 had another tanker version, the Tu-16N *Badger-A* equipped with the KAZ hose drum unit. However, the Tu-16N prototype was unavailable. Therefore, the abovementioned IL-28 '01 Red' (c/n 2402101) was converted into a makeshift 'tanker trainer' deploying a drogue on a steel cable.

The MiG-19 would make contact with the tanker at 7,000 m (22,965 ft) and 450-470 km/h (250-261 kts) IAS, approaching from a stand-by position 10-20 m (32-65 ft) behind the drogue. Contact was usually made in a climb, with or without side slip. Approach speed varied from 0.3 to 12 m/sec (1-39 ft/sec) or 1-30 km/h (0.54-16.2 kts) IAS. After making contact the MiG-19 stayed locked into the drogue for three to five seconds, then slowed down and broke contact. For safety's sake the drogue lock was set at an unlocking force of 60-80 kgf (132-176 lbf). Usually the fighter

carried drop tanks to increase mission time.

As mentioned earlier, Mashkovskiy, Kaz'min and Anokhin made ten 'refuelling' flights in the MiG-19; see table on the previous page for details.

As mentioned earlier, the tenth flight had to be terminated prematurely when the drogue entered the fighter's air intake and disintegrated, the debris damaging one of the engines.

The chances of making contact with the tanker depended mainly on the stability of the drogue – which left much to be desired. The nose probe allowed the pilot to make contact relatively easily, providing approach speed was 0.3-0.5 m/sec (1-1.6 ft/sec). Making contact with the other probes was difficult if approach speed was less than 2 m/sec (6.5 ft/sec) and was largely a matter of chance because the drogue swayed 2.03 m (6.5-10 ft) from side to side. The odds were better if the fighter approached at 7-12 m/sec (23-39 ft/sec) with sideslip, but this called for considerable flying skills. The test results were preliminary, since the drogue was attached to a cable and would probably behave differently when attached to a less flexible hose.

Kaz'min reported that *'making contact with the wing-mounted probes is difficult and impracticable because of the considerable asymmetry and wing sweep. The pilot has to look sideways and backwards, which is difficult and dangerous.*

Making contact with the nose-mounted probe is easier on the MiG-15 which allows the pilot to do it on the first try after two or three training flights, with an approach speed of 4-6 km/h [2.16-3.24 kts]. On the MiG-19, drogue deflection is greater; so is the approach speed – 6-8 km/h [3.24-4.32 kts] or 1.5-2 m/sec [5-6.5 ft/sec].

The MiG-15's engine is less sensitive to turbulence. On the MiG-19, missing the drogue

MiG-23UB '19' was used by LII in several test programmes. Here it is depicted as an in-flight refuelling system testbed with a fixed IFR probe.

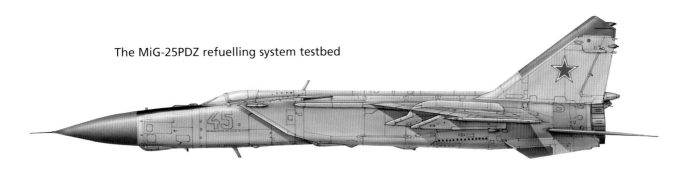

The MiG-25PDZ refuelling system testbed

can cause the latter to be ingested by the engines and disintegrate, which is exactly what happened on the tenth flight. The *MiG-19 is less responsive to stick inputs; when using the side probe* (No. 1 – Auth.) *the drogue is deflected by the shock wave generated by the fuselage.'*

By late 1956 it became clear that IFR capability was more important for long-range bombers than for fighters. Besides, the Mikoyan OKB was already working on new-generation fighters which would outperform the MiG-19 by far; thus the idea of giving the MiG-19 IFR capability was abandoned. The KAZ hose drum unit, on the other hand, entered production; apart from the Tu-16N, it was fitted to M-4 and 3MS/3MN bombers converted for the tanker role (these were redesignated M-4-2, 3MS-2 and 3MN-2 respectively).

Mikoyan MiG-23UB refuelling system testbed

A MiG-23UB *Flogger-C* trainer coded '19' was converted by LII into a testbed for a probe-and-drogue IFR system. The fixed L-shaped refuelling probe was installed on the starboard side of the nose immediately ahead of the windshield and canted to starboard. Apparently this is the same 'MiG-23UB No. 919' (c/n ...90...919) as described below.

Mikoyan MiG-25PDZ refuelling system testbed

In the late 1980s, Soviet aircraft designers returned to the issue they had been trying intermittently to resolve for the previous forty years – providing tactical aircraft with in-flight refuelling capability. The first experiments involving Yakovlev Yak-15 *Feather*, MiG-15 and MiG-17 fighters dated back to the late 1940s and early 1950s. A second series followed in the late 1960s with the abovementioned MiG-19 (SM-10). The advent of the Ilyushin IL-78 *Midas-A* tanker (a derivative of the IL-76MD *Candid-B* transport), the UPAZ-A podded HDU (*oonifit-**see**rovannyy podves**noy** agre**gaht** za**prahv**ki* – 'standardised suspended refuelling unit') developed by OKB-918 and retractable refuelling probes meant that tactical fighters and interceptors could finally enjoy IFR capability.

In the late 'eighties the Mikoyan OKB fitted three MiG-25s experimentally with IFR probes and launched a trials programme. One of the aircraft in question, converted from a production MiG-25PD *Foxbat-E* interceptor coded '45 Blue', served for assessing the possibility of extending the intercept range and operating jointly with MiG-31 interceptors. The modified aircraft was designated MiG-25PDZ, the Z standing for za**prahv**ka (refuelling).

The conversion was performed by one of the Air Force's aircraft repair plants in 1985 and the MiG-25PDZ entered flight test in early

'45 Blue', the MiG-25PDZ. The retracted IFR probe is just visible ahead of the cockpit windshield.

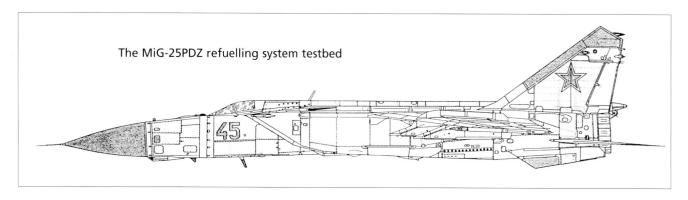

The MiG-25PDZ refuelling system testbed

1986. The retractable L-shaped IFR probe was located ahead of the cockpit windshield and slightly offset to starboard. A plug had to be inserted in the nose to accommodate the probe and associated equipment. After the first few test flights the probe was shortened.

At about the same time, Mikoyan and Sukhoi were working on fitting IFR probes to

The MiG-25RBVDZ IFR system testbed ('68 Red') on final approach to Zhukovskiy.

The MiG-25RBVDZ and Su-24M '29 Blue' refuel from the second prototype IL-78, CCCP-76556.

fourth-generation fighters (the MiG-29 light tactical fighter and the Su-27 interceptor). For commonality reasons, the probes on both aircraft were offset to port; therefore, the designers considered moving the probe to the port side on the MiG-25PDZ as well. This called for

major modifications, including more pipelines, changes to the fuel metering system, additional short-range radio navigation (SHORAN) equipment ensuring rendezvous with the tanker and special lights for illuminating the probe and the tanker's drogue during night refuelling.

The flight tests were a complex and dangerous affair, since the hose and drogue could hit the fighter's cockpit if the fighter pilot misjudged his position; a broken hose could douse the fighter with fuel, with an ensuing fire more than probable. To simplify flying the MiG-25PDZ and make refuelling easy enough for average service pilots, a micro-control system was proposed and tested successfully. It involved engine thrust vectoring by slightly moving the petals of the variable nozzles.

Despite its complexity, the IFR system tested on the MiG-25PDZ in its definitive form was clearly efficient enough, and a proposal was drafted to retrofit existing MiG-25s with it. However, the Air Force did not have enough IL-78 tankers to fill the PVO's needs, since the type was primarily intended to work with Tu-95MS *Bear-H* and Tu-160 *Blackjack* strate-

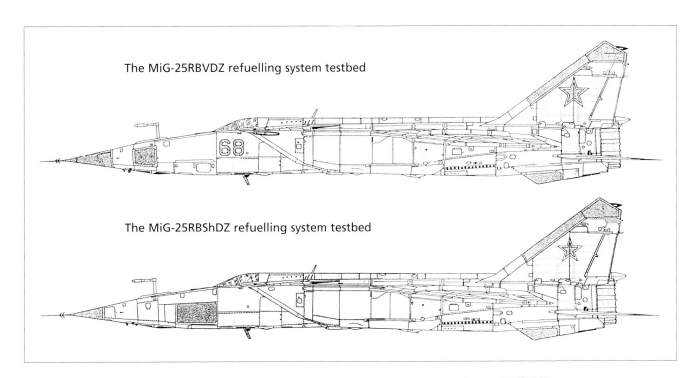

The MiG-25RBVDZ refuelling system testbed

The MiG-25RBShDZ refuelling system testbed

gic cruise missile carriers and the Ilyushin/ Beriyev A-50 *Mainstay-A/B* AWACS aircraft (another spin-off of the IL-76MD). Hence the PVO dropped the idea of refitting MiG-25s with the IFR system, deciding to fit it only to the more modern MiG-31 *Foxhound*.

Mikoyan MiG-25RBVDZ and MiG-25RBShDZ refuelling system testbeds

As mentioned earlier, the in-flight refuelling system fitted to the MiG-25PDZ interceptor was also tested on two reconnaissance/strike versions of the *Foxbat*. On these aircraft the IFR probe was offset to starboard (as originally on the interceptor) but was located much further forward and was non-retractable (this was dictated by the mission equipment installed in the nose). As on the MiG-25PDZ, a slight stretch of the nose section was necessary to accommodate the probe and associated equipment.

Two aircraft – a MiG-25RBV *Foxbat-B* coded '68 Red' and a MiG-25RBSh *Foxbat-D* coded '34' – were modified and redesignated MiG-25RBVDZ and MiG-25RBShDZ respectively (DZ = *dora**bot**annyy dlya za**prahv**ki* – modified for refuelling). The IL-78 tanker prototypes, CCCP-76501 and CCCP-76556, were used during the tests; however, the aircraft could also receive fuel from other tankers, including Sukhoi Su-24M *Fencer-D* tactical bombers carrying a UPAZ-1A refuelling pod on the centreline as 'buddy' tankers.

The MiG-25RBShDZ ('34') makes contact with the IL-78's starboard drogue.

The same scene from the MiG driver's perspective.

The MiG-25RBShDZ is seen from the IL-78's refuelling systems operator station as it approaches the starboard drogue.

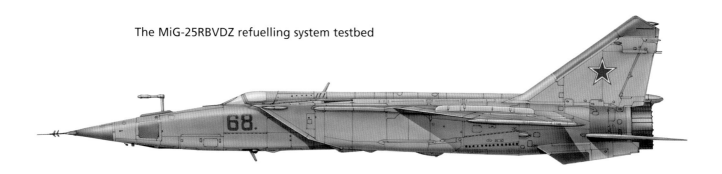

The MiG-25RBVDZ refuelling system testbed

At that point, however, the Su-24MR *Fencer-E* tactical reconnaissance aircraft entered flight test. Like the basic Su-24M, it had IFR capability as standard and was considered more advanced. The VVS quickly dropped the idea of extending the MiG-25RB's range and, as with the interceptor version, the proposed upgrade programme never materialised.

Mikoyan MiG-25 refuelling system testbed

A further *Foxbat* referred to as 'MiG-25 No. 304' was also involved in IFR system trials, being used by LII in 1974-76 to develop testing techniques and assess the conditions in which contact with the tanker was possible. The version is unknown, and 304 could be the last three digits of the c/n or the f/n (0304).

Here, the MiG-25RBShDZ makes contact with the first prototype IL-78, CCCP-76501 – singly and in company with a camouflaged Su-24M.

Mikoyan MiG-29 refuelling system testbed

A standard production MiG-29 (*izdeliye* 9.12) *Fulcrum-A* was converted into a testbed for the optional IFR probe which was to be fitted to the MiG-29SD, MiG-29SE and MiG-29SM export versions at the customer's demand. Unlike the MiG-29K and MiG-29M, both of which had a fully retractable L-shaped probe offset to port ahead of the windscreen, this probe was semi-retractable and detachable. The largish elongated fairing, rather like that of the Panavia Tornado IDS, was located on the port side of the forward fuselage (starboard-side installation was impossible because the No. 2 pitot was located there). The probe itself weighed 65 kg (143 lb); the additional fuel lines, valves and other fuel system components added by another 30 kg (66 lb) to the aircraft's empty weight. The system made it possible to fill both internal tanks and drop tanks. A contract with Malaysia signed on 7th June 1994 stated that 14 of the 16 MiG-29s ordered by the Royal Malaysian Air Force (the MiG-29N *Fulcrum-A* single-seaters) were to be retrofitted with refuelling probes upon completion of deliveries.

In developing the system Mikoyan relied on experience with the retractable probe tested on the MiG-25PDZ testbed (and later fitted to production MiG-31B *Foxhound* interceptors). The MiG-29's probe was suitable for working both with the IL-78/IL-78M *Midas-A/B* three-point tanker and with Western hose-and-drogue tankers, such as the Lockheed C-130K Hercules and Vickers VC10 C.1K.

The only problem was how to squeeze the new components into the very limited space available. Originally the engineers wanted to place the probe immediately below the cockpit canopy but later moved it downward a little. The fuel line could not be stowed completely in the fuselage, and part of it adjacent to the probe had to be covered by a fairing.

The probe, complete with the mounting frame and the protruding part of the fuel line, could be removed or reinstalled within an hour if need arose. The system could be retrofitted to any version of the *Fulcrum* with minimum modifications and at minimum cost.

In-flight refuelling is a complex procedure demanding great concentration and skill, so changes were made to the fighter's semi-automatic control system to facilitate approach to and contact with the tanker. A SHORAN system ensured rendezvous with the tanker. After extending the refuelling probe the pilot switched the control system to 'refuelling stabilisation' mode.

The modified MiG-29 (c/n 2960536034, f/n 4808) originally flew in primer finish without markings; in this guise it was used for refuelling trials. On 16th November 1995 MAPO-MiG chief test pilot Roman P. Taskayev made the first successful refuelling from an IL-78 *sans suffixe* registered CCCP-78782. (Many Soviet Air Force transport aircraft ostensibly wore civil registrations and the colours of the national airline Aeroflot, and a few retained the old CCCP- prefix and the Soviet flag as late as 1997.) Mikoyan test pilot Marat A. Alykov and GNIKI VVS test pilots Vladimir D. Shooshoonov and Aleksandr A. Goncharov also flew the aircraft. Mikoyan engineers and test pilots claimed that the improved flight control/stabilisation system was better than all others used on Russian combat aircraft with IFR capability. Refilling one fuselage tank and three drop tanks extended the MiG-29's ferry range to 6,000 km (3,726 miles). The IFR probe had no adverse effect on the fighter's performance and handling.

Contact with the IL-78 was made at altitudes up to 8,000 m (26,250 ft) and speeds between 400 and 600 km/h (248-372 mph). Refuelling was also performed at 350-500 km/h (217-310 mph) to simulate working with the slower KC-130 turboprop tanker. This was because the Malaysian Air Force had acquired

MiG-29 c/n 2960536034 makes contact with the drogue during trials.

six C-130 transports which could be readily converted into tankers.

The trials programme was completed in January 1996. The fighter was then painted in the standard two-tone grey camouflage and coded '357 Blue'. That year it appeared at two major airshows – ILA'96 at Berlin-Schönefeld and Farnborough International '96. In June 1997 '357 Blue' participated in the 42nd Paris Air Show; two months later it appeared in the static park of the MAKS-97 airshow in Zhukovskiy (19-24th August).

After ILA'96 Western aviation specialists tended to refer to '357 Blue' as a 'MiG-29SM' (although the data placard at MAKS-97 read 'MiG-29SD'!). However, both designations are hardly applicable to this particular aircraft, since it was merely a testbed. Curiously, a stencil on the nose immediately aft of the radome

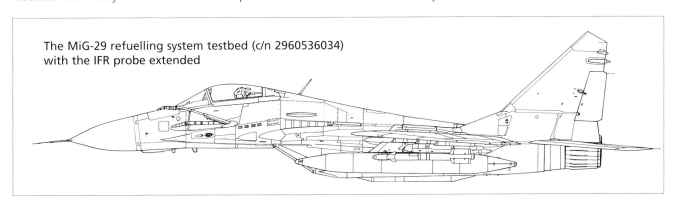

The MiG-29 refuelling system testbed (c/n 2960536034) with the IFR probe extended

read 'N019EM' – which is probably a mispaint, as the radar is designated N019ME. Indeed, there is no positive proof that '357 Blue' is really equipped with this radar!

Mikoyan MiG-31 refuelling system testbeds

The original production version of the MiG-31 (*izdeliye* 01) lacked IFR capability. The first *Foxhound* to feature a refuelling probe was MiG-31 f/n 3603; this was actually a dummy installation and fuel transfer was impossible. The aircraft was possibly coded '77 Red'; however, some sources list MiG-31 '77 Red' based at Zhukovskiy as c/n N69700104695 and f/n 0304.

The modified MiG-31 served for testing the probe actuation mechanism and for 'dry runs' with the refuelling tanker. The MiG-25PDZ, MiG-25RBVDZ and MiG-25RBShDZ testbeds had featured a micro-throttle system enabling minute engine power adjustments to facilitate contacts with the tanker's drogue. Initial tests showed that the IFR-capable MiG-31 did not need this system, as the standard throttles allowed contact to be made without any diffi-

culty; nevertheless, the micro-throttle system was fitted after all to enhance flight safety and make the pilots fell more comfortable.

V. S. Yegorov was appointed engineer in charge of the tests. The greater part of the test programme was performed by OKB pilots in Zhukovskiy, followed by further testing in Akhtoobinsk at the hands of GNIKI VVS pilots.

Trials continued with another Batch 36 MiG-31 ('368 Blue', f/n 3608). This aircraft had a fully functional IFR probe and fuel transfer system; however, it lacked the system automatically monitoring the quantity of fuel transferred. The successful trials allowed the IFR-capable MiG-31B (and *izdeliye* 01DZ mid-life upgrade) to enter service.

Polikarpov R-5 refuelling system testbed

The R-5 reconnaissance biplane was the first Soviet aircraft to be used in IFR experiments. The work was undertaken in 1931 by a NII VVS team under A. K. Zapanovannyy. The R-5 tanker (called *benzinovoz* – 'petrol carrier' – in the day's terminology) had a hand-driven hose drum, the hose being fitted with a 3-kg (6.6-lb) stabilising weight at the end. The tanker would take up a position directly above the receiver aircraft and deploy the hose which the technician of the receiver aircraft would catch and connect to the fuel system. To increase the fuel supply the tanker carried additional fuel tanks under the lower wings.

Initially another R-5 acted as the receiver aircraft; later, in June-July 1932, a Tupolev TB-1 bomber was refuelled, followed in September-November 1933 by a Tupolev TB-3. Water was used at first for safety's sake; later, actual transfer of petrol took place. However, the system was extremely inconvenient to use (on the TB-3 the two dorsal gunners could

Front view of MiG-31 '77 Red' with the probe extended. Note the ciné camera 'egg' on top of the port air intake to capture the refuelling sequence.

Bottom: Another view of the same aircraft, showing the deployed IFR probe. The object on the port forward fuselage hardpoint may be a test equipment pod.

barely mange to catch the thrashing hose, and it was all too easy to get hit by the weight at the end!). Also, the extra tanks and fuel transfer lines were notoriously leaky and the initial version was not even flight tested. A version of the tanker with a 500-litre (110 Imp gal) belly tank made of steel entered test in June 1932; the whole set of IFR equipment weighed 120 kg (264 lb). Maximum transferable fuel was 1,040 litres (228.8 Imp gal).

Sukhoi Su-7 refuelling system testbed

In the late 1970s and early 1980s NPP Zvezda developed an in-flight refuelling (IFR) system for Soviet tactical aircraft (see next entry for more details). The result was the UPAZ-1 podded HDU used both as a 'buddy' refuelling pack by tactical aircraft and as a permanent installation on the IL-78. In order to carry out flight tests and development of the new system, work out IFR techniques and find the optimum flight modes for refuelling various types of aircraft, a series of testbeds were produced at LII, including a modified Su-7BM *Fitter-A* fighter-bomber (c/n 5106) used in this capacity in 1968-70.

Sukhoi Su-15 refuelling system testbeds (Sakhalin programme)

The Sukhoi OKB's first involvement with in-flight refuelling dates back to 1971. The need to master IFR techniques was caused by the necessity to increase the range and combat radius of the Su-24 tactical bomber. By the end of the year the OKB prepared a set of project documents envisaging the installation of an IFR system on what was then known as the T-58M (the manufacturer's designation of the Su-24 was soon changed to T-6).

Unlike the Sukhoi OKB, which was breaking new ground with in-flight refuelling, the Flight Research Institute and the NPP Zvezda design bureau (formerly OKB-918) under Guy Il'yich Severin were old hands at this technique. By the early 1970s these two establishments had accumulated a wealth of experience in developing, testing and using various IFR systems. In the 1950s and 1960s such systems were created in the Soviet Union both for heavy bombers and for tactical aircraft, although it would be a while before the latter would actually benefit from in-flight refuelling.

NPP Zvezda had been working on IFR systems since the late 1960s and the programme was code-named *Sakhalin* after an island in

'01 Red' (formerly '34 Red', c/n 0015301), the first Su-15 converted into a 'buddy' tanker equipped with a UPAZ-1A podded HDU. The large size of the pod is readily apparent. Note the photo calibration markings on the fuselage and tail and the AKS-5 camera pod on the starboard wing pylon.

Bottom: '37 Red' (c/n 1115337), the second Su-15 modified to take the UPAZ-1A HDU. Note the absence of the missile launch rails.

An air-to-air of Su-15T '37 Red' as it streams the drogue.

the Soviet Far East. (Perhaps the implication was that 'with this system our aircraft will be able to reach Sakhalin non-stop'!) In 1971-73 the OKB's specialists designed and tested the principal components of a hose-and-drogue refuelling system.

The preliminary design project of the T-58M strike aircraft featuring a retractable IFR probe was assessed at a session of MAP's Scientific & Technical Council. In keeping with the latter's recommendations the Sukhoi OKB decided to hold a series of tests in advance so that the system would be fully mastered by the time the bomber was ready to take it. The Su-15 was selected for conversion into IFR system testbeds, one aircraft acting as the tanker and the other as the receiver. In October 1973 the VVS issued a specification in which three Sukhoi types – the Su-15TM, the Su-17M/Su-17M2 and the T-6 (by then allocated the service designation Su-24) – were stated as possible tanker aircraft fitted with 'buddy' refuelling pods. The pod itself, or hose drum unit (HDU), was designated UPAZ-1.

Pursuant to a joint MAP/Air Force ruling the Sukhoi OKB allocated two Su-15s – the first pre-production aircraft (c/n 0015301) and an early production aircraft (c/n 0215306) – for testing the Sakhalin-1A IFR system. Both aircraft by then had a long history as 'dogships', having been used for testing various systems and equipment. Additionally, LII allocated one of its Su-15s ('37 Red', c/n 1115337) for testing the Sakhalin-1A system and evolving IFR techniques. Since each of these aircraft served a different purpose, they are described separately hereunder, followed by an account of the trials.

Su-15 'buddy' IFR tanker testbeds

By December 1972 the first pre-production Su-15, '34 Red' (c/n 0015301), had been modified as the first of two examples fitted with a dummy UPAZ-1A (Sakhalin-1A) pod for aerodynamic testing (actually the pod featured a functional hose drum allowing the drogue to be deployed and stowed). On 19th December Sukhoi OKB test pilot Vladimir A. Krechetov performed the first test flight for the purpose of assessing the aircraft's stability and handling with this rather bulky external store.

In 1973 the OKB issued a set of documents for the adaptation of the production-standard Su-15's fuel system to accept the UPAZ-1A pod. This involved installation of additional fuel pumps in the Nos. 1, 2 and 3 fuselage tanks, modifications to the electric system and replacement of the standard TRV1-1A fuel metering kit with a new TRK1-1 kit. The display of the *Oryol* (Eagle) fire control radar was removed to make room for the HDU's control panel.

The UPAZ-1A had a 26-m (85 ft 3 in) hose and a flexible 'basket' drogue. The hose drum was powered by a ram air turbine (RAT) with an intake scoop on the port side which was normally closed flush with the skin. A second air intake at the front closed by a movable cone was for an RAT driving a generator for the electric transfer pump. Normal delivery rate was 1,000 litres (220 Imp gal) per minute but this could be increased to 2,200 litres (484 Imp gal) in case of need. 'Traffic lights' were installed at the rear end of the pod to indicate fuel transfer status to the pilot of the receiver aircraft. The pod was attached to the airframe via a so-called standardised piping/electrics connector (UURK – *oonifitseerovannyy oozel razyoma kommunikatsiy*) and could be jettisoned pyrotechnically in an emergency. A test equipment suite was fitted.

Outwardly the aircraft, which received a new tactical code ('01 Red') after the conversion, differed from its standard sister ships in having an egg-shaped pod housing an AKS-5

Su-15 (c/n 0215306) refuelling systems testbed with IFR probe

ciné camera fitted to the starboard PU-1-8 missile launch rail. The aft-looking camera served for filming the moments when the receiver aircraft made and broke contact.

The conversion was performed in April-July 1974 by the OKB's Novosibirsk branch. On 4th July 1974 Sukhoi OKB test pilot Aleksandr S. Komarov made the first checkout flight from Novosibirsk-Yel'tsovka, whereupon the modified aircraft was flown back to Zhukovskiy for testing.

The second aircraft to be fitted with a dummy UPAZ-1A pod was the aforementioned Su-15 coded '37 Red' (c/n 1115337). Manufactured on 13th February 1970 and delivered new to LII for special tests, this aircraft featured double-delta wings and an operational BLC system; also, it could be fitted with both old R11F2SU-300 and new R13-300 engines. In late 1973 after completing a spinning test programme the fighter was modified in house by LII for installation of an operational UPAZ-1A. Su-15 c/n 1115337 was intended for testing the pod proper and for verifying the approach and contact technique. Contact would be made in 'dry' mode (that is, without actual fuel transfer), hence the scope of the modification work on this aircraft was much smaller as compared to Su-15 c/n 0015301.

The conversion involved removing the radar, which was replaced by ballast, and installing the HDU control panel, the UURK 'wet' centreline pylon and the appropriate data recording equipment. The missile launch rails were removed for the duration of the Sakhalin test programme.

In 1974 LII test pilots held a series of tests on this aircraft, checking the operation of the HDU (that is, hose deployment/retraction and the operation of the pod's other systems). Afterwards '37 Red' continued in use with the institute, participating in several research programmes. Unfortunately on 24th December 1976 the aircraft crashed near Lookhovitsy, Moscow Region, killing LII test pilot Leonid D. Rybikov. The exact cause was never determined. Five (some sources say three) minutes after taking of from Tret'yakovo airfield (the factory airfield in Lookhovitsy) on a positioning flight to Zhukovskiy, which is about 90 km (56 miles) away as the crow flies, the aircraft inexplicably entered a shallow dive with considerable bank, impacting in a field 39.5 km (24.5 miles) from the point of origin and disintegrating utterly. Examination of the crash site and analysis of the wreckage showed that the aircraft had been perfectly serviceable up to the point of impact. At the final section of the

Two views of Su-15T c/n 0215306 following conversion as the receiver aircraft for the Sakhalin test programme with a fixed IFR probe. Note the camera fairing on top of the fin, the photo calibration markings and the non-standard comms radio aerial on the centre fuselage.

flight path the aircraft had been levelled out but continued losing altitude due to the high sink rate prior to that, impacting in a wings-level attitude; the pilot did not eject and was killed. Temporary pilot incapacitation was eventually cited as the cause of the crash.

Shortly after the completion of the Sakhalin IFR system's trials Su-15 c/n 0015301 was returned to the PVO as time-expired. Actually the aircraft could have been returned to service after an overhaul and a service life extension, but in view of the extensive conversion it had undergone, using it as a combat aircraft was inadvisable. Hence '01 Red' was transferred to a PVO Junior Aviation Specialists School (ShMAS – **Shko**la **mlahd**shikh aviat-si**on**nykh spetsia**lis**tov) located in Solntsevo just outside the Moscow city limits as a ground instructional airframe. After the closure of the school in 1991 the Su-15 was returned to the OKB for preservation and, after cosmetic repairs, joined the collection of the open-air aviation museum at Moscow-Khodynka established in 1994. Regrettably the museum is now closed and the aircraft, like all the other exhibits, is in serious danger.

Sukhoi Su-15T IFR system testbed (receiver aircraft)

In late 1973 the Sukhoi OKB started converting the abovementioned Su-15T c/n 0215306, another long-serving 'dogship', into the receiver aircraft under the Sakhalin programme. Again the conversion was quite extensive, as the aircraft had to be equipped with a completely new IFR system ensuring the correct fuel transfer sequence in order to keep the CG within the allowed limits.

The uncoded aircraft was equipped with a new TRK1-1 fuel metering kit and a fixed L-shaped refuelling probe offset to starboard ahead of the cockpit windshield; the tip of the probe was located 4.25 m (13 ft 11⅜ in) aft of the tip of the radome, 0.9 m (2 ft 11½ in) to the right of the fuselage centreline and 0.8 m (2 ft 7½ in) from the fuselage surface. A forward-looking AKS-5 camera was installed in a cigar-shaped fairing replacing the dielectric fin cap of the communications radio antenna for filming the contact with the tanker's drogue; hence a non-standard ASM-1 blade aerial serving the R-832M radio was fitted to the centre fuselage. An MSL-2 (aka KhS-62) flashing beacon was installed on the upper centre fuselage for synchronising the operation of all photo and movie cameras capturing the refuelling sequence.

Since the refuelling operation imposed considerable stress on the pilot of the receiver aircraft, Su-15T c/n 0215306 was fitted with special *Koovshinka* (Water lily) medical equipment recording the pilot's physiological parameters (pulse and breath) and recording the distribution of his concentration during various stages of the process. Finally, a PAU-467 ciné camera was fitted.

The course of the Sakhalin test programme

Apart from the three aircraft described above, the tests involved a MiG-21U owned by LII which acted as a chase plane and camera ship. According to plan the first stage of the tests (performed jointly by the Sukhoi OKB and LII) was devoted to evolving and perfecting the optimum approach and contact technique. Yevgeniy S. Solov'yov was the OKB's project test pilot, with M. L. Belen'kiy as engineer in charge; LII appointed V. D. Koorbesov and Yu. N. Goonin respectively.

Before the joint flights and attempted contacts with the tanker could begin, the probe-equipped Su-15T made a series of test flights to see if and how the refuelling probe affected the aircraft's stability and handling. The actual flight test programme commenced on 31st May 1974 when the first two flights were made. The approach to the tanker and attempted contacts were made at 8,200 m (26,900 ft) on the first occasion and about 6,000 m (19,680 ft) on the second occasion; the speed was 550 km/h (340 mph) in both cases. Both attempts ended in failure; moreover, on the second try the fuel transfer hose got entangled with the refuelling probe and snapped as the receiver aircraft manoeuvred.

A pause was then called while the engineers made corrections to the piloting technique during the final approach phase. The pause turned out to be a long one, the flights resuming only on 24th December 1974. During this period the Zvezda OKB revised the UPAZ-1A pod, increasing the length of the hose to 27.5 m (90 ft 2 in). In the meantime Yevgeniy S. Solov'yov took special training on a purpose-built simulator at the OKB and in the actual aircraft (the approach to the tanker was simulated on the ground and special reference lines were applied to the cockpit canopy to mark the drogue's position at different ranges). It was decided that the tanker aircraft would deploy the flaps and that the dampers in the receiver aircraft's control circuits would be switched on. The flight altitude was reduced to 4,000 m (13,120 ft) and the tanker approach technique was revised accordingly.

The effect of the additional training taken by the pilots was felt immediately; as early as 14th January the modified Su-15T made the first stable contact with the tanker, maintaining refuelling formation for a while. This flight marked the end of the phase involving LII's 'buddy tanker' (Su-15 c/n 1115337), as the OKB's own 'buddy tanker' (Su-15 c/n 0015301) had been completed by then.

Solov'yov remained the receiver aircraft's project test pilot; the other aircraft was flown by Aleksandr S. Komarov and Vladimir A. Krechetov. M. L. Belen'kiy likewise continued in his capacity as engineer in charge.

Duly prepared and endorsed at all levels, the test schedule for the summer of 1974 envisioned 70 flights, 29 of which would be made by the two aircraft together. Yet the beginning of the programme's main phase involving contacts with the tanker was delayed until 21st January 1975. The first contacts were made in 'dry' mode (without fuel transfer). The missions were flown at altitudes of 2,000-7,500 m (6,560-24,600 ft) and speeds

of 480-660 km/h (300-410 mph); each mission involved two to seven attempts at making contact. This stage continued until the end of February, after which another lengthy hiatus followed due to the need to analyse the results obtained and prepare for the next phase. This was devoted to testing and perfecting the UPAZ-1A HDU proper and the associated equipment of the carrier aircraft during refuelling missions.

The work resumed in June 1975. Again, initially contacts were made in 'dry' mode; in early July, however, the plug closing the orifice of the HDU's drogue was removed, allowing fuel to be transferred. An immediate problem arose: on 2nd, 4th and 18th July fuel was seen to be leaking from the drogue after the receiver aircraft's probe had locked into position. It turned out that the drogue had been manufactured poorly, making it impossible to seal

The receiver aircraft has made stable contact with the drogue; fuel transfer can begin.

This picture taken by the camera on the receiver aircraft's tail illustrates the offset of the IFR probe and the design of the drogue.

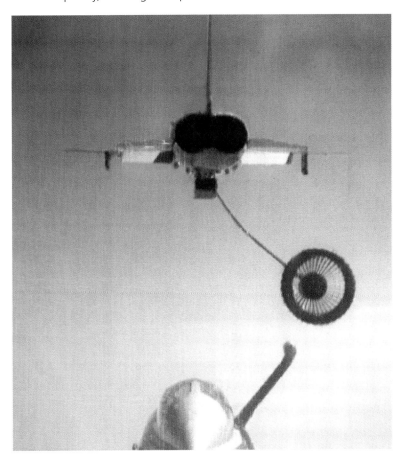

the joint well; a new, properly made drogue took care of the problem. The first successful transfer of 250 kg (550 lb) of fuel took place on 30th July 1975; the following day Su-15T c/n 0215306 made two more fuel top-ups, receiving 500 and 1,000 kg (550 and 1,100 lb) of fuel respectively.

By early December 1975 the Sukhoi OKB and the Zvezda design bureau had ironed out all the bugs discovered in the course of the preceding tests. After that, another four flights involving three fuel top-ups were made between 10th and 23rd December; on the latter date the receiver aircraft 'hit the tanker' twice in a single mission, taking on 400 and 250 kg (880 and 550 lb) of fuel respectively. Thus the test programme was successfully completed in full.

The test report said, '...The IFR system and technique verified [...] can be recommended for use primarily on the Su-24 aircraft'. Further tests of the Sakhalin-1A system were done, using Su-24 bombers.

Sukhoi T10U-2 refuelling systems testbed ('aircraft 02-01')

In 1985 the Sukhoi OKB began exploring ways of increasing the Su-27's range and endurance. Hence the second prototype Su-27UB *Flanker-C* combat trainer (T10U-2) coded '02 Blue' (c/n 49021002...01, f/n 02-01) was fitted experimentally with a fully retractable refuelling probe. The objective was to test the Su-27K shipboard fighter's IFR system, perfect refuelling techniques and investigate crew workloads and physical condition in long-range missions. The L-shaped probe was installed ahead of the windscreen, offset to port, with appropriate changes to the fuel system and some other systems. Huge black and white phototheodolite calibration markings were added to the fins and

The 'aircraft 02-01' testbed on the ground at Zhukovskiy. Note the retracted IFR probe ahead of the windshield.

'Aircraft 02-01' was used as a 'buddy' tanker during the tests of the Su-27K (Su-33). Here it refuels the first prototype ('37 Blue').

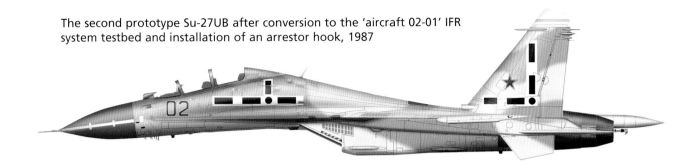

The second prototype Su-27UB after conversion to the 'aircraft 02-01' IFR system testbed and installation of an arrestor hook, 1987

nose. The testbed was known in-house by its fuselage number as 'aircraft 02-01'.

The carefully-designed cockpit of the Su-27UB with conveniently located controls, together with the flight refuelling capability, greatly increased mission time. In June 1987 'aircraft 02-01' piloted by Nikolay F. Sadovnikov and Igor' V. Votintsev made an unprecedented (for a Soviet fighter) non-stop flight across the entire country from Moscow to Komsomol'sk-on-Amur. In March 1988 the same aircraft with the same crew flew from Moscow to Komsomol'sk-on-Amur and back, covering 13,440 km (8,347 miles) in 15 hrs 42 min. The flight included four top-ups from an IL-78 tanker near Novosibirsk and Chita; the fairly frequent contacts with the tanker were not caused by the aircraft running low on fuel but by the need to train in IFR techniques.

Soon afterwards, 'aircraft 02-01' escorted by operational Su-27Ps belonging to the PVO's 941st IAP (istre**bit**el'nyy a**via**polk – fighter regiment) paid a visit to the world's northern-most airfield, Graham-Bell AB on the Zemlya Frantsa-Iosifa (Franz Josef Land) archipelago in the Barents Sea. This flight demonstrated that the type could operate successfully in the harsh climate of the Far North.

When the refuelling system had been successfully tested the aircraft was modified again under the Su-27K programme. This involved the addition of an arrestor hook under the tail 'stinger', just like on the T10-25. Further, it served as an avionics testbed for the Resistor-K42 automatic carrier approach system and a

tanker for the first prototype Su-27K (T10K-1, '37 Blue'), transferring fuel by means of an UPAZ-1A 'buddy' refuelling pod.

Tupolev TB-1 refuelling system testbeds

The first Soviet all-metal bomber, the Tupolev TB-1 (ANT-4), was used in several test and research programmes. In particular, in-flight refuelling experiments were carried out between 1933 and 1935, using one TB-1 (or an R-5 biplane) as a tanker and a second TB-1 as the receiver aircraft. The latter assumed a position below the tanker, which paid out a hose with a weight at the end; the hose was grabbed by a crewman and connected to the receiver aircraft's fuel system, whereupon fuel was transferred by gravity. This method was not adopted for service due to its complexity and operational inconvenience.

Tupolev Tu-2 refuelling system testbeds

In 1949 LII modified a pair of Tupolev Tu-2 Bat piston-engined bombers for testing the above-mentioned wingtip-to-wingtip IFR system developed by Igor' I. Shelest and Viktor S. Vasyanin. One aircraft was the tanker with a hose drum unit in the starboard wing, while the other was fitted with a receptacle under the port wingtip. Flight tests with a dummy installation on these aircraft began in the summer of 1949.

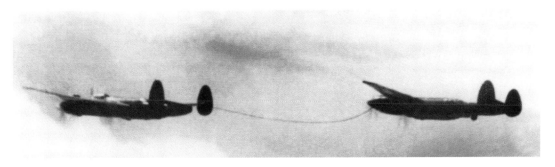

A pair of Tu-2s modified for in-flight refuelling experiments at LII. The tanker is on the left.

Yak-15 '47 Yellow' served as a testbed for the Shelest/Vasyanin wing-to-wing IFR system. The tanker's hose can just be seen beneath the tail.

Tupolev Tu-22 refuelling system testbed

In 1970-71 the Tu-22 bomber mentioned in Chapter 1 (c/n ...41605...) was used by LII for defining the parameters of the tanker's refuelling drogue. The results were used in developing the IFR-capable versions of the *Blinder*.

The Yak-15 formates with the Tu-2 'tanker' during a simulated refuelling. The IFR system was non-functional.

Yakovlev Yak-15 refuelling system testbed

A production Yak-15 *Feather* jet fighter serialled '47 Yellow' was used by LII as a refuelling system testbed in 1949. It was similarly outfitted with a dummy receptacle for the Shelest/Vasyanin wingtip-to-wingtip system, making contact with the Tu-2 'tanker'.

2. Landing gear testbeds

Alekseyev I-215D bicycle landing gear testbed

In 1946-47 the Gor'kiy-based OKB-21 headed by Semyon M. Alekseyev brought out a number of combat aircraft projects, of which only two – the I-211 and I-215 twin-turbojet fighters – were destined to materialise. Both types shared the same airframe, differing only in the powerplant, and featured a tricycle landing gear. However, the second prototype I-215 bearing the designation I-215D (*doob**lyor*** –

lit. 'understudy', the then-current Soviet term for second prototype) differed considerably from the first I-215, being custom-built to fill a special order placed by OKB-1. The latter design bureau was staffed with captive German engineers supervised by Dr. Brunolf Baade, developing heavy aircraft, including the 'aircraft 150' twinjet medium bomber. The I-215D was to verify the bicycle landing gear arrangement envisaged for the '150'.

The I-215D's nose and main gear units were placed fore and aft of the aircraft's CG; the rear strut incorporated a 'kneeling' feature to increase the angle of attack on take-off by 3°. The outrigger struts retracted into the engine nacelles; when the landing gear was extended the outrigger wheels were almost in line with the CG. The wings had an incidence of 3°30'; together with the 'kneeling' feature this provided an adequate angle of attack and sufficient lift for take-off. Hence, unlike aircraft with a conventional tricycle landing gear, the I-215D could become airborne without rotation, not requiring the pilot to haul back on the control stick.

Many of the Soviet aircraft engineers and airmen were opposed to the bicycle landing gear at the time, claiming it was inherently unsafe. However, Semyon M. Alekseyev graphically demonstrated with the help of the I-215D that these fears were unfounded. The I-215D completed by the Gor'kiy aircraft factory No. 21 was perfectly airworthy and ready for its first flight. Test pilot Ivan Ye. Fyodorov, Hero of the Soviet Union, agreed to ferry the aircraft to the flight test facility of OKB-1 (to which Alekseyev had been transferred by then). The flight, which involved a refuelling stop at LII's airfield (then referred to as Ramenskoye airfield), took place in October 1949 (the exact date is unknown) and went perfectly. On arrival the pilot put on a brief aerobatics display with steep banking turns; after landing he did several sharp turns and figure-eights at maximum taxying speed, showing that an aircraft with a bicycle landing gear could manoeuvre safely on the ground with no tendency to tip over on one wing.

Antonov An-2TD high-flotation landing gear testbed

Outstanding though the field performance of the An-2 *Colt* utility biplane was, an attempt was made to enhance it further by improving the aircraft's soft-field capability. In 1950 MAP plant No. 279 developed and built a special

An-2TD c/n 10547302 (with star insignia in eight places as per the pre-1955 standard) was converted in to a testbed for a high-flotation landing gear designed for operation from soft surfaces.

high-flotation landing gear designed for operation from soft and soggy runways. The main units were provided with special six-wheel bogies consisting of swivelling frames with two small 470 x 210 mm (18.5 x 8.26 in) wheels – the same type as fitted to the An-2's tail gear – positioned fore and aft of the existing 800 x 260 mm (31.49 x 10.23 in) mainwheels. The tail gear unit was provided with a skid fitting around the tailwheel. As a result, the landing gear footprint was doubled as compared to the standard aircraft.

The experimental landing gear was tested on an unserialled Kiev-built An-2TD paradrop aircraft in standard Soviet Air Force olive drab camouflage (c/n 10547302). Tests held in the spring of 1952 showed that the aircraft could operate from ploughed fields, sand and cross ditches up to 50 cm (19 in) wide without nosing over. However, the landing gear impaired the An-2's ground manoeuvrability, which is why it was not recommended for production.

Close-up of the main gear bogies of An-2TD c/n 10547302. Each bogie is a frame with two stock An-2 tailwheels fitted over the existing main-wheels.

The tailwheel of An-2TD c/n 10547302 was fitted with a skid.

en pumps. Each module weighed 28 kg (61.73 lb). The aircraft was designated An-714; the 7 referred to MAP's 7th Main Directorate to which Berezhnov's OKB was subordinated.

The 'centipede' An-2TD c/n 10547302 takes off from a dirt strip during trials in 1950 with flaps and slats fully deployed.

Antonov An-714 landing gear testbed

In an attempt to enhance the ability of the An-14 **Pchol**ka (Bee; NATO reporting name Clod) utility aircraft to operate from soft and soggy ground, the Antonov OKB fitted a production An-14A with an air cushion undercarriage developed by the specialised landing gear design bureau under I. Berezhnov in Kuibyshev. The three air cushion modules were identical, featuring three inflatable rubber rings and hydraulically actuated centrifugal blowers mounted on platforms atop the skirts; hydraulic power was provided by engine-driv-

Two views of the An-714 air cushion landing gear testbed taxying at Gostomel'. The nose titles actually read 'An-714'.

The logo of the Antonov OKB which developed the An-714 and An-14Sh testbeds.

Antonov An-14Sh landing gear testbed

In 1971 the Antonov OKB made another attempt to equip the An-14A with an air cushion landing gear; the resulting aircraft was designated An-14Sh (for *shassee* – landing gear). GSOKB-473 Chief Designer Oleg K. Antonov was the motive power behind the project; the design effort was performed by an eight-man team under B. M. Kolomiyets in accordance with a plan which MAP endorsed on 20th August 1975.

This time redesign was much more extensive. The single large inflatable skirt located under the fuselage had the shape of an inverted rubber dinghy; its overall area was equal to the An-14's wing area. Inside it were four pairs of smaller inflatable rings at the points where compressed air was fed into the skirt. This design automatically created a levelling force if the aircraft banked while taxying on uneven ground, preventing it from tipping over.

Carrying no registration (only the CCCP nationality prefix), the An-714 made its first flight on 20th October 1970 with Antonov OKB test pilot Vladimir A. Kalinin at the controls. However, it soon became apparent that the air pressure in the air cushion modules (850 kg/m^2) was too high and the landing gear footprint was too small. The air jets eroded the soft ground to such an extent that pieces of soil flew every which way and were kicked up into the engine and blowers. This led all further work on the An-714 to be abandoned.

The An-714 becomes airborne. Note that the aircraft carried no registration, only the CCCP prefix.

The bizarre An-14Sh testbed at rest. The air cushion skirt was to retract into the huge ventral pod and be closed by twin doors. Note the exhaust of the TA-6A-1 APU providing air for the air cushion; the air intake is on the port side.

Here the An-14Sh is seen taxying. Note that the air cushion pod has obstructed the entry door normally located at the rear. The standard main gear units have been retained.

Compressed air for the air cushion was supplied by a TA-6A-1 gas turbine engine (used as an APU on several Soviet airliners and transports); it was located in the cabin, breathing via a lateral air intake. This meant that the cabin could no longer be used for passengers or cargo; the An-14Sh was purely a technology demonstrator. The intention was to develop a similar system (twice as large) for the An-12.

The carefully chosen placement of the air feed points and the wear-resistant liners along the perimeter of the skirt minimising the air leak made sure that the escaping air had a relatively low speed, minimising ground erosion. After take-off the skirt was to deflate, retracting into a massive elliptical-section fairing; the aperture was closed by clamshell doors.

In 1980 a Soviet Air Force An-14A built in 1965 ('95 Red', c/n 500404) was converted into the An-14Sh. Taxi tests commenced in

December 1981; in the winter of 1983 the aircraft finally took to the air, flown by OKB test pilot Vladimir G. Lysenko and flight engineer V. Mareyev. In taxying mode the skirt's ground clearance was 6-10 mm ($0^{15}\!/_{64}$ to $0^{25}\!/_{64}$ in). The aircraft was stable at taxying speeds up to 100 km/h (62 mph) and exhibited adequate directional stability in crosswinds up to 12 m/sec (24 kts).

One more aspect of the parked An-14Sh.

A rare in-flight shot of the An-14Sh. Note the new entry door on the port side.

In the course of the trials the An-14Sh covered a total distance of more than 700 km (435 miles) along paved and dirt runways. The tests had to be interrupted frequently for modifications and repairs to the experimental landing gear, causing the test programme to continue until April 1986. The skirt was never retracted in the test flights.

Ilyushin IL-28 landing gear testbeds

a) IL-28LSh: In 1958 a Moscow-built IL-28 coded '12 Red' (c/n 53005112) was converted into the IL-28LSh testbed (LSh = *lyzhnoye shassee* – ski landing gear) for testing the efficiency and durability of aircraft skis designed for dirt strips. The aircraft was fitted with a semi-retractable sprung skid under the centre fuselage. The skid was equipped with pressure sensors and mounted on a hollow box which could be filled with ballast to test it for various loads; the whole assembly could be raised and lowered by hydraulic rams. The nose gear unit was fitted with larger wheels and the main units had widely-spaced twin wheels rather

than the usual single ones. This modified undercarriage could not be retracted, so the mainwheel well doors were deleted to avoid making contact with the wheels. The skid was tested on airstrips with various soil density; the aircraft made high-speed runs but did not become airborne.

Interestingly, one Russian publication asserts that the IL-28 used for skid landing gear tests in 1953-57 and 1978-79 (!) was 'No. 710' (that is, '10 Blue', c/n 53005710). However, this cannot be the case, as *this* aircraft was used for ejection seat tests in the early 1950s!

b) tracked landing gear testbed: To enhance the *Beagle*'s ability to operate from tactical airfields a special tracked (caterpillar) landing gear was designed, built and tested on an IL-28 pursuant to a Council of Ministers directive of 11th January 1951. It allowed the bomber to operate from soft, wet, soggy or snow-covered airfields which rendered take-off with a conventional wheeled landing gear was very difficult or utterly impossible. The tests were considered successful but, owing to the extra weight and complexity of the experimental landing gear, it was not retrofitted to production aircraft.

Lisunov Li-2 (PS-84) with tracked landing gear

In 1942-43 an attempt was made to equip the Lisunov Li-2 *Cab* transport (originally called PS-84) with a tracked undercarriage designed by S. A. Mostovoy; this held the promise of considerably improving the aircraft's ability to taxi on soggy ground or slush. The undercarriage made use of the so-called 'Chechubalin caterpillars' (designed by N. A. Chechubalin, an employee of the North Sea Route's Inventors Section, in 1937); the tracks had a

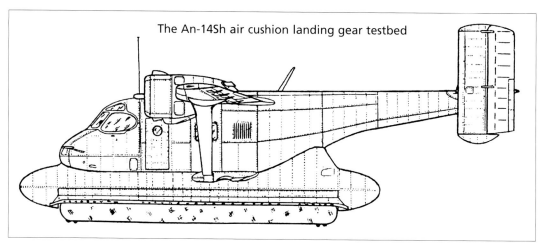

The An-14Sh air cushion landing gear testbed

The IL-28LSh ('12 Red') was used to investigate the efficacy of a skid landing gear that would enable tactical aircraft to operate from soggy strips.

The IL-28LSh had a special frame with an experimental skid that was lowered from the bomb bay on parallel arms. Note the fixed landing gear with twin mainwheels.

width of 300 mm (11¹³⁄₁₆ in) and were composed of Textolite composite cylinders having a diameter of 50 mm (1³¹⁄₃₂ in). Earlier a similar device had been successfully tested on the Polikarpov U-2 and R-5 biplanes.

Fitted with a fixed caterpillar undercarriage, the aircraft (an early Moscow-built PS-84, c/n 1841314) commenced manufacturer's tests in September 1942. Taxi runs, take-offs and landings were made on a normal airfield, on a ploughed field (a potato field) and on a marsh covered with a water layer up to 70 cm (27 in) deep. In the concluding part of the test report it was noted that a caterpillar undercarriage could be utilised on all types of combat aircraft, enabling them to operate from a ploughed field, an airfield with a soggy surface and even marshland which were unsuitable for aircraft with a normal wheeled undercarriage. The flight and field performance of the aircraft equipped with a caterpillar undercarriage, as

Close-up of the frame. The boxes could be filled with ballast to simulate various loads. Note the skid's shock absorber.

well as its handling, were no worse than those of a standard aircraft with an extended wheel undercarriage. The designers of the undercarriage considered that minor modifications to the Li-2 would make it possible to retract the caterpillar undercarriage. The prototype installation weighed twice as much as the normal wheels and 1.25-1.3 times more than the skis used on the Li-2. The test commission noted that the experience accumulated in the process of developing and testing the caterpil-

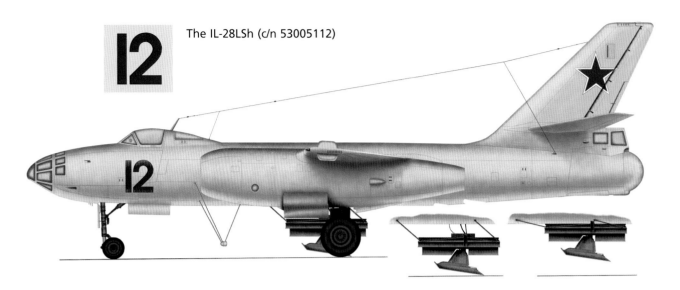

The IL-28LSh (c/n 53005112)

One of the two PS-84s fitted experimentally with a caterpillar undercarriage for soft-field operations.

lar undercarriage for the Li-2 made it possible to start the work on a similar undercarriage for combat aircraft immediately.

In April 1943 manufacturer's tests were conducted of another Li-2 (c/n 1846310) fitted with an improved version of the caterpillar undercarriage. Generally the report on the testing of this aircraft repeated the conclusions made after testing the first version of the caterpillars. It was stated that the undercarriage had passed manufacturer's tests successfully and the aircraft ought to be submitted to state acceptance tests.

No sooner said than done. NII VVS project test engineer Engineer-Capt. Shoobralov and test pilot Maj. Aleksey M. Khripkov were assigned to conduct the state acceptance tests of the Li-2 fitted with the caterpillar undercarriage. The aircraft's flight performance was determined with wheels only, as both sets of the caterpillar undercarriage had suffered damage in the course of testing. The trials showed that the caterpillars caused a marked reduction of ground manoeuvrability, the take-off and landing techniques became considerably more complicated, and the take-off run was increased. It was noted that the caterpillar design was not fully developed and its operation was unreliable. Even in its fixed version the caterpillar undercarriage was 290 kg (640 lb) heavier than the standard one.

The trials report stated that the flights had confirmed the possibility of utilising a caterpillar undercarriage for single special flights, providing it was well manufactured, but the design was unsuitable as a standard undercarriage type for mass operation. The caterpillar undercarriage failed to pass state acceptance tests, and it was not recommended for series manufacture. Despite the fact that the trials report was a 'thumbs-down', it was nevertheless confirmed that N. A. Chechubalin's concept was feasible. The state acceptance trials protocol was endorsed on 2nd July 1943.

Mikoyan/Gurevich SL-19 skid landing gear testbed

A MiG-19 was converted at LII into a research vehicle designated SL-19 (samo**lyot**-labora-**tor**iya – lit. 'laboratory aircraft', that is, research aircraft or testbed). It was fitted with a skid landing gear whose track and wheelbase (or rather 'skidbase') could be varied widely.

Mikoyan/Gurevich Ye-5/2 development aircraft with skid landing gear

In the late 1950s the Soviet Air Force attached much importance to the ability of its aircraft to operate from unpaved tactical airstrips that could be set up near the frontlines. Therefore at that time the Mikoyan OKB's landing gear design group evolved a unique skid landing gear design as part of the effort to refine the MiG-21. The 'skids' were quite unlike the devices traditionally described by this term – they were circular, functioning as metal wheels on paved runways and as skids on dirt strips; the idea was pioneered by M. N. Voronov.

The circular skids were tested in 1960 on the specially modified Ye-5/2 (the second prototype of the initial MiG-21 sans suffixe), which had by then received the tactical code '12 Red'. The aircraft was flown by Mikoyan OKB test pilots Gheorgiy K. Mosolov and Konstantin K. Kokkinaki; Ye. F. Nashchokin was the engineer in charge. The aircraft was light, with an all-up weight of 5,500 kg (12,125 lb), including 910 kg (2,010 lb) of fuel. The test programme started off in the autumn/winter season of 1959 with towing and taxying on a dirt strip at the LII airfield in Zhukovskiy. From 4th May 1960 onwards the tests proceeded at a specially prepared dirt airfield (Tret'yakovo airfield in Lookhovitsy, Moscow Region), continuing until 29th July.

The aircraft was fitted with non-retractable levered-suspension main gear units making use of stock half-forks and shock absorbers from the MiG-19. Two types of skids – the novel circular ones and traditional rectangular ones – could be fitted; for ground handling they were replaced with dummy wheels. The nose gear unit was patterned on that of the production MiG-21F, featuring a 500 x 180 mm (19.68 x 7.08 in) KT-38 nosewheel and an MRK-10 steering mechanism (mekha**nizm** razvo**rot**a kole**sa**).

The rectangular skids were AL-19 aluminium alloy castings measuring 1,168 x 240 mm (45⁶³⁄₆₄ x 9²⁹⁄₆₄ in); the surface resting on the ground measured 900 x 210 mm (35⁷⁄₁₆ x 8¹⁷⁄₆₄ in). The skids featured detachable upturned nose sections preventing them from digging into soft or soggy ground. Additional trailing arms were attached to the rear ends of the skids, moving in parallel with the half-forks to ensure that the skids remained parallel to the ground, regardless of oleo compression. To stabilise the aircraft's movement on soft and soggy ground the skids were fitted with removable stabilising keels on the underside.

The circular skids were cast discs made of AL-19 alloy (alyu**min**iy **li**teynyy – aluminium optimised for casting) faced with EI-659 steel sheet on the underside. Each disc was 724 mm (2 ft 4½ in) in diameter and 126 mm (4³¹⁄₃₂ in) thick; the junction between the undersurface and the sides of the discs was curved, with a 300-mm (11¹³⁄₁₆ in) radius. The discs were mounted on vertical axles with roller bearings, rotating freely in the horizontal plane; the suspension mechanism with parallel arms was similar to that of the first version. The horizontal position (in which the discs rested fully on the ground) was for landing; for taxying and take-off the axles of the discs were tilted 15° by hydraulic rams so that the discs touched the ground obliquely, acting as wheels. With a take-off weight of 5,500 kg, the aircraft's runway loading was 1 kg/cm² (14.28 lb/sq in) with the skids in landing position and 2.5 kg/cm² (35.7 lb/sq in) with the skids in take-off position. A keel 30 mm (1³⁄₁₆ in) high and 115 mm (4³³⁄₆₄ in) long was located centrally on the underside of each disc.

With rectangular skids the aircraft commenced its take-off run on a grass surface with a bearing strength of 7-15 kg/cm² (100-214 lb/sq in) when the engine thrust amounted to 1,300-2,100 kgp (2,865-4,630 lbst); with the turf removed and a bearing strength of 10-12 kg/cm² (142-171 lb/sq in) the aircraft started moving with the engine in afterburner mode delivering 3,800 kgp (8,380 lbst). On ploughed and then rolled ground with a bearing strength of 7 kg/cm² the take-off run likewise began when the engine thrust reached 3,800 kgp; on snow-covered ground with a bearing strength of 15 kg/cm² a thrust of 1,700 kgp (3,750 lbst) was enough.

The Ye-5/2's directional stability and controllability during take-off and landing were adequate on all types of runway surface; however, taxying at speeds up to 40-60 km/h (25-37 mph) on soggy ground and frozen earth revealed inadequate stability. Taking off from a dirt strip was somewhat more complicated as

The Ye-6T/2 after conversion as a landing gear testbed with special levered suspension main units fitted with skids.

This sequence of stills from a ciné film shows the modified Ye-6T/2 taking off from the dirt strip at Zhukovskiy.

compared to operation from a paved runway – the nosewheel was reluctant to lift, but when it did the aircraft became airborne immediately. The landing procedure was also more complicated, as the aircraft dropped its nose immediately on touchdown instead of the usual nose-up run.

On a grass strip with tall grass and a bearing strength of 12-15 kg/cm^2 the aircraft had a take-off run of 800-1,000 m (2,640-3,280 ft), becoming airborne at 310 km/h (192 mph); on landing the fighter touched down at 320 km/h (198 mph) and came to a standstill after a landing run of 1,760 m (5,770 ft). The testers noted that during the dry season the jet blast created a tremendous pall of dust, making formation take-offs impossible. The shock absorption of the landing gear also proved to be inadequate. With rectangular skids the Ye-5/2 had a normal take-off weight of 5,535 kg (12,200 lb).

The circular skids created somewhat higher drag when moving on the ground; on the other hand, with these skids the aircraft was more manoeuvrable on the ground, making tighter turns when one of the skids was set horizontal. At 60 km/h (37.2 mph) the aircraft tended to yaw on take-off and landing.

The concluding part of the 'Report on the taxi tests on unpaved strips undertaken with the Ye-5 aircraft equipped with an experimental skid landing gear' stated:

'1. Rectangular skids: It is advisable to equip the MiG-21F with skids, taking account of the changes made to the aircraft's design. (as compared to the Ye-5 – Auth.)

2. The circular skids undoubtedly offer the advantage of higher manoeuvrability and easier setting in motion on hard-packed earth surfaces.'

The report was signed by the MiG-21's project chief Anatoliy G. Broonov and his deputy Gheorgiy A. Sedov; General Designer Artyom I. Mikoyan endorsed the document on 28th September 1960.

Mikoyan/Gurevich Ye-6T/2 skid landing gear testbed

When the trials of the K-13 weapons system had been completed, the Ye-6T/2 (the second prototype of the missile-armed MiG-21F-13) found further use as a test vehicle. In 1961 the aircraft was fitted with a non-retractable skid landing gear with levered-suspension main units for operation from dirt strips; rectangular and circular skids were tried out.

The tests took place at Lookhovitsy-Tret'yakovo airfield, Moscow Region, involving 18 taxi runs and 11 flights. Perhaps it is this aircraft that was referred to in one source as 'MiG-21 No. 105 tested in 1968-72'.

Mikoyan/Gurevich MiG-21F-13 unpaved strip operations testbed

A production MiG-21F-13 coded '01 Red' (c/n unknown) was modified under a research programme concerned with tactical fighter operations from semi-prepared airstrips; thus it can, to a certain degree, be regarded as a landing gear testbed. The rear transparency of the cockpit canopy was replaced with sheet metal, as on Czechoslovak-built MiG-21F-13s (Aero S-106s), and the pitot boom was located dorsally on the centreline instead of ventrally to minimise the risk of foreign object damage. Retaining the normal wheel undercarriage, the aircraft made a series of test flights from dirt and snow-covered airstrips. Shortly after the beginning of these trials, phototheodolite calibration markings in the form of sets of vertical stripes were applied to the airframe to assist measurements and calculations.

The dorsally mounted pitot became a feature of all subsequent MiG-21 variants.

MiG-21F-13 PSP strip operations testbed

As part of the same research programme, in the early 1960s another production MiG-21F-13 coded '12 Red' (c/n unknown) was used to explore the possibility of fighter operations from *ad hoc* tactical airstrips paved with perforated steel plate (PSP).

Mikoyan/Gurevich MiG-21PF wheel/skid landing gear testbed

A single production MiG-21PF *Fishbed-D* interceptor coded '02 Red' (c/n N76210820) was modified in 1963 with a view to improving its ability to operate from unpaved tactical airstrips. The aircraft was equipped with a wheel/skid landing gear similar to that tested on the Ye-7/4 prototype in 1962. The skids were fitted to the main units only; they were made of titanium and had a surface area of 600 cm^2 (93 sq in), reducing the runway loading by 50%. The wheel/skid landing gear allowed the fighter to operate from soggy airstrips with a bearing strength of no more than 4 kg/cm^2 (57 lb/sq in). The skids were lowered into position by hydraulic rams, being almost level with the bottoms of the mainwheels; for operation from normal (firm) runways they were raised clear of the ground. The nose gear unit was unaltered.

Like the modified Ye-7/4, MiG-21PF c/n N76210820 was tested with two dummy R-3S missiles on the wing pylons. Curiously, the pylons proper and the missile launch rails attached to them were borrowed from another MiG-21PF (c/n N76210725). Ciné films shot during the tests clearly show that the wing pylons of MiG-21PF '02 Red' were marked 0725.

MiG-21PF '02 Red' (c/n N76210820) was fitted experimentally with a wheel/skid landing gear.

MiG-21PF '02 Red' during tests on a snow-covered airstrip.

The starboard hydraulically powered skid of MiG-21 PF '02 Red' in standby (far left) and operating position.

Boris A. Orlov was project test pilot for this programme. The following is an extract from his memoirs:

'With the skids lowered into position the aircraft turned into a real all-road vehicle, if not exactly a cross-country vehicle. On occasion a lorry would have trouble moving across the field, whereas my MiG would just whizz through all that mud… Yet we took care to keep out of the worst mud because the nose-wheel tended to sink into the soft ground almost up to the axle.

The flights proper presented no major problems – I would take off and land again

This series of stills from a ciné film shows MiG-21PF '02 Red' taxying and taking off during soft-field tests.

Far right:
The ruts left by MiG-21PF '02 Red' after landing on a soggy dirt airstrip.

MiG-21PF c/n N76210820 with a wheel/skid landing gear

some time later after burning off the fuel. However, I had to touch down exactly at the specified spot where the bearing strength of the strip had been measured – in other words, on a very limited stretch of runway.'

Even though the results were good, the Mikoyan OKB decided against using the wheel/skid landing gear on the MiG-21. The only Soviet tactical aircraft to utilise this type of landing gear in service was the Su-7BKL fighter-bomber (the KL suffix denoted *ko**lyos**no-**lyzh**noye shas**see*** – wheel/skid landing gear).

Petlyakov Pe-2 landing gear testbed

Following successful tests of the Yakovlev UT-2N (see below), a production Pe-2 dive-bomber was converted into a testbed for an air cushion undercarriage conducted by engineers Nikolay Ivanovich Yefremov and Aleksandr Davidovich Nadiradze at TsAGI and LII in 1939-41. After take-off the inflatable rubber skirts with blowers creating the air cushion were to be deflated, allowing the landing gear to be retracted. In 1941 the modified Pe-2 began initial taxiing runs, but the outbreak of the Great Patriotic War led further work on the project to be halted.

Polikarpov U-2 with tracked landing gear

As noted above, the tracked undercarriage designed by N. A. Chechubalin was fitted experimentally to an example of the U-2 primary trainer/utility aircraft.

Polikarpov R-5 with tracked landing gear

At about the same time the tracked undercarriage designed by Chechubalin was successfully tested on a Polikarpov R-5 reconnaissance aircraft.

A Polikarpov U-2 trainer fitted experimentally with a tracked landing gear.

A Polikarpov R-5 with similar tracked main gear units. Note the bracing wires at the front.

Sukhoi S22-4 testbed

To explain the reasons leading to the appearance of the Su-7BKL – a derivative of the Su-7BM with a wheel-skid landing gear – in 1965, we have to go back in time to 1960. It was then that the Sukhoi OKB began a new phase of work in solving the problem of operating the Su-7B from unpaved airstrips. Now, after fitting and testing an all-skid landing gear on the S-23, it was proposed to test a combined version by fitting very small supporting skids, so-called 'skilets' (*lyzhonki*), to the standard main gear units in order to increase the landing gear footprint and reduce the pressure.

This provided the option of using either the main wheeled undercarriage or the auxiliary wheel-skid variant, depending on the

Loaded with twenty-eight S-3K unguided rockets on four APU-14U launchers, the S22-4 landing gear testbed based on the Su-7BM undergoes trials with skids on the main gear units. Note the aft-facing ciné camera under the intake.

conditions. For take-off and landing on normal concrete runways the 'skilets' remained retracted, but when operating from unpaved airstrips where the ground was insufficiently firm for a conventional wheeled undercarriage, the 'skilets' were lowered, reducing the surface loading. Importantly, the aircraft's field performance remained acceptable for service use. It was also decided to test on this aircraft the new twin-canopy braking parachute system and the attachment of two SPRD-110 jet-assisted take-off (JATO) boosters (***star**tovyy porokho**voy** ra**ket**nyy **dvig**atel'* – solid-fuel take-off rocket [booster]).

The S22-4 experimental aircraft was used in this programme, which was performed in various climatic zones, at various times of the year, operating from concrete and earth runways with and without JATO boosters. The design of the wheel-ski undercarriage and the whole set of measures for improving field performance tested on the S22-4, including the brake parachute container relocated to the base of the fin and the installation of the SPRD-110 boosters, was recommended for

series production on the Su-7BKL version which replaced the Su-7BM on the production line in 1965.

Sukhoi S-23 landing gear testbed

In addition to the work on the Su-7B fighter-bomber, the Sukhoi design bureau (OKB-51) was engaged in other development programmes at the same time. One of these, based on the Su-7, was an experimental programme to improve the field performance of tactical aircraft.

OKB-51 was at that time effectively the only design bureau in the USSR to take the problem of operating aircraft from unpaved airfields seriously, and it had considerable success in finding ways to solve it. The reason why this problem came into the picture was that, in the event of war, most airfields with paved runways might be put out of action in the first hours of the hostilities. A possible answer appeared in fitting tactical aviation aircraft with either removable skids or a composite wheel-skid undercarriage to enable them to operate from unpaved (dirt or grass) tactical airstrips.

To explore these ideas, in 1958 OKB-51 devised the S-23 which had various types of undercarriage: all-skid (with skids on all three units) or composite (with a nosewheel and skids fitted to the main gear units). Tests of the S-23 lasted from April 1959 to August 1960, at different times of the year, in various parts of the country under different weather and climatic conditions, and on different surfaces, including snow and earth with differing degrees of firmness. The results of these tests yielded a great deal of material for further work in this field which was continued later on the S22-4 and S-26.

The S-23 prototype served as a testbed for the radical (for a jet fighter) ski-type main landing gear. These tests proved surprisingly successful.

Here the S-23 development aircraft based on the Su-7 fighter is shown at GK NII VVS with small skids on the main gear units and a ciné camera fairing under the nose.

The S-23 seen taking off from a dirt strip. Note the short nose and the bulky fairing under the nose accommodating ciné cameras trained on the main gear units.

Sukhoi S-26 (S26-1 and S26-2) testbeds

Practical research by OKB-51 in the field of improving the aircraft's performance when operating from unpaved airstrips was carried out in the course of several years. Tests of two modified production Su-7Bs, the S26-1 (c/n 3601) and S26-2 (c/n 3608), on which testing of the skid-type undercarriage continued, were a new phase in the test programme carried out on the S-23. Unlike the latter aircraft, however, it was only the composite version with a nosewheel and skids on the main gear units that was tested. The two S-26 testbeds were converted at OKB-51's prototype construction facility (MMZ No. 51 at Moscow-Khodynka) in 1963. The changes included:

• fitment of two skids, each with a surface area of 0.25 m² (2.68 sq ft), to the main gear units;

• fitment of a 660 x 200 mm (25.98 x 7.87 in) K2-106 non-braking nosewheel instead of the 510 x 140 mm (20.0 x 5.51 in) KN-100 braking nosewheel;

• provision of a new twin-canopy brake parachute housed in a fairing at the base of the rudder (on the S26-1 only);

• attachments under the rear fuselage for two SPRD-110 JATO boosters.

As in the case of the S22-4, the two S-26 machines were tested in various seasons of the year, under various climatic conditions and on earth surfaces with varying degrees of hardness. The results showed a very high degree of improvement.

The uncoded S26-2 is now preserved in the Central Russian Air Force Museum in Monino in an all-skid configuration (!).

The S26-1 with twin-wheel handling dollies on the skid- equipped main gear units. The aircraft carries FAB-250M62 low-drag bombs, drop tanks and an aft-facing ciné camera is fitted ahead of nosewheel well to record the main gear's operation.

The S26-1 (note the dorsal brake parachute housing which was not fitted to the S26-2) blasts off with SPRD-110 JATO boosters. The ski main landing gear is readily visible. The JATO units shortened the take-off run appreciably.

Close-ups of the starboard ground handling dolly. There is no way these bulky assemblies could have fitted into the Su-7's main-wheel wells.

The S-26's main skid, complete with rocking dampers and attachment.

The S-26's steerable nose gear unit with a skid.

Sukhoi Su-17 wheel/skid landing gear testbed

A Su-17 coded '30 Red' was used for investigating the possibility of operating the aircraft from unpaved tactical strips. To this end the normal main gear units were replaced with modified ones similar to those of the S-26, which could be fitted with staggered-tandem wheel dollies for taxying or towing. A cylindrical pod with a pointed nose housing ciné cameras trained on the main gear units was mounted on a short pylon under the nose, and photo calibration markings were applied to the fuselage and fin.

In this guise the aircraft operated from unpaved runways, carrying various payloads which included a *Metel'-A* (Blizzard-A) radar detector/guidance system pod on the starboard inner wing pylon. This pod worked with the Raduga Kh-28 (NATO AS-9 *Kyle*) anti-radar missile used for suppression of enemy air defences.

Su-17M2D experimental fighter-bomber (S-32M2D)

The Su-17M2D fighter-bomber (in-house designation S-32M2D) was the prototype of the Su-17M2's export version, Su-22 *Fitter-F* (S-32M2K). The D stood for ***dvig**atel'* (engine), since the Lyul'ka AL-21F-3 afterburning turbojet (which was in short supply) was replaced with the Tumanskiy R29BS-300 afterburning turbojet and the rear fuselage and tail unit were redesigned accordingly. Appropriate changes were made to the fuel system, as well as the hydraulics, electrics, and fire suppression system.

The Su-17M2D first flew in January 1975. In June it was delivered to GK NII VVS for special trials; apart from performance testing, these included operations from unpaved and soggy airstrips, for which purpose the aircraft was fitted with a modified all-skid landing gear.

Sukhoi T-58L development aircraft

In the early 1960s, when the Cuban Missile Crisis put the two superpowers on the brink of an all-out war, the Soviet Armed Forces paid much attention to dispersing troops (including Air Force units) in order to make them less vulnerable to enemy strikes. In the case of military aviation this meant operations from unpaved tactical and reserve airstrips – for which, as it turned out, the units of the Air Force and the PVO fighter arm were totally unprepared. This led the military to demand insistently that all

Su-17 '30 Red' takes off from a dirt strip, carrying external stores, including a Metel'-A air-to-surface missile guidance pod on the starboard inboard wing pylon.

'10 Blue', the Su-17M2D (S-32M2D), which was the prototype of the Su-22 export version, was flown both with a normal landing gear and with a skid landing gear for operations from dirt strips as shown here.

tactical aircraft types should be capable of operating from semi-prepared dirt strips.

After the state acceptance trials the Sukhoi OKB followed the recommendations of the State commission, developing a special skid landing gear for the second prototype Su-15 *sans suffixe* (T58D-2) with a view to exploring the possibility of operating the new interceptor from unpaved strips. By then the OKB had accumulated a wealth of experience with the S22-4, S-23 and S-26 development aircraft; thus the optimum layout could be chosen quickly and the design work completed within a short time frame. The engineers selected a mixed arrangement with skids on the main units and a wheeled nose unit, as on the S-26.

Modification work on the T58D-2 was completed in the first half of 1965, whereupon the aircraft was redesignated T-58L (the designation is sometimes rendered as T58-L), the L standing for *lyzhnoye shassee* – skid landing gear. The conversion involved installing new main gear units which could be quickly reconfigured from wheels to skids and back again, with appropriate modifications to the mainwheel wells and main gear doors; the skids were provided with a lubrication system to facilitate movement on grass and packed

earth surfaces. The standard castoring nose gear unit was replaced by a steerable unit, as on the S-26 (normally the T-58 was steered on the ground by differential braking, which was impossible with skids). Provisions were made for installing solid-fuel JATO boosters, changes were made to the forward fuselage and cockpit, and a new Sukhoi KS-4 ejection seat was fitted.

Vladimir S. Ilyushin performed the T-58L's first flight on 6th September 1965. From 1966 until the mid-1970s the aircraft underwent extensive testing on various semi-prepared grass, dirt and snow strips in various climatic zones; GK NII VVS also participated in these tests. Apart from Ilyushin, the aircraft was flown by Sukhoi OKB test pilots Yevgeniy K.

The Su-17M2D pictured during wet-runway trials. Note that the main gear doors have been removed, showing to advantage the landing gear oleo compression links.

The T-58L in late configuration. It was converted from the T58D-2, retaining the tactical code '32 Blue'. The new twin-wheel nose gear unit giving the aircraft a nose-up attitude is clearly visible.

Kukushev, Yevgeniy S. Solov'yov, Vladimir A. Krechetov *et al*, as well as GNIKI VVS pilots.

In one of the test flights with Kukushev at the controls the aircraft banked at the moment of rotation and the anti-flutter boom on one of the stabilators dug into the ground, ripping away together with a portion of the skin. To prevent a repetition of this incident the anti-flutter booms were angled up 15° on all Su-15s, production aircraft and prototypes alike. Shortly afterwards the T-58L was refitted with a new, taller nose gear unit featuring twin 620 x 180 mm (24.4 x 7.0 in) KN-9 non-braking wheels; the purpose of this modification was to increase the angle of attack on take-off (thereby increasing lift and shortening the take-off run), raise the air intakes higher above the ground (thereby reducing the risk of foreign object ingestion) and improve ground manoeuvrability.

The skid landing gear was not introduced on the production model because the tests of the T-58L revealed major operational problems associated with operations from unpaved strips. The vibrations experienced on uneven runways subjected the avionics and armament to augmented loads which could ruin them; also, the missiles were liberally spattered with dirt, which likewise could put them out of action. It should be noted that production

Su-15s with a standard landing gear also underwent tests on unpaved runways; these tests ultimately led to the introduction of a twin-wheel steerable nose gear unit on the Su-15TM.

Upon completion of the test programme in 1974 the T-58L was donated to the Air Force Academy named after Nikolay Ye. Zhukovskiy as an instructional airframe. Fortunately it was not 'vivisected', as ground instructional airframes often are, and subsequently moved to the Soviet Air Force Museum (now Central Russian Air Force Museum) in Monino, where it resides to this day.

Tupolev SB-2M-103 tricycle landing gear testbed

Like everywhere else, Soviet aircraft of the pre-war period were 'taildraggers' (that is, had a tailwheel landing gear). However, at the end of the 1930s the Soviet aircraft designers were already looking at the more advanced tricycle (nosewheel) landing gear layout. Before the new arrangement could be used on a new design, however, it needed to be verified. Therefore LII converted a late-production Tupolev SB **bomber** powered by Klimov M-103 engines (SB-2M-103) into a landing gear testbed. The standard short glazed nose accommodating the navigator/bomb-aimer was replaced by an elongated 'solid' nose mounting the nose gear unit; the main units were also new. The experimental gear was non-retractable, all three units being connected to a truss mounted externally under the fuselage. The standard tailwheel was retained, acting as a tail bumper in the event of overrotation.

The modified bomber was tested in 1940-41. The outbreak of the Great Patriotic War on 22nd June 1941 interrupted these experiments.

Tupolev/Myasishchev ShR-1 and ShR-2 bicycle landing gear testbeds

When the newly organised OKB-23 led by Vladimir Mikhaïlovich Myasishchev began development of the M-4 (*izdeliye* M) *Bison-A* strategic bomber in March 1951, the designers chose the bicycle landing gear arrangement from the outset. The twin-wheel nose unit envisaged originally and the main unit fitted with a four-wheel bogie absorbed 30% and 70% of the aircraft's weight respectively. This arrangement was not totally new to Soviet aircraft designers by then, as the Alekseyev I-215D had successfully flown in October 1949. However, there was as yet no experience of using the bicycle landing gear on a heavy aircraft (the 'aircraft 150' bomber did not make its first flight until 5th September 1952), and the need arose to verify the novel arrangement on a suitably converted bomber.

Hence OKB-23 developed a testbed version of the Tupolev Tu-4 for studying the behaviour of a heavy aircraft with a bicycle landing gear in take-off, landing and taxying modes and for

training flight crews. The aircraft was designated ShR-1, the Sh denoting *shassee* (landing gear) and the R referring to the Tu-4's in-house product code (*izdeliye* R).

A brand-new *Bull* manufactured by the co-located production plant No. 23 at Moscow-Fili (no serial, c/n 230322) was delivered to the Myasishchev OKB for conversion into the ShR-1 testbed. The standard nose gear unit was replaced by a new twin-wheel levered suspension strut with larger 1,450 x 520 mm (57 x 20.4 in) wheels. The scratchbuilt main gear unit consisted of two stock Tu-4 main gear oleos mounted in tandem and rigidly connected; to these was hinged a four-wheel bogie with 1,450 x 520 mm wheels. The entire assembly was attached to a hefty steel frame installed near the aft bomb bay; it could be installed in three different ways so that the main gear absorbed 72%, 85% or 90% of the total weight. This unusual design feature was introduced to see how the changing weight distribution and wheelbase affected the aircraft's field performance and ground manoeuvrability (the reason was that different versions of the M-4's PD project featured an aft-retracting or forward-retracting nose gear unit, with

This extensively modified SB-2M-103 was fitted with a fixed tricycle landing gear to explore the techniques of operating tricycle-gear aircraft. Note the thicket of main gear bracing struts.

Tu-4 c/n 230322 was converted into the ShR-1 bicycle landing gear testbed for the M-4 bomber. The fixed experimental landing gear is clearly visible here.

The main gear unit of the ShR-1; note the tandem shock absorbers.

began in April, lasting until June. The ShR-1 made 50 taxi runs and 34 flights to check the aircraft's stability and controllability, the operation of the nose gear steering mechanism and the optimum position of the main gear and outrigger struts. As the main gear was moved aft the load on the nose gear gradually increased from 10.6% to 20.8% to 28.6%. Test data were recorded automatically.

Meanwhile the M-4's twin nosewheels had been rejected in favour of a four-wheel bogie with electrically steerable nosewheels. Hence the Tu-4 testbed was modified accordingly and redesignated ShR-2. In this guise the aircraft underwent further tests in 1953, making 24 taxi runs and 17 flights; the nose gear bogie absorbed 20% of the weight versus 40% on the real *Bison-A*. The tests showed that the bicycle landing gear offered excellent ground handling and simplified the take-off technique – the M-4 required no rotation to become airborne, taking off almost of its own will.

attendant changes in wheelbase and weight distribution). The nosewheels and mainwheels were equipped with hydraulic brakes.

The outrigger struts installed a short distance outboard of the Nos. 1 and 4 engines consisted of stock Tu-4 nose gear units mated to special truss-type mountings which allowed the length of the outriggers to be adjusted. Interestingly, the non-retractable experimental landing gear could be removed and the normal gear reinstalled for positioning flights.

The conversion job was completed in January 1952; Stage 1 of the test programme

Tupolev Tu-16 landing gear testbed

In the late 1950s a single Tu-16 coded '56 Red' (c/n unknown) was adapted for ground tests of the 'jump strut' nose landing gear mechanism devised for the Myasishchev M-50 *Bounder* supersonic heavy bomber. An extensible extra twin-wheel strut was mounted immediately aft of the nose gear unit, emulating the M-50's four-wheel nose gear bogie which could be tilted to increase the angle of attack on take-off. This feature was necessary because the M-50 had a bicycle landing gear.

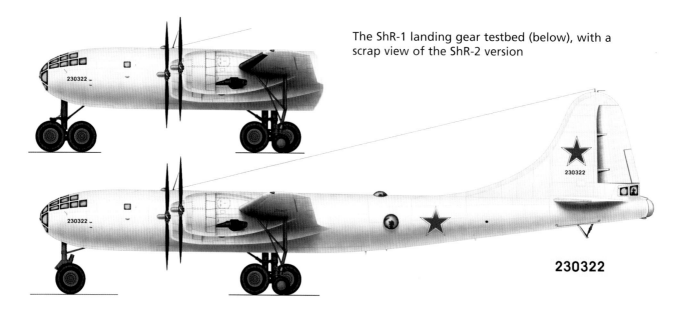

The ShR-1 landing gear testbed (below), with a scrap view of the ShR-2 version

230322

Left and below: '56 Red', a Tu-16 bomber equipped experimentally with a second nose gear unit featuring a jump strut. The arrangement emulated the nose bogie tilting mechanism of the Myasishchev M-50 bomber.

Above: The tail bumper of the Tu-16 testbed had a pair of wheels added as an extra protection in the event of overrotation.

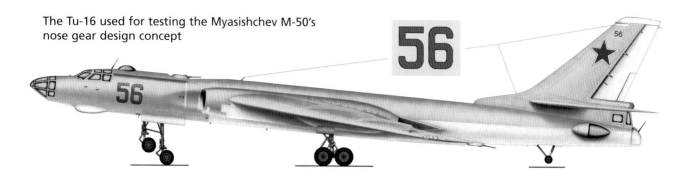

The Tu-16 used for testing the Myasishchev M-50's nose gear design concept

Tupolev Tu-134A landing gear testbed

In the early 1970s the ninth production Tu-134A airliner, CCCP-65653 (c/n 0351009), was outfitted by the Tupolev OKB and the Siberian Aviation Research Institute (SibNIA – *Sibeerskiy naoochno-issledovatel'skiy institoot aviahtsii*) in Novosibirsk for measuring the structural loads applied to the landing gear on different runways. This instrumented test aircraft made a round of all Soviet airports into which the Tu-134 could operate. As a result of these tests the main gear bogies were reinforced and the original KT-81/3 wheels replaced by lighter KT-153 wheels to preclude fatigue cracking of the bogies – a problem encountered on the first Tu-134As.

The logo of SibNIA which undertook tests with a modified Tu-134A.

Yakovlev UT-2N (SEN) landing gear testbed

A production Yakovlev UT-2 primary trainer was converted into a testbed, or rather technology demonstrator, in experiments with an air cushion landing gear designed by Nikolay I. Yefremov and Aleksandr D. Nadiradze at TsAGI and LII in 1939-41. The aircraft, designated UT-2N, was also known as SEN (or, more exactly, SYeN, *samolyot Yefremova i Nadiradze* – Yefremov's and Nadiradze's aircraft). Nadiradze had previously worked on air cushion vehicles (ACVs) at a different OKB headed by V. I. Levkov.

The air cushion undercarriage was intended to replace the normal wheel or ski undercarriage and enable the aircraft to take off from any surface along a downhill path with next to no take-off run. Mounted under the aircraft on a rear centreline pylon and lateral N-struts was a big annular rubber balloon or skirt resembling an inflatable dinghy turned upside down. The size of the skirt was calculated to ensure the aircraft's stability in the event of sideways movement during the take-off. A fan driven by a cowled 25-hp motorcycle engine mounted on a platform atop the skirt supplied the pressure for the air cushion.

During the tests, which were conducted in 1940 by LII test pilots Igor' I. Shelest, Mikhail

The UT-2N (alias SEN) was a testbed for an experimental air cushion landing gear developed by Yefremov and Nadiradze. The fan forcing air into the cushion was driven by a 25-hp motor in the housing on top of the skirt.

The Yakovlev UT-2N (SEN)

The Yak-27LSh development aircraft based on the production Yak-27R. The small skid on the main gear unit between the wheels is just visible in this view.

M. Gromov, Andrey B. Yumashev and Aleksandr P. Chernavskiy, the aircraft behaved quite normally, demonstrating the practicability of the concept. Encouraged by the results, NKAP tasked the two engineers with developing a retractable air cushion undercarriage for the Petlyakov Pe-2 dive-bomber. This was duly done in 1941, as mentioned earlier.

Yakovlev Yak-25 landing gear testbed

The third (or first?) prototype of the single-seat, single-engined, straight-wing Yak-25 fighter of 1947 (the first aircraft to be thus designated) was fitted experimentally with a non-retractable bicycle landing gear. The uncertainty as to which aircraft was involved is due to the fact that the first (c/n 115001) and third (c/n 115003) Yak-25s wore the same tactical code, '25 Yellow', in order to fool hypothetical spies! This test programme gave the Yakovlev OKB the necessary experience to develop the Yak-120 two-seat twinjet swept-wing interceptor – the second Yak-25 (*Flashlight-A*).

Yakovlev Yak-27LSh wheel/ski landing gear testbed

A production Yak-27R *Mangrove* reconnaissance aircraft was used to test an experimental heavy-duty landing gear designed for operation from semi-prepared snow-covered airstrips. The twin mainwheels were augmented with a large skid that could be raised by a hydraulic ram when not in use, while the nose unit was reinforced and equipped with twin 600 x 125 mm (23.62 x 4.92 in) K-262 wheels replacing the usual single size 600 x 155 mm (23.62 x 6.10 in) nosewheel in a fork. The landing gear was non-retractable because the twin nosewheels were too wide to fit into the standard wheel well. The aircraft was appro-

priately designated Yak-27LSh (***lyzh***noye shas***see*** – ski undercarriage).

Trials began in late 1963 and continued until the spring of 1964; at one time the aircraft was also flown with a standard single-wheel nose gear unit. The aircraft was flown by LII test pilots Boris V. Polovnikov, Vladimir I. Kryzhanovskiy, Oleg V. Droozhinin and Anatoliy N. Presnov. Similar test programmes were undertaken by the Sukhoi and Mikoyan design bureaux.

3. Control system testbeds

Ilyushin DB-3 in experiments with remote-controlled aircraft

A research programme undertaken between 1935 and 1941 focused on remote control of aircraft. This research was done by the Ostekhbyuro (***O***sob***oye tekh***nich***eskoye byu***ro po vo***yen***nym izobre***ten***iyam spetsi***ahl'***novo nazna***chen***iya – Special Technical Bureau for Special-Purpose Military Inventions) headed by Vladimir I. Bekauri in Leningrad, an establishment responsible for developing advanced weapons systems. The experiments with several types of 'telemechanical' aircraft (the Soviet term of the time for remote-controlled aircraft) were aimed at turning them into radio-controlled flying bombs or into reconnaissance drones for use against well-protected targets where the risk of being shot down was especially high. In addition to the obsolescent Tupolev TB-3 (ANT-6) four-engined heavy bomber and Tupolev SB (ANT-40) fast bomber, these experiments involved Ilyushin DB-3 bombers. The latter were used primarily in the role of drone director aircraft for the pilotless TB-3s (references to conversion of the DB-3 into flying bombs are rather vague).

Ilyushin IL-28 control system testbed

Between 1953 and 1954 an unidentified IL-28 was used to verify the automatic take-off and landing approach system developed for the IL-28M remote-controlled target drone (*mishen'*).

Mikoyan/Gurevich MiG-15 control system testbeds

Similarly, in 1954 an unidentified MiG-15 served for verifying an automatic take-off system – apparently also intended for a target drone. One more testbed of the same sort was used in 1966 for testing the control system of the MiG-15MNV low-altitude target drone (*mishen' nizkovysotnaya*).

Mikoyan/Gurevich MiG-15U autopilot testbed

Also in 1953-54, LII used an unidentified MiG-15 for testing an autopilot developed for fighters. The aircraft was referred to in a Russian publication as the MiG-15U. Some Western authors have used this designation in error for the UTI-MiG-15 *Midget* trainer; however, Russian authors should know better than to make such blunders. Therefore, it is possible that the U stands for *oopravleniye* – control.

'02 Blue', the MiG-25PU modified for verifying the Buran space shuttle's control system. It is seen here at Zhukovskiy, its home base.

Mikoyan/Gurevich MiG-17 control system testbeds

In 1952 an early-production MiG-17 *sans suffixe* (*Fresco-A*) was used to test a BU-1A hydraulic actuator in the elevator control circuit which reduced the stick forces threefold. The actuator was housed in the fin and accessed via a removable panel on the starboard side of the fin.

Three years later an altogether different MiG-17 testbed served for verifying an automatic control system that found use on the MiG-17M (M-17) target drone.

LL-MiG-17 reaction control testbed

As Aram Nazarovich Rafaelyants developed his Toorbolyot vertical take-off and landing (VTOL) technology demonstrator vehicle, LII tested the efficiency of wingtip reaction control jets, or 'jet ailerons', on a modified *Fresco-A* designated LL-MiG-17. The work proceeded in 1956-59 under the supervision of V. V. Matveyev. Engineer Ye. N. Toropchenko and test pilots Yakov I. Vernikov and Sergey N. Anokhin were involved with the LL-MiG-17, while engineers A. I. Kvashnin, A. M. Lapshin, Yuriy I. Sneshko and test pilot Yuriy A. Garnayev participated in the development and trials of the Toorbolyot.

The initial research results obtained on the LL-MiG-17 were described in two reports prepared by LII in 1956. These research pro-

grammes provided valuable data for designing future VTOL aircraft – the Yak-36 *Freehand* technology demonstrator, the Yak-38 *Forger* shipboard attack aircraft and ultimately the Yak-41 *Freestyle* shipboard fighter.

Mikoyan MiG-23UB DLC system testbed

In 1976-82 a MiG-23UB *Flogger-C* trainer referred to in the press as 'MiG-23UB No. 919' (these are the last three digits of the c/n – *90**919) was converted by LII into a direct lift control (DLC) technology testbed. A stick-shaker was undergoing tests on the same aircraft during the same time frame. Before that, in 1972-75, 'MiG-23UB No. 919' had been used for stall/spinning tests and was suitably equipped with spin recovery rockets under the wings.

Mikoyan MiG-25PU control system testbed

A MiG-25PU *Foxbat-C* interceptor trainer coded '02 Blue' (c/n N22044011) was also used as a research aircraft under the *Buran* (Snowstorm) space shuttle programme – specifically, for evolving methods of checking the operation of the automatic flight control system and training Buran crews.

Mil' Mi-1 autopilot testbed

In 1951-53 an unidentified Mi-1 *Hare* light utility helicopter operated by LII was used for testing an autopilot developed for helicopters.

Mil' Mi-6 control system testbed

In 1975-1990 LII operated a Rostov-built Mi-6 *Hook* heavy-lift helicopter coded '76 Red' (c/n 4681710V) which had been modified for investigating the flight dynamics and verifying the control systems of advanced helicopters. The helicopter was thus a total in-flight simulator (TIFS). For the first time in Soviet helicopter design practice this testbed had a side-stick cyclic pitch controller instead of the usual centrally mounted cyclic pitch stick.

The results obtained with Mi-6 '76 Red' were used in developing and testing the Mi-26 *Halo* heavy-lift helicopter and the Mi-28 *Havoc* attack helicopter. They also allowed recommendations for standardising the stability and handling characteristics of helicopters to be worked out.

Mil' Mi-8T fly-by-wire control system testbed

In the mid-1980s LII began research on fly-by-wire (FBW) control systems for advanced helicopters. Accordingly in 1986 a Mi-8T *Hip-C* utility helicopter (identity unknown) also became part of a TIFS complex which also included a ground-based computing system and a data link system. The helicopter's control system included a VUAP-1 standardised helicopter autopilot (*vertolyotnyy oonifitseerovannyy avtopilot*).

In flight the on-board test equipment measured the main flight parameters, as well as cyclic pitch stick/pedal movements, stick forces and autopilot inputs, transmitting them to the ground control centre via data link. The data was processed in accordance with preset algorithms and monitored by test engineers; the ground control centre would then send corrective commands to the helicopter which were decoded and fed into the autopilot. This method allowed the characteristics of stability augmentation systems and FBW artificial-feel units to be developed and optimised.

Mil' Mi-8T side-stick controller testbed

In 1989 the same Mi-8T was further modified for testing the feasibility of using a side-stick cyclic pitch controller. Two such controllers (for the left and right hands) were installed at the co-pilot's workstation (on the starboard side). The tests took place in November and December 1990 with LII test pilots Vladimir L. Teben'kov, Boris Yu. Barsukov, Mikhail V. Pavlenko and Vener M. Mukhametgareyev flying the modified chopper. These pilots had different skill levels and different experience with helicopters, so one of the objectives was to see how demanding it is for pilots of different skill levels to fly a helicopter with a side-stick. Test flights were made with the chopper's experimental stability augmentation system (SAS) activated or deactivated.

The tests revealed that the side-stick controllers allowed the pilot to assume a more natural and relaxed posture when controlling the aircraft, as both arms lying on armrests were positioned at the same level. As a result, pilot fatigue was reduced dramatically. Moreover, with the SAS activated, flying the helicopter became much easier, allowing the pilot to concentrate on the mission and pay more attention to relevant inputs.

The Toorbolyot VTOL technology test rig on the ground.

Rafaelyants Toorbolyot research vehicle

In 1955-56 LII conducted several research programmes associated with VTOL. Among other things, a Mikoyan/Gurevich MiG-15 fighter was used as a research aircraft to investigate the low-speed control characteristics of an aircraft in vertical climb mode. The forward fuselage and Klimov VK-1A engine of a time-expired MiG-17 fighter was mounted vertically on a ground test rig to check the erosion effect of a VTOL aircraft's exhaust jets on paved (concrete) and unpaved runways.

Apart from the key issue of providing a thrust/weight ratio in excess of 1, the designers and engineers creating VTOL aircraft had numerous other problems to tackle. One of them was the need to ensure the aircraft's stability and controllability in the hover and at low speeds; conventional control surfaces were useless in these flight modes due to the insufficient dynamic pressure. Hence in 1955 the LII's design bureau developed a VTOL technology demonstrator named *Toorbolyot* (lit. 'Turbo-Flyer'). This was a unique aerial vehicle intended for exploring the behaviour of a VTOL aircraft during vertical take-off, hover, the transition to forward flight and back to the hover, and vertical landing, as well as to forward speeds close to zero. The Toorbolyot was the Soviet counterpart of the British TMR (Thrust Measurement Rig), better known as the Flying Bedstead, though the two were totally different in layout. The programme was supervised by V. N. Matveyev, with A. I. Kvashnin as engineer in charge; Aram N. Rafaelyants headed the actual design effort, with the participation of G. N. Lapshin. Yuriy A. Garnayev was appointed the Toorbolyot's project test pilot, with G. I. Kobets as operations engineer.

The Toorbolyot was powered by a single Mikulin RD-9BP axial-flow turbojet; this was a special non-afterburning version of the RD-9B powering the MiG-19 fighter, with the afterburner and associated fuel system compo-

Behold the flying octopus! The Toorbolyot becomes airborne in a cloud of black exhaust smoke.

nents replaced by a simple fixed-area nozzle to save weight. In order to check how the engine would run in the vertical position (for which it was not designed, after all), a LII team under O. Konstantinov built a test rig representing a tetrahedral pyramid constructed of rolled steel beams on a concrete foundation. The engine was mounted vertically in the middle, and the base of the pyramid was rigged with temperature sensors. A test team headed by S. Shcherbakov undertook a series of experiments with this rig.

At first, a Klimov RD-45F centrifugal-flow turbojet from a MiG-15 was installed, the engine speed being controlled by a throttle via a long cable, which was later replaced with a hydraulic drive. For safety's sake the fuel tank was placed well away from the rig. Later, in the summer of 1956, the engine was substituted by the intended RD-9BP.

The tests confirmed that the turbojet could run in the unaccustomed vertical position without any trouble. It was established that the exhaust jet fanned out uniformly in all directions like a thin veil upon hitting the ground rather than rising up like a billowing cloud. Thus the conditions were favourable for a vertical take-off.

The Toorbolyot had neither fuselage nor wings. The engine was perched on a welded steel truss with four vertical struts terminating in small castoring wheels and with four long outrigger booms of equal length at right angles to each other carrying reaction control nozzles. (If the TMR was the Flying Bedstead, then the Toorbolyot could easily be the Flying Octopus!) A small angular cockpit looking almost like the cab of a bulldozer was mounted on one side of the engine, featuring the usual flight control – a stick, rudder pedals (which belied their name because there was no rudder) and a throttle quadrant. The fuel was accommodated in two 200-litre (44 Imp gal) tanks flanking the engine.

The vehicle fitted into a 10 x 10 m (3 ft 3 in x 3 ft 3 in) square and was 3.8 m (12 ft 5³⁹⁄₆₄ in) tall when parked. The take-off weight was 2,340 kg (5,160 lb), which was just a little less than the RD-9BP's take-off thrust of 2,835 kgp (6,250 lbst).

On the whole the contraption looked like the least likely thing to fly. Yet fly it did. Predictably, controlling the Toorbolyot wasn't easy at all. The engineers at LII had to invent a special lift control device. Two petals controlled by the pilot were located below the edge of the engine nozzle at right angles to

the engine axis. They were made of heat-resistant alloy and moved symmetrically in the horizontal plane to enter the exhaust jet, deflecting part of the efflux and thereby reducing lift; the engine ran at constant rpm all the while. This unusual device ensured sufficiently precise control in the vertical plane.

Yuriy Garnayev made dozens of test flights in this unique aerial vehicle. He later recalled that the tests were mainly concerned with determining whether jet vanes or reaction control nozzles (puffers) were best for controlling an aircraft in VTOL mode. The jet vanes were positioned in the engine efflux, so their efficiency depended on the engine's operating mode and thrust at the moment. The puffers, on the other hand, used compressed air bled from the engine. It turned out that a combination of both types provided the best results, allowing the VTOL aircraft to move in all

Upon completion of the trials the Toorbolyot was preserved at the Soviet Air Force Museum (now Central Russian Air Force Museum) in Monino.

directions while manoeuvring at low speeds.

The machine's stability and controllability were rather poor, requiring a steady hand and causing some concern on the part of the test crew; one false move could cause the Toorbolyot to flip over. Added to that, the flights were made at low altitude; therefore the pilot was not equipped with a parachute, which would have been useless anyway.

Little by little Garnayev mastered the unusual flying machine, concurrently preparing to demonstrate the Toorbolyot to the Soviet government and the public at the 1958 Aviation Day display at Moscow-Tushino airfield (such displays were held there annually). This sure was a sight to behold! Emitting an ear-splitting roar and belching a stream of hot exhaust, the strange contraption slowly rose from a cloud of dust and hovered above the ground; it tilted slowly here and there and did a full 360° turn around its vertical axis, as if waltzing in the air. Next, it tilted forward and moved towards the far end of the airfield, picking up speed. The journalists reporting on the event immediately dubbed the machine *letayushchiy stol* (flying table).

In good weather with no wind, the Toorbolyot was easy enough to fly. In a wind of up to 12 m/sec (24 kts) the take-off and landing procedure became somewhat complicated, as the vehicle drifted and there were no

control surfaces with which to parry the drift. Yet the problem was solved by tilting the vehicle in the direction opposite to the drift. Garnayev's conclusion was that, providing the pilot was proficient enough, the Toorbolyot presented no great problems even in windy conditions. Usually the thing took off from and landed on a large sheet of metal, but on one occasion Garnayev managed to land the Toorbolyot successfully on an even grass surface at the abovementioned Tushino flying display.

The Toorbolyot featured an automatic flight control system – the first of its kind in the Soviet Union; however, in Garnayev's opinion, the system did little to improve the vehicle's control characteristics and could just as well be excluded. Apart from Garnayev, in 1957 the machine was flown by other LII test pilots – Fyodor I. Boortsev, G. N. Zakharov and Sergey N. Anokhin. The report on the VTOL tests of this vehicle was endorsed in September 1957.

The tests of the Toorbolyot showed the need for automatic stabilisation systems to be used by VTOL aircraft in take-off, landing, hover and transition modes. They also allowed the designers to determine the required efficiency of the reaction control nozzles, verify the hovering altitude control system and find the optimum seat incline and control locations. The results of this research proved valuable when the Soviet Union's first true VTOL aircraft, the Yak-36, was being designed.

Sukhoi Su-7U stability augmentation system testbed

A Su-7U trainer (identity unknown) was used for developing and testing stability augmentation systems (SAS) in 1965. This cannot be the 100LDU CCV because the latter was developed and tested later.

Sukhoi Su-9U control system testbed

In 1962-64 LII used a modified Su-9U (identity unknown) to test an automatic stability augmentation system.

Sukhoi Su-15 CCV with a side-stick

In 1980 the Sukhoi OKB converted a production Su-15 *sans suffixe* (c/n 1115328) into a control-configured vehicle with variable in-flight stability and control parameters. For the first time in Soviet aircraft design practice the aircraft featured a side-stick; the standard centrally mounted stick was retained and the pilot was able to switch the control system from one stick to the other as required.

The CCV underwent tests at LII in 1981-82 (one source says 1973-75!); it was flown by LII test pilots Vladislav I. Loychikov, Rimas-Antanas A. Stankiavicius, Anatoliy S. Levchenko, Igor' P. Volk, Aleksandr S. Shchookin, Viktor V. Zabolotskiy and Yuriy A. Oosikov. Unfortunately on 11th November 1982 the aircraft crashed and was destroyed before the test programme could be completed. Oosikov ejected and survived but sustained serious injuries which forced him to give up flying.

Su-15TM control system testbed

Service pilots kept complaining that that the Su-15's lateral stability was poor, especially during the landing approach. The Sukhoi OKB worked in several directions, trying to eliminate this shortcoming. Eventually the second production Su-15TM *Flagon-F* ('75', c/n 0315303) was turned over to GK NII VVS for testing in 1974; this aircraft was retrofitted with a trim mechanism in the aileron control circuit and a lateral stability augmentation system. The new features received a positive appraisal and were introduced on the production line.

Later the same aircraft was used to test an increased-area horizontal tail. The test programme included 29 flights; the new horizontal tail was likewise recommended for production – too late, as Su-15TM production had ended by then.

Sukhoi Su-27 (T10-30) development aircraft

An uncoded early-production Su-27 *Flanker-B* (c/n 36911009406, f/n 09-06) known at the Sukhoi OKB as the T10-30 served as a testbed for an automatic stall/spin recovery system ('panic button' feature) and research into stall/spin modes and fighter superagility. As an insurance policy in case the system didn't work, the T10-30 was equipped with spin recovery rockets on the inboard wing hardpoints.

Sukhoi LMK-2405 CCV

LII used several Su-27s as research aircraft and systems testbeds. One such aircraft was a *Flanker-B* coded '05 Red' (c/n 36911024205, f/n 24-05). The aircraft was a CCV, a part of the LMK-2405 active flight safety research complex developed for testing control techniques for tomorrow's agile fighters (*lyotno-modeleeruyuschchiy kompleks* – 'flight modelling complex' or total in-flight simulator). The other part of the complex was a ground processing unit which analysed the aircraft's trajectory and systems operation and formulated control algorithms.

The cockpit of the LMK-2405 CCV. Note the side-stick controller and associated armrest on the starboard console.

Su-27 '05 Red' (c/n 36911024205) was converted into the LMK-2405 CCV. Here it is seen in its original colours at Zhukovskiy during one of the MAKS airshows.

The aircraft featured an advanced reprogrammable digital FBW control system with a pressure-sensing side-stick (the standard control stick was retained as a back-up) and full-authority digital engine control (FADEC). Special multi-faceted angle reflectors were fitted under the wingtip missile rails to give a better radar signature, assisting trajectory measurements. Flight and systems data were transmitted to the ground in real time by an omnidirectional data link system with aerials aft of the cockpit and beneath the port air intake. These, and the angle reflectors, were the LMK-2405's only outward distinguishing features. For high-alpha/low-speed handling trials the air-

craft could be fitted with spin recovery rockets.

The LMK-2405 was unveiled at MosAeroShow '92 in Zhukovskiy on 11-16th August 1992 where it was in the static park. It was displayed again at the MAKS-93 (31st August – 5th September 1993), MAKS-95 (22nd-27th August 1995), MAKS-97 (19-24th August) and MAKS-99 (17th-22nd August) airshows in Zhukovskiy in standard camouflage colours. At the MAKS-2001 (14-19th August) and MAKS-2003 (19-24th August) the aircraft was presented in a striking grey/blue colour scheme worn by some of LII's flight test aircraft, wearing 'ACE – Advanced Control Experiment' titles and a tactical code amended to '05 Blue outline'.

The LMK-2405 CCV (c/n 36911024205) in early colours
Zhukovskiy, 1992

The same aircraft in late colours
Zhukovskiy, 2003

Tupolev TB-1 – TMS remote control system testbed/flying bomb

The aforementioned Ostekhbyuro team experimented with a radio-controlled TB-1 bomber known as the TMS (*telemekhanicheskiy samolyot* – 'telemechanical aircraft'). This was to be the flying bomb.

An attempt was made to use the TMS operationally in the initial period of the Great Patriotic War. Once a pilot had taken the explosive-laden TB-3 aloft, engaged the remote control system and bailed out, the operator aboard the DB-3 drone director aircraft took over. However, as the two approached the target, the Germans put up a tremendous anti-aircraft artillery barrage. A flak shell exploded next to the DB-3 and put the transmitting aerial out of action, rendering the TMS uncontrollable; the latter disappeared in the clouds and flew on until it finally crashed.

Tupolev/Myasishchev UR-1/UR-2 control system testbed

The bicycle landing gear was not the only novel feature of the Myasishchev M-4 bomber. For the first time in Soviet aircraft design practice the *Bison* incorporated fully powered controls with reversible and irreversible hydraulic actuators in the aileron, rudder and elevator control circuits, plus an artificial feel mechanism.

Of course, testing these features on a ground rig ('iron bird', to use Boeing terminology) was not enough and the actuators had to be put to the test in flight. Hence a Tu-4 (identity unknown) was converted into the UR control system testbed by the Myasishchev OKB. The U stood for [*sistema*] *oopravleniya* – control system and the R was again a reference to the Tu-4 (*izdeliye* R).

The first version designated UR-1 featured reversible actuators; tests in this configuration were completed in March 1952, involving 12 flights totalling about 20 hours. The test flights showed that the powered controls worked acceptably, even though a few failures did occur; incidentally, the powered controls improved the *Bull's* handling dramatically. Later the aircraft was refitted with irreversible actuators and tested in April 1952 as the UR-2. The test results obtained on this aircraft enabled OKB-23 to simplify the M-4's control system, utilising a simple spring-loaded artificial feel mechanism instead of a complex automatic device. Nevertheless, further tests with a pneumatically-operated automatic artificial feel system were conducted on the UR-2 in September 1952.

Tupolev Tu-16 control system testbeds

In 1954-55 an unidentified Tu-16 served for verifying certain features of the control system developed for the ill-fated *Boorya* (Storm) hypersonic intercontinental ground-launched cruise missile (GLCM), aka *izdeliye* 350 – a product of the OKB-301 fighter design bureau led by Semyon A. Lavochkin.

An altogether different testbed served for verifying an automatic take-off system developed for the Tu-16M (M-16) remote-controlled target drone.

Tupolev Tu-154LL CCV/avionics testbed

Development of the Buran space shuttle involved selecting the optimum automatic landing system (ALS), testing and perfecting it. Also, future space shuttle pilots had to be trained in approach and landing techniques – a point all the more important because the Buran would land in gliding mode, with no chance to make a go-around if it missed the runway. Therefore, several aircraft were extensively modified or even purpose-built for testing and calibrating the Buran's ALS; these included the three examples of the Tu-154LL.

These aircraft can be regarded both as avionics testbeds and as control-configured vehicles (which is why they are included here). Creating such an aircraft turned out to be a major challenge. The machine had to be outfitted with automatic and manual control/landing systems integrated in such a way as to emulate the Buran's control parameters as closely as possible. The Tu-154 was selected for this role because it came closest to the space shuttle in size and weight.

In order to give the Tu-154LL the required 'Buranesque' handling, Soviet flight dynamics specialists designed a special stability and handling alteration system which acted as an interface between the airliner's existing systems and the ALS. This allowed the aircraft to emulate the Buran's flight pattern from an altitude of 10,000 m (32,810 ft) right to the point of touchdown.

Several other structural and systems changes had to be made. In order to provide the required steep descent from high altitude, the thrust reversers would be engaged in flight (this necessitated reinforcement of the rear fuselage to absorb the augmented loads and installation of revised thrust reverser grilles) and the flight and ground spoilers would be deployed all together.

The logo of the M. M. Gromov Flight Research Institute (LII), the leading Soviet/Russian flight test establishment which operated a huge fleet of testbeds.

CCCP-85024, the first Tu-154LL; note the instrumented nose probe, the fairing with a blade aerial on the flight deck roof and dorsal strake aerial.

The flight deck was modified considerably, featuring a Buran pilot's workstation with a fighter-type control stick and appropriate instruments instead of the normal first officer's station. The captain's workstation was retained, yet a pilot selected to fly the Buran would take the captain's seat. A second, complete version of the orbiter's flight deck was installed in the forward cabin.

Three second-hand Tu-154s were converted to Tu-154LL CCVs. The first of these was CCCP-85024 (c/n 72A024). Built as a Tu-154 *sans suffixe*, it had been upgraded to Tu-154B standard and had served with the Aeroflot's Central Directorate of International Air Services/63rd Flight and the East Siberian Civil Aviation Directorate/1st Irkutsk United Air Detachment/201st Flight before being transferred to LII.

The aircraft's most obvious identification feature was a long pointed air data boom mounted on a scratchbuilt metal nosecone replacing the regular radome; this nosecone incorporated small dielectric sections at the front and underneath. Other non-standard details included a small teardrop blister on the flight deck roof incorporating a round antenna (identical antennas were built into the sides of the nose just ahead of the nose gear unit) and

small oval antennas built into the sides of the nose immediately aft of the ex-radome; the abovementioned blister roof carried the blade aerial usually located there. The forward over-wing emergency exit covers were modified to incorporate low-set circular sensor windows, a tall dorsal strake aerial offset to starboard was mounted ahead of the wings, a small thimble-shaped ventral fairing housing a data link antenna was fitted aft of the wings, and a small sensor sat on the centre engine's air intake trunk. The 14th pair of cabin windows had the glazing replaced by emergency decompression valves with air vent grilles allowing the cabin to be depressurised within two or three seconds prior to bailing out, should the need arise.

The other two examples, CCCP-85083 (c/n 74A083) and CCCP-85108 (c/n 75A108), were sourced from the West Siberian CAD and the Ukrainian CAD/Borispol' UAD/222nd Flight respectively. Both were originally Tu-154As but had been updated to Tu-154B and Tu-154B-1 standard respectively by then. Tu-154LLs CCCP-85083 and CCCP-85108 lacked the distinctive air data boom of the first machine, retaining the standard radome, because the aircraft's handling in the unconventional flight modes had been more or less studied by then;

One more view of CCCP-85024. The aircraft was displayed at Zhukovskiy in 1991 along with other LII test aircraft on occasion of LII's 50th anniversary.

the emergency exits were also standard. Moreover, CCCP-85083 had a flat-bottomed rectangular ventral fairing with dielectric portions ahead of the wings and two additional blade aerials located on the centre engine's air intake trunk (aft of the abovementioned sensor) and at the top of the fin; the tall dorsal strake aerial was offset to port ahead of the wings on CCCP-85083 and omitted altogether on CCCP-85108.

When the programme was completed the three Tu-154LLs had made more than 200 test flights between them. No fewer than 16 versions of the ALS were developed and tested consecutively on the Tu-154LL, and the three aircraft had different equipment fits at any one time; thus, CCCP-85024 and CCCP-85108 could land in fully automatic mode but CCCP-85083 could not. Seven of these 16 versions were selected for further development on the BTS-002 research vehicle, or GLI-Buran (CCCP-3501002) – an example of the space shuttle fitted with four Lyul'ka AL-31 turbofans that enabled it to take off and fly like a normal aeroplane (GLI = gorizon**tahl'**nyye **lyot**nyye ispy**tah**niya – 'horizontal flight tests'). Later, the definitive version of the ALS underwent an additional series of tests on the Tu-154LL involving 80 flights.

The outside world became aware of the Tu-154LL's existence in 1992 when all three examples were seen at Zhukovskiy during MosAeroShow '92. CCCP-85083 was in the static park, while CCCP-85108 took part in the flying display on the public days, accompanied by the Mikoyan MiG-25PU-SOTN chase plane ('22 Blue', c/n N22040578). The performance staged by the Tu-154LL was truly hair-raising; viewed from the spectator area, the aircraft appeared to be dropping like a stone and guaranteed to crash – which of course it did not.

By 1995 the Tu-154LLs had been withdrawn from use. It appears that all three were ultimately scrapped at Zhukovskiy.

Tupolev Tu-154M-LL (FACT) CCV

By 1985 the first prototype Tu-154M (CCCP-85317, c/n 78A317) had been transferred to LII and converted into a CCV that would best be called a total in-flight simulator (TIFS) for verifying new control and flight data presentation techniques evolved for future transport aircraft and airliners. Outwardly the only signs of the conversion were a small sensor on the centre engine's air intake trunk and

a small ventral data link antenna 'thimble' aft of the wings, just like on the Tu-154LL CCV. The internal changes were far more extensive: CCCP-85317 was fitted with a digital fly-by-wire control system and digital ('power-by-wire') engine controls based on a computer that could be reprogrammed in flight. The flight deck featured alternative pilot's controllers (left and right side-sticks, mini-wheels for roll control and new throttles) with variable-in-flight parameters, plus new data presentation systems – a fighter-style monochrome head-up display (HUD) and a colour cathode-ray tube (CRT) head-down display – which could likewise be reprogrammed in flight. Traditional electromechanical instruments were retained as a back-up. The cabins were crammed with test and recording equipment.

The aircraft communicated in real time with a ground data processing facility via a data link system functioning in two-way (uplink/downlink) mode. Its applications were as follows:
- in-flight simulation and assessment of flight dynamics of future transports in normal and abnormal situations;
- verification of different types of pilot's controllers;
- in-flight optimisation of the aircraft's handling qualities, development of handling quality criteria and certification standards;
- development of the philosophy and standard graphic formats for HUDs and head-down displays;
- pilot training;
- flight deck ergonomics studies and flight safety research (the onboard equipment permitted analysis of the pilot workload and the pilots' vital signs.

In this guise CCCP-85317 was first displayed at the MosAeroShow '92. It has been an invariable participant of the Moscow airshows ever since, gaining the RA- prefix by August 1995. The experimental equipment changed from time to time.

At the MAKS-2001 airshow (14-19th August 2001) RA-85317 was first shown in LII's new grey/blue livery with titles and logo. By 2003 the aircraft had gained huge FACT titles; the acronym stood for Future Aircraft Control Testbed. The aircraft's equipment had been considerably altered by then (among other things, it included ambient air water content/humidity sensors). The Tu-154LL's onboard navigation suite included the indigenous BIMS-T, BSPN-2-01, A-737D, A-737I, A-737PL, I-21 and K-161 systems, as well as

The Tu-154M-LL (FACT) CCV, RA-85317, at one of the MAKS air-shows.

the GG-12 and JGG-30 (the latter designations were rendered in Roman letters, indicating the systems' foreign origin). Part of the associated data analysis equipment was housed in a Volkswagen LT 35D van outfitted as a mobile laboratory which was displayed alongside.

In addition to the applications stated above, the aircraft now fulfilled the following missions:

• research of aircraft condensation trail (contrail) formation conditions;
• development and research of communication, navigation, surveillance/air traffic management (CNS/ATM) concepts;
• flight test of high-accuracy inertial and inertial/satellite navigation systems.

The Tu-154M-LL (FACT) was still active in mid-2010 but due for retirement as time-expired.

Yakovlev UT-2 remote control system testbed

The UT-2 trainer was one of the aircraft types involved in the experiments with remote-controlled ('telemechanical') aircraft. In January 1940 the State Defence Committee issued a special directive tasking NKAP with the development and testing of prototype 'telemechanical' versions of the TB-3, the SB and the UT-2. In September 1940 People's Commissar of the Aircraft Industry Aleksey I. Shakhoorin asked the State Defence Committee to postpone the deadlines for the completion of this work; in particular, he requested five more months for the development of a UT-2 version capable of radio-controlled take-off and landing. The request was granted.

This time it was not Ostekhbyuro but the NII-22 research institute that did the work; a team led by engineer Nikol'skiy was responsible for was entrusted with the task). The remote-controlled UT-2 was undergoing manufacturer's tests in Leningrad in early 1941, and there were plans for submitting it for state acceptance trials in July-August 1941.

Yakovlev Yak-25 remote control system testbeds

Several Yak-25s served with LII as testbeds of various kinds. One of them was used in 1956-57 to develop a remote control system for target drones; the results obtained were incorporated into the design of the Yak-25MSh (*mishen'*) target drone version. Another Yak-25 testbed was used in 1957-58 for developing remote control systems for cruise missiles.

Similarly, an example of the Yak-25RV *Mandrake* single-seat high-altitude reconnaissance aircraft was used to develop the control system of the Yak-25RV-II target drone.

4. Ejection seat testbeds

Antonov An-12BK ejection seat/APU/recovery systems testbed (An-12M LL)

LII's test aircraft fleet included an An-12BK in Soviet Air Force colours coded '43 Red' (c/n 8345902) which was converted into a multi-purpose testbed known as An-12M LL. The second mission fulfilled by this aircraft was the testing of new models of ejection seats in 1987-97. To this end the DB-65U tail turret was replaced by a detachable elongated pod

emulating the cockpit of a combat aircraft. This 'cockpit' was attached by multiple bolts and could be installed at any angle from upright to inverted (0° to 180°) at 30° increments to emulate different attitudes of the stricken aircraft at the moment of ejection. This was because state-of-the-art ejection seats, such as the famous Zvezda K-36, were designed to ensure safe ejection even in inverted flight at low altitude – the worst possible

combination. Two video cameras in orange egg-shaped pods were mounted under the wingtips to capture the ejection sequence; test equipment heat exchangers in characteristic white-painted teardrop fairings were installed high on the fuselage sides immediately aft of the wings.

The existence of An-12BK '43 Red' in this form was revealed at the Konversiya '91 trade fair which took place at the VDNKh fairground

The An-12M LL taxies with the ejection seat module fitted with 60° right bank. Note the test equipment heat exchangers aft of the wings, the video cameras under the wingtips and the LII logo on the nose.

The An-12M LL fires a K-36D-3,5 ejection seat in an inverted position during the MosAeroShow '92.

The same aircraft makes a smoky landing approach at Zhukovskiy.

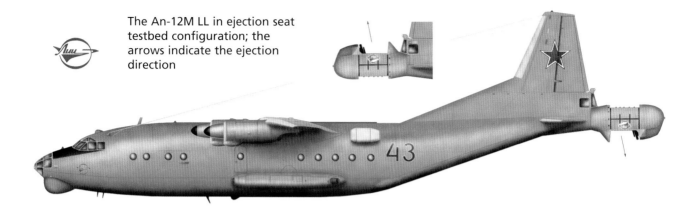

The An-12M LL in ejection seat testbed configuration; the arrows indicate the ejection direction

Right and far right: Tests of the Zvezda K-37 ejection seat created for the Kamov Ka-50 Black Shark attack helicopter. The seat is extracted by a squib attached to a telescopic boom. At the time the aircraft still had the hinged booms of an aerial recovery system under the rear fuselage.

(**Vys**tavka dosti**zhen**iy na**rod**novo kho**ziay**stva – National Economy Achievements Exhibition) in Moscow in December 1991. (The word

*kon**ver**siya* (conversion) means the adaptation of military technologies and defence industry enterprises to civilian needs.) At that time the aircraft still had the 'grabbing booms' left over from the previous programme but these were later removed. The testbed made demo flights involving live ejections of a seat with a dummy at the MosAeroShow '92 (11-16th August 1992) and MAKS-97 (19-24th August 1997) airshows in Zhukovskiy; on the former occasion (15th August) a K-36DM seat was fired from an inverted position, while at MAKS-97 the latest K-36DM-3.5 (featuring a reduced 3.5-G load limit to prevent pilot injuries) was fired with the cockpit set at 60° right bank.

Ilyushin IL-28LL ejection seat testbed

In the early 1960s several IL-28 bombers were converted into testbeds for various systems of the *Vostok* (East) manned spacecraft under development by Sergey P. Korolyov's OKB. One of these was a Moscow-built example coded '10 Blue' (c/n 53005710), an ejection seat testbed used to test, among other things, the ejection seat of the Vostok's re-entry vehicle. This aircraft has been referred to as the IL-28LL.

Port side view of IL-28LL c/n 53005710 with the bomb bay doors open for firing the Vostok ejection seat from the forward cabin

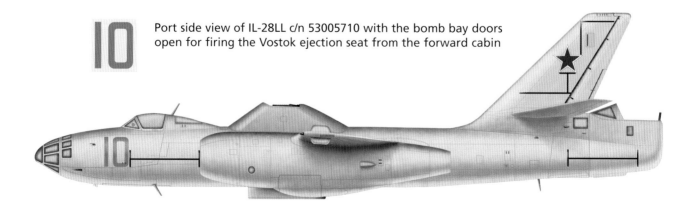

The IL-28LL fires the ejection seat developed for the Vostok space-craft. Note the open blast doors below the dorsal 'hump', the open top of the rear test cabin (with a small slipstream deflector) and the photo calibration markings.

A Zvezda K-36 ejection seat is fired through the canopy of the IL-28LL's rear cabin. Note the stabilising booms extending from the headrest.

Far left: The Vostok ejection seat spouts terrific flames as it leaves the aircraft.

The bulky Vostok ejection seat was installed in the faired-over bomb bay immediately ahead of the wing torsion box and protruded above the upper fuselage; hence a large teardrop fairing with flattened sides had to be installed aft of the pilot's cockpit to protect the test pilot sitting in the seat from the slipstream. Additionally, the tail gunner's compartment was replaced by a large slab-sided fairing extending much further aft, from which another ejection seat could be fired both upwards and downwards. Ciné cameras were mounted in teardrop fairings above and below the wingtips to capture the ejection sequence.

The Vostok ejection seat was tested successfully by future cosmonaut Gherman S. Titov. The Mikoyan SM-50 fighter (aka MiG-19SU, an experimental version of the

53005710

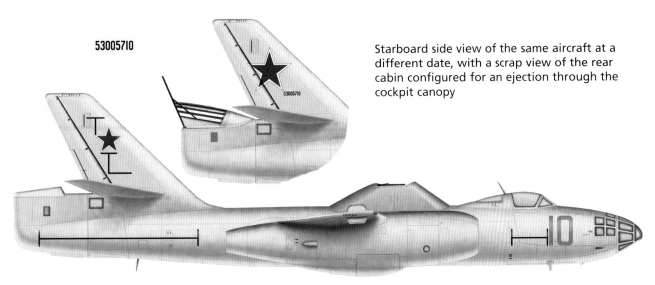

Starboard side view of the same aircraft at a different date, with a scrap view of the rear cabin configured for an ejection through the cockpit canopy

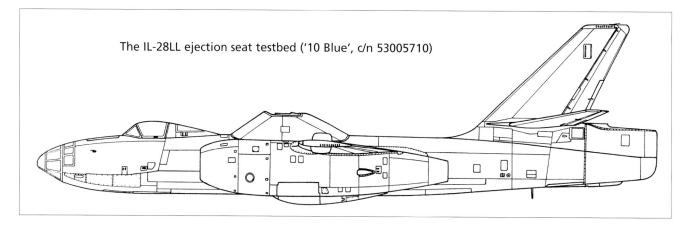

The IL-28LL ejection seat testbed ('10 Blue', c/n 53005710)

MiG-19SF *Farmer-C* with a ventral U-19 liquid-propellant rocket booster) acted as a chase plane.

IL-28LL '10 Blue' was used for testing many models of ejection seats, serving with LII from 1953 all the way to 1978.

Mikoyan/Gurevich UTI-MiG-9 ejection seat testbed

OKB-155 was required to install ejection seats in the UTI-MiG-9 (*izdeliye* FT, the trainer version of the *Fargo*) and present the aircraft for state acceptance trials in this configuration. The work was mostly conducted in 1948. The seat was developed by the Mikoyan OKB and weighed 128.5 kg (283.3 lb). The front and rear ejection seats of the second prototype (FT-2) were installed at an angle of 22.5° and 18.5° respectively. Ejection was supposed to proceed as follows: the forward canopy would be jettisoned, followed by the rear one, after which the seats would fire – the rear seat first.

Manufacturer's flight tests of the ejection seat were conducted on the FT-2 at LII in July and August 1948. The seat was fired from the rear cockpit (from which the canopy had been removed); parachute tester Gavriil A. Kondrashov performed the role of the ejectee.

On 29th September 1948 the FT-2 equipped with ejection seats was turned over to GK NII VVS for state acceptance trials. On the whole the ejection system and the ejection seat received a positive assessment from the military who recommended them for use on production aircraft.

In addition to the ejection seat, wing-mounted airbrakes and auxiliary fuel tanks underwent state acceptance trials at GK NII VVS on the UTI-MiG-9. However, the trainer itself was not put into production. By mid-1948 the MiG-9 was already outmoded and there was no point in building its two-seat version, as the swept-wing MiG-15 was due to enter flight test at the end of the year.

Mikoyan/Gurevich UTI-MiG-15 (*izdeliye* ST-10) ejection seat testbeds

Experience with first-generation ejection seats showed that pilots were often injured in high-speed ejections because the seat offered no protection from the slipstream. Besides, the seats were not stabilised and tumbled head over heels after separation from the aircraft, increasing the risk of injury. The engineers examined several means of increasing the chances of safe egress at high speed, including detachable crew modules (as on the General Dynamics F-111 fighter-bomber). Eventually, however, it was decided that improved ejection seats were the cheapest and simplest solution to the problem. In particular, the second-generation seat was to be fired together with the forward-opening canopy, which would protect the pilot from the slipstream.

As a first step, a ground rig was built to test the seats, ejecting dummies at speeds up to 1,000 km/h (621 mph). This installation veri-

A tester is ejected from the rear cockpit of the UTI-MiG-9 (FT-2) during tests of the ejection seat at LII.

fied the seat's stabilisation system ensuring safe ejection at high speed. Another ground rig was used to study the effects of G loads on human pilots during ejection. The efficiency of the canopy doubling as a protective visor was tested on a Tu-2 bomber converted into an ejection system testbed.

In 1954 the first production Kuibyshev-built UTI-MiG-15 *Midget* trainer serialled '101U Blue' (c/n 10101) was converted by LII for conducting ejection seat tests with both dummies and pilots. The aircraft was designated *izdeliye* ST-10. The aircraft was flown in two configurations with the experimental ejection seat fired from the front and rear cockpits; it is hard to say which came first.

In the former configuration '101U Blue' was a nearly-standard *Midget* with the prototype seat installed in the forward cockpit. Special ejection seats were manufactured; the hinged forward canopy was replaced by a metal fairing with a large opening through which the seat was ejected. This was an attempt to kill two birds with one stone (to avoid jettisoning the canopy and facilitate installation and removal of the ejection seat

while reducing turbulence around the forward cockpit). High-speed ciné cameras were installed in small teardrop fairings on the wing upper surface near the wingtips to record the ejection sequence. The aircraft had photo-theodolite calibration markings on the forward/rear fuselage and fin.

In its other configuration the aircraft had the sliding rear canopy replaced by a large shallow metal fairing extending almost all the way to the fin.

Initially the ST-10 fired a standard MiG-15 ejection seat suitably modified for the experiment, featuring canopy retention locks and a totally new seat belt system. The rear cockpit had a non-standard canopy inside the above-mentioned fairing. During rotation two clamps on the seat headrest engaged two lugs on the canopy, causing it to rotate up and forward. The front end of the canopy slid aft along guide rails until it locked into position on the seat pan, disengaging itself from the guide rails in so doing; the canopy now offered protection for the pilot. The seat complete with canopy weighed 225 kg (496 lb) and was ejected by a telescopic ejection gun at an

101U Blue (c/n 10101), the first of LII's three UTI-MiG-15s converted into ST-10 ejection seat testbeds. The open front cockpit is clearly visible. Note the photo calibration markings.

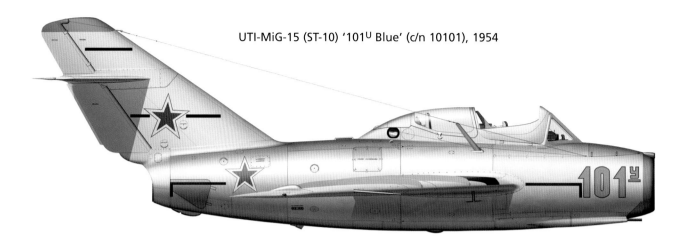

UTI-MiG-15 (ST-10) '101ᵁ Blue' (c/n 10101), 1954

initial speed of 18.5-19.0 m/sec (60.7-62.3 ft/sec).

Tests began in the second half of the 1950s. Several stabilising systems were tried, including four drogue parachutes on telescopic booms extending aft of the seat. Several versions of the canopy-to-seat retention locks were tested; the canopy separation system was verified, as the canopy obviously needed to be discarded before the pilot used his own parachute. The programme was conducted by LII test pilot Eduard V. Yelyan and test parachutist V. Golovin. As a result, an ejection mount with a sliding canopy for pilot protec-

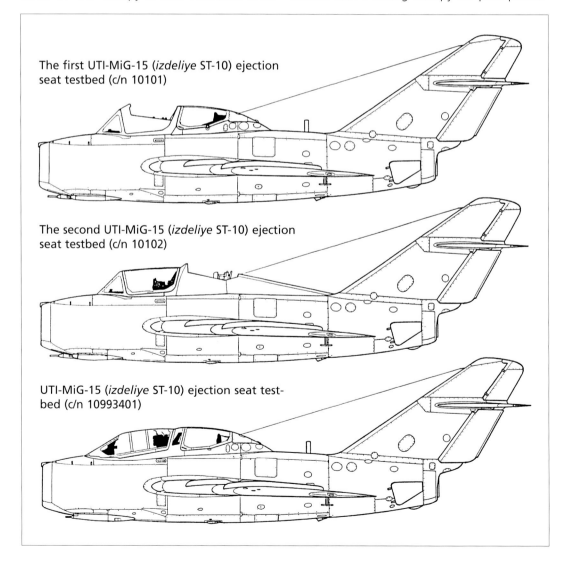

The first UTI-MiG-15 (*izdeliye* ST-10) ejection seat testbed (c/n 10101)

The second UTI-MiG-15 (*izdeliye* ST-10) ejection seat testbed (c/n 10102)

UTI-MiG-15 (*izdeliye* ST-10) ejection seat testbed (c/n 10993401)

tion from the airflow was recommended for all Soviet fighters capable of exceeding 1,000 km/h – such as the Mikoyan Ye-2 *Faceplate* swept-wing fighter prototype, the delta-wing Ye-5, the Ye-50 mixed-power fighter based on the Ye-2 and the Mikoyan I-3U swept-wing fighter prototype.

The designation ST-10 has also been quoted for two other UTI-MiG-15s converted into ejection seat testbeds. The second aircraft, '102U Blue' (c/n 10102), had the prototype seat in the rear cockpit. The sliding rear canopy was replaced by a large shallow metal fairing extending almost all the way to the fin.

As part of the MiG-21 development programme, the third UTI-MiG-15 (Ulan-Ude-built '401U Blue', c/n 10993401) was converted into a testbed for its parachute-stabilised SK ejection seat (see **den'ye katapool'tee**ruye-moye – ejection seat). The forward canopy section complete with fixed windshield was replaced by a one-piece forward-hinged canopy developed for the MiG-21F *Fishbed-C* (or perhaps, more precisely, its precursor – the Ye-2A). This was rather lower than the rest of the *Midget*'s canopy, resulting in a pronounced step between it and the sliding rear canopy. '102U Blue' and '401U Blue' had similar photo-theodolite calibration markings.

Curiously, the MiG-21F's bulletproof windscreen was not part of the hinged canopy, being attached to the airframe under the canopy, and was left behind after ejection. On '401U Blue' it obviously played no part in the ejection sequence and was very probably fitted in lieu of a windshield. With the forward canopy gone, the turbulent airflow around the forward fuselage might otherwise have complicated flying excessively for the pilot who stayed behind to land the aircraft!

'401U Blue' also had a different configuration with a standard forward canopy (including windshield) and a non-standard rear canopy resembling that of the Sukhoi Su-7; it was longer than the *Midget*'s and a new fairing had to be installed between it and the

Above left: The same aircraft reconfigured for firing the experimental seat from the rear cockpit, which is provided with an extended fairing.

Above: ST-10 '101U Blue' fires an experimental seat from the rear cockpit. Note how the canopy protects the ejectee (most probably a dummy) from the slipstream.

Below: '102U Blue', the second ST-10, with a similarly modified rear cockpit.

Bottom: The third ST-10, '401U Blue', with a MiG-21 ejection seat and canopy in the front cockpit. Note the separate windshield which remains when the one-piece canopy is jettisoned.

fuselage. The rear canopy likewise acted as a protective visor and the seat was fired by a long telescopic ejection gun. Curiously, the canopy sported something like a large blade aerial, though this may have been a sort of stabilising surface. It is not known which configuration came first.

LII's MiG-25RU ejection seat testbed, '01 Blue', fires a Zvezda K-36RB seat from the rear cockpit. The seat's distinctive stabilising booms tipped with drogue parachutes are deployed. Note the LII logo on the nose.

Mikoyan MiG-25RU ejection seat testbed

The prototype of the MiG-25RU *Foxbat-C* reconnaissance trainer (c/n 390SA01, f/n 0101) was turned over to LII by the Mikoyan OKB after completing its flight test programme. The new owner converted it into a testbed for the Zvezda K-36RB zero-zero ejection seat – a version of the standard Soviet ejection seat modified for use in the Buran space shuttle. The seat (referred to in some sources as the K-36M11F35) was fired from the rear cockpit, which was suitably modified with a cutaway metal fairing replacing the standard canopy. A ciné camera was installed in a dorsal fairing on the nose to record ejection seat separation, and photo calibration markings were applied to the air intake trunks and the outer faces of the fins.

Initially the aircraft was coded '46 Red';

later it was recoded '01 Blue'. When the Buran programme was terminated the MiG-25RU was used for testing other ejection seats – the Zvezda K-93 zero-zero ejection seat developed for the MiG-AT advanced trainer and the K-36D-3,5 seat. The aircraft was in service from 1979 to 1997.

On 22nd-27th August 1995 the MiG-25RU ejection seat testbed was in the static display at the MAKS-95 airshow in Zhukovskiy.

Mikoyan MiG-31LL ejection seat testbed

A non-IFR-capable MiG-31 coded '79 Red' (c/n N69700115548) was modified for testing ejection seats and training flight crews in ejection procedures. This aircraft was known as the MiG-31LL. The aircraft belonged to GNIKI VVS, which was renamed the Russian Air

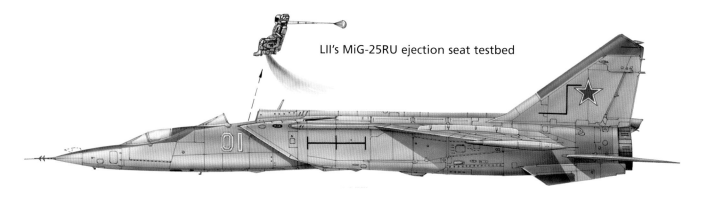

LII's MiG-25RU ejection seat testbed

Force's 929th State Flight Test Centre named after Valeriy P. Chkalov (GLITs – *Gosudarstvennyy lyotno-ispytahtel'nyy tsentr*) in the early 1990s.

Outwardly the MiG-31LL differed from standard machines in lacking the glazing on the inter-canopy crash frame and having small camera pods at the wingtips to record the ejection sequence. On 11-16th August 1992 the aircraft was displayed statically at the MosAeroShow-92 in Zhukovskiy with a standard canopy over the rear cockpit. Curiously, the MiG-31LL sports nose art in the shape of a definitely moth-eaten lion's head. This is obviously a pun on the aircraft's testbed role: the Russian for 'bald lion' is *lysyy lev*, which abbreviates as LL (standing for *letayushchaya laboratoriya*, of course). In September 1995 the aircraft had another show performance, demonstrating the ejection of a dummy from the rear cockpit during the take-off run at an 'open house' at Akhtoobinsk on occasion of the former GNIKI VVS's 75th anniversary.

Petlyakov Pe-2LL ejection seat testbed

A Pe-2 dive-bomber coded '45 Yellow' was converted into a testbed for the first Soviet ejection seats at LII in the late 1940s. The seat was fired from a suitably modified waist gunner's station which had a cutout in the roof.

Sukhoi Su-7U ejection seat testbeds

At least three Su-7U trainers coded '10 Red', '20 Red' and '21 Red' (one of them was c/n 1603) were converted into testbeds for the new family of Zvezda K-36 ejection seats. The rear cockpit was specially modified. Everything was removed from its interior which was covered in steel plate and the canopy was deleted. Cameras in streamlined housings were fixed to the wings to record the ejection sequence. Su-7U c/n 1603, which was operated in 1975-91, served for check-up tests of production models of ejection seats and manufacturer's tests of K-36D family seats.

One of these testbeds ('20 Red') performed a demonstration flight at the MosAeroShow '92 on 15th August 1992, firing a K-36DM seat with a dummy during the take-off run. Minutes later, however, a fire broke out aboard the aircraft and test pilot Aleksandr A. Murav'yov was forced to eject, the aircraft

The MiG-31LL at the Akhtoobinsk test centre. The cylindrical wingtip pods house video or ciné cameras recording the ejection sequence. Note the rear canopy has been replaced by a special open-top fairing.

The first Soviet ejection seats were tested on LII's Pe-2LL testbed ('45 Yellow'), being fired from the heavily modified ventral gunner's station.

125

being totally destroyed. The ejection took place well out of sight of the spectators, who did not suspect anything amiss.

Sukhoi L-43 (*izdeliye* 94) and other Su-9U ejection seat testbeds

Upon completion of the state acceptance trials the Su-9U trainer prototype (U-43; according to some sources, the first prototype) was converted into a testbed for verifying new ejection seats developed by the Sukhoi OKB. The modified aircraft received the designation L-43 or *izdeliye* 94. The tests confirmed that the latest seats in the KS series permitted safe ejection at speeds up to Mach 1.8 and altitudes of 150 to 15,000 m (490 to 49,210 ft).

According to some sources, in 1962 the first two production Su-9Us (c/ns 111000101 and 111000102?) were converted for testing new ejection seats and high-altitude pilot gear, the modified rear cockpits permitting installa-

tion of seats developed by various manufacturers. One aircraft was delivered to LII, operating from Zhukovskiy, while the other belonged to GK NII VVS and was home-based at Vladimirovka AB. On the other hand, and rather confusingly, there is evidence that in 1967-75 LII operated 'Su-9U c/n 1018' (that is, c/n 112001018?) converted into an ejection seat testbed which was used for state acceptance trials of the Sukhoi KS-4, Mikoyan KM-1 and Yakovlev KYa-1 seats, as well as the Czech VS-1BRI and VS-2 zero-zero ejection seat developed for the Aero L-39 Albatros advanced trainer (VS = *vystřelovací sedačka* – ejection seat) and the later VS-2 model. The exact number and identities of the Su-9U ejection seat testbeds is unknown (different sources give conflicting data), but one aircraft was coded '10 Blue'.

The changes made to the aircraft were as follows. The pressurised cockpit was divided into two separate pressurised compartments, making sure that the forward cockpit

'20 Red' was one of three Su-7Us modified as ejection seat testbeds by LII. Here an experimental seat is fired from the rear cockpit. Note the photo calibration markings.

Another Su-7U ejection seat testbed, '21 Red', fires an experimental seat while still on the runway; note the squib deployed from the seat headrest to extract the seat.

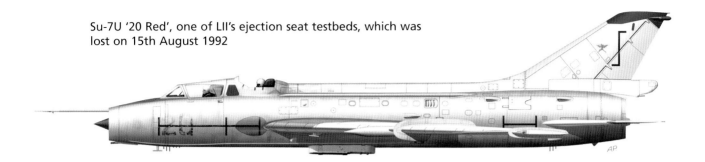

Su-7U '20 Red', one of LII's ejection seat testbeds, which was lost on 15th August 1992

remained pressurised and the pilot thus remained able to fly the aircraft after the rear seat had been fired. The flight controls, engine controls and instruments were deleted from the rear cockpit; the normal canopy could be fitted for positioning flights but was removed for test flights and a special open-top fairing was fitted instead. The testbed was equipped with three high-speed ciné cameras to capture the ejection sequence – two in fairings on the wings and one in the cockpit. Performance

and handling were almost identical to those of the standard trainer.

Ejections were performed during the take-off run, at high altitude and at supersonic speeds; initially test dummies were used, of course, but these were later joined by parachutists made of flesh and blood. Later the Su-9U ejection seat testbeds were used for testing the Zvezda K-36 seat which became a standard fit on all Soviet/Russian combat aircraft developed from the late 1970s onwards.

Su-9U '10 Blue' converted into an ejection seat testbed fires a K-36 seat.

The same aircraft performs a ground ejection of an experimental seat.

127

The Su-29KS testbed, RA-01485/ '02 Black', demonstrates the SKS-94 ejection system in action. The seat headrest is catapulted by the ejection gun and extracts the pilot (in this case, a dummy) by means of a harness before his parachute opens.

The logo of NPP Zvezda, the ejection seat manufacturer in whose interests several testbeds were operated.

Sukhoi Su-29KS ejection seat testbed

Unlike combat aircraft pilots, the pilots of competition aerobatic aircraft and primary trainers did not enjoy the luxury of ejection seats – until the 1990s at least. Since aerobatics are performed at low altitude, pilot error – or catastrophic structural failure if the airframe was overstressed during a high-G manoeuvre (which happens occasionally) – would almost certainly be fatal for the pilot(s), as there was not enough altitude or time for bailing out safely. After several fatal crashes involving Sukhoi Su-26M aerobatic aircraft had occurred, in 1991 the Sukhoi OKB tasked NPP Zvezda with developing a lightweight ejection seat for the Su-31 (a refined derivative of the Su-26M). After four years of intensive work Zvezda brought out the SKS-94 ejection system.

Hence in 1994 a production Su-29 aerobatic aircraft – the two-seat trainer version of the Su-26M with mid-set wings and fixed tailwheel landing gear – was converted into a testbed for the SKS-94 ejection system. The cockpits were modified to permit installation of the ejection seats, and the lateral glazing panels in the rear cockpit (just aft of the wing training edge) were eliminated. For positioning flights a standard cockpit canopy (with a fixed windshield and a large section hinged to starboard over both seats) was used; in operational configuration the hinged portion was removed and a second windshield was installed for the rear seat occupied by the test pilot. The seat was normally fired from the front cockpit. Unlike most production Su-29s, this aircraft had only one aerodynamic balance on each aileron instead of two.

Registered RA-01485 and wearing the tactical code '02 Black', the Su-29KS was used to perform the manufacturer's flight tests and then the state acceptance trials of the SKS-94. In the course of testing some 20 ejections of dummies were performed at altitudes ranging from 50 to 2,000 m (164 to 6,560 ft) and at speeds between 180 and 400 km/h (112 and 249 mph), reportedly including an ejection in inverted flight at 50 m. In August 1995 the aircraft made its public debut at the MAKS-95 airshow in Zhukovskiy, participating in the flying display which included the ejection of a dummy pilot. In 2000 the Su-29KS was among the Russian aircraft participating in the Air China 2000 show in Zhuhai where it demonstrated the rescue system in action again; a similar demonstration was performed in Berlin at an ILA airshow.

As for the SKS-94 system, it is fitted as standard to the single-seat Su-31M and the Su-29AR trainer (the version supplied to the Argentinean Air Force). It was also envisaged for the radically reworked Su-29T primary trainer, a low-wing aircraft with a retractable tricycle landing gear and stepped-tandem seats, which subsequently became the Su-32 and then the Su-49. Yet this trainer intended to supersede the venerable Yakovlev Yak-52 has not materialised as of this writing.

Tupolev/Myasishchev KR crew rescue system testbed

The aforementioned UR-2 control system testbed was later modified for testing the M-4 bomber's crew rescue system and redesignated KR, the K denoting *katapool'**tee**ruyemoye **kres**lo* (ejection seat). Its *raison d'être* was the *Bison*'s unconventional crew rescue system with movable ejection seats, the captain and co-pilot ejecting consecutively through the same hatch.

The KR was equipped with ejection seats for the pilots, navigator, flight engineer and dorsal gunner. Between October 1952 and January 1953 it performed 12 test flights in which 20 ejections were made, including seven with live parachutists. As a result, some changes had to be introduced into the ejection seat's design.

Tupolev Tu-22M3 crew escape system testbed

By way of experiment, one Tu-22M3 *Backfire-C* was fitted with Zvezda K-36D zero-zero ejection seats. Still, Tu-22Ms continued rolling off the production line with the old Tupolev KT-1 ejection seats.

Yakovlev Yak-25L ejection seat testbeds

LII converted an early-production Yak-25 *sans suffixe* two-seat interceptor coded '01 Red' into a testbed for new ejection seats, ejection guns and pressure suits designated Yak-25L. The L suffix denoted [*letayushchaya*] *laboratoriya* – 'flying laboratory'.

The cockpit section of the Yak-25L was extensively modified, featuring two separate pressurised cockpits with individual canopies. The experimental seat was installed in the rear cockpit. For safety reasons the rear canopy

was often removed on test missions and a fairing was fitted over the rear cockpit to shield it from the slipstream. The cannons and some equipment items were deleted to save weight, and high-speed ciné cameras were installed at the wingtips for capturing the ejection sequence. Numerous test ejections were performed with this aircraft; the ejection seat was invariably occupied by a dummy.

Another *Flashlight-A*, a Yak-25M coded '27 Red', was also converted into an ejection seat testbed with a different cockpit design. The front cockpit was enclosed by an abbreviated

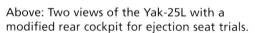

Above: Two views of the Yak-25L with a modified rear cockpit for ejection seat trials.

Above right: The Yak-25L in action: the experimental seat is seen leaving the rear cockpit. This example carries neither a tactical code nor photo calibration markings.

Right: Here, Yak-25L '01 Red' performs an ejection from ground level.

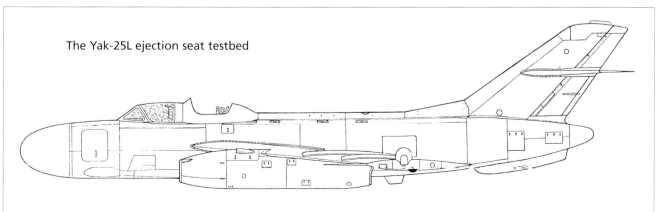

The Yak-25L ejection seat testbed

One more
Yak-25L, '16 Red',
performs a test
ejection.

version of the standard canopy, apparently opening sideways, while the rear cockpit was completely open. One more example was coded '16 Red'.

The Yak-25L testbeds were in use in 1960-65, serving for testing the K-5, K-7, KS-3, KS-4 and KM-1 ejection seats.

Yak-27L '27 Red'
on the ground
with the experi-
mental seat
partly exposed.

5. De-icing system testbeds

Antonov An-12BK 'Tanker' de-icing systems testbed

LII's large fleet of test and research aircraft included an An-12BK in ex-Air Force grey colours which initially wore the non-standard civil registration CCCP-48974 (c/n 6344510) inherited from one of MAP's Li-2T transports. At first this machine was used as a 'spray tanker' for testing the de-icing systems of other aircraft. Developed in 1981, the modification involved the fitment of an 8,700-litre (1,914 Imp gal) water tank and a drum from which a 47-m (154-ft) hose terminating in a circular sprinkler grid was deployed through the cargo door, the grid acting as a stabilising drogue. The whole thing looked so much like a single-point hose-and-drogue refuelling tanker that the aircraft was immediately

dubbed 'Tanker'. This fairly complex arrangement was needed to keep the grid and the aircraft being tested out of the tanker's wake vortex.

The water was presumably fed by gravity; the delivery rate could vary up to 4 litres (0.88 Imp gal) per second. The aircraft created a cloud of water mist measuring 3-5 m (10-16 ft) in diameter. Depending on the flight speed and delivery rate, the mission time could be anything between 30 minutes and six hours.

CCCP-48974 No. 2 is known to have been used in the trials of the Antonov An-72 *Coaler* twin-turbofan STOL tactical transport. Using the 'tanker' allowed the time required for verifying a new aircraft's de-icing system to be reduced by a factor of four; it also enhanced flight safety and allowed icing tests to be performed at realistic speeds (that is, the ones at which the aircraft was likely to operate). Curiously, the An-12/An-72 combination was displayed in model form at the 1983 Paris Air Show; the *Cub* featured its actual registration but was painted in 1973-standard blue/white Aeroflot livery. This puzzled Western observers completely, leading to the misconception that the model represented an IFR tanker akin to the RAF's C-130K Hercules C.1K and that the new tanker was perhaps intended to have some kind of civilian role (!).

Later the An-12 was extensively modified. A test article with a symmetrical airfoil and a leading-edge de-icer was installed vertically aft of the wing centre section, requiring the fin fillet to be cropped slightly. A large circular sprinkler grid with bracing struts was mounted ahead of the wing centre section. The freight hold accommodated test equipment consoles and a water tank; a big 'elephant's ear' air intake was provided on each side in line with the wing trailing edge for pressurising this tank and feeding water to the nozzles. The first cabin window to port was blanked off with sheet metal; cameras were installed in special fairings on the wing upper surface near the inboard engines, allowing the icing process to be filmed. Later, a large observation blister was added high on each side of the forward fuselage, providing a view of the grid and the test article. A small sprinkler was installed on the port side of the nose for testing air data sensors, icing detectors, aerials and such in icing conditions.

Test flights could be made at altitudes up to 8,000 m (26,250 ft) and speeds of 300-400 km/h (186-248 mph). The trail of droplets

An-12BK '10 Red' (c/n 6344510) in its final test configuration with a dorsally mounted de-icer test article, sprinkler grid and observation blisters. Note the additional sprinkler (for testing sensors and pitots) below the said blister.

generated by the sprinkler grids had a water content of up to 2 g/m³; maximum continuous spraying time with this water content was two hours, the droplets having an average diameter of 15-30 microns. The dorsally mounted test article could be up to 5.5 m (18 ft 0¹⁷⁄₃₂ in) tall, with a maximum chord of 2.5 m (8 ft 2²⁷⁄₆₄ in). Depending on the type of de-icer under test, hot air with a temperature up to 190°C (374°F) could be supplied at a rate of up to 0.8 kg/sec (1.76 lb/sec); for electric de-icers, 27V DC power with an output of up to 20 kW or 115 V/400 Hz AC power with an output of up to 12 kVA could be supplied.

By 1987 the aircraft had lost its civil identity for some reason, gaining Air Force insignia and the tactical code '10 Red'; the registration CCCP-48974 passed to a brand-new Antonov An-32A *Cline* transport (c/n 1407) in late 1987. In early 1992 the An-12 testbed was stripped of all non-standard features except the blisters and the abbreviated fin fillet, receiving another non-standard regis-

tration, RA-13331. Later it was sold to a Russian airline called Start and repainted in basic 1973-standard Aeroflot colours; concurrently an ROZ-1 Lotsiya radar in an An-12B/An-12BP-style small radome was fitted instead of the An-12BK's Initsiativa radar and associated large radome.

Close-up of the test article and sprinkler grid. The black stripes applied to the leading edge are for icing visualisation.

An-12BK CCCP-48974 No. 2 in its second icing testbed configuration

Mission equipment placement on the An-12BK de-icing system testbed (CCCP-48974):

1. Main water spraying grid.

2. Pumping station.

3. 8,700-litre water tank.

4. Test article (airfoil with leading-edge de-icer).

5. Water droplet cloud parameter measuring suite.

6. Small water spraying grid for testing air data sensors.

7. Mounting platform for air data sensors under test.

8. Test engineers' workstation.

9. Ciné cameras

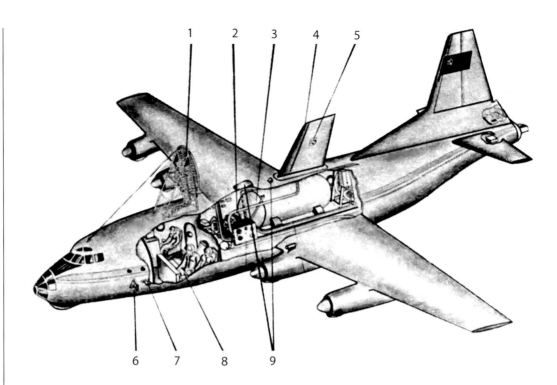

An-12BP icing research aircraft

In the early 1960s a series of fatal crashes involving the Antonov An-10 *Cat* airliner occurred. Investigation revealed that the accidents had been caused by stabiliser icing leading to loss of control – the aircraft would dive unexpectedly of its own accord. Hence GosNII GA at Moscow-Sheremet'yevo undertook an icing research programme with its own An-12BP, CCCP-11101 (c/n 01347703). The aircraft had the wing and stabiliser leading edges painted in black stripes for icing visualisation.

Ilyushin IL-12T de-icing system testbed

A Soviet Air Force IL-12T *Coach* transport was used for developing advanced de-icing systems. Unfortunately no details are known.

Ilyushin IL-18LL de-icing systems test aircraft

In the early 1960s the sixth production IL-18A *Coot* medium-haul airliner, CCCP-Л5821 (that is, SSSR-L5821, c/n 188000201), was transferred to LII which converted it for testing the de-icing systems of other aircraft. In this form the aircraft was known as the IL-18LL (not to be confused with the identically designated engine testbed converted from an IL-18V in Czechoslovakia, which in not included here).

The changes to the IL-18LL's airframe were quite extensive. A flat-topped superstructure was fitted to the centre fuselage for mounting the test article – a section of airframe incorporating de-icing system elements. To create artificial icing conditions a large circular sprinkler grid with bracing struts at the front was mounted ahead of the wing centre section;

The An-12BK 'Tanker' de-icing system testbed in its original guise as CCCP-48974 and minus blisters and other appendages on top. The sprinkler grid in the cargo door aperture is stowed but seems guaranteed to scrape the runway on rotation.

the cabin accommodated a water tank and test equipment consoles. Ciné cameras installed in angular fairings on top of the outer engine nacelles captured the ice formation on the test article.

In 1965 the IL-18LL was used to verify the de-icing systems developed for the wings of the IL-62 *Classic* long-haul airliner and the engine air intakes of various high-speed combat aircraft. According to test pilot Yakov I. Vernikov who captained the aircraft, the added area above the centre of gravity (CG) made the testbed sensitive to crosswinds, complicating flying, especially at low speeds. Towards the end of the decade the IL-18 (by then reregistered CCCP-75637 under the new system) was written off as time-expired, yielding its test mission to An-12BK CCCP-48974 No. 2 described above.

Ilyushin IL-28 de-icing system testbed

In 1953 a Soviet Air Force IL-28 was used for testing a de-icing system that found use on the An-12 and IL-18.

Lisunov Li-2 de-icing system testbed

In 1949 a Li-2 transport served as a testbed for a hot air de-icing system developed for the An-10 and Tu-104 airliners.

Mil' Mi-6VR Vodoley de-icing systems testbed

In 1976 LII commissioned the Mi-6VR test aircraft (**vod**oraspy**lite**l' – water sprayer), which served for testing the de-icing systems of other helicopters in the conditions of artificially induced icing. In similar manner to the An-12

The An-12BK 'Tanker' in military guise as '10 Red'. The sprinkler grid is stowed; note the observation blister on the forward fuselage.

The grid and hose are deployed. Note the dorsal test equipment heat exchanger and the lateral air scoops for water tank pressurisation.

'Tanker', the cargo cabin housed a water tank. In this case, however, there was no deployable grid – the water was fed to ventral spraybars installed immediately ahead of the cargo door threshold; the spraybars were forward-swept and were attached to the underside of the stub wings by long bracing struts. Because of its operational mode the machine was dubbed *Vodo**ley*** (Aquarius).

Tupolev Tu-16 de-icing system testbed

A Tu-16 coded '44 Red' (c/n 4200404) was used for aerodynamic tests and de-icing system trials, featuring an aerofoil-shaped test article installed atop the centre fuselage.

Seen here at its base, Moscow/Sheremet'yevo-1, An-12BP CCCP-11101 in pre-1973 Aeroflot colours was used by GosNII GA for icing investigation, hence the black-striped wing and tail unit leading edges.

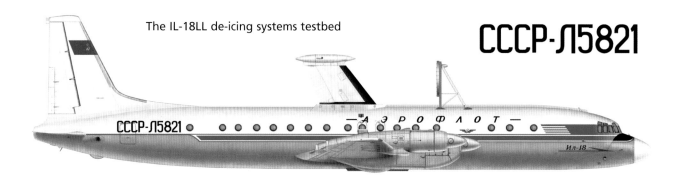

The IL-18LL de-icing systems testbed

CCCP-Л5821

IL-18A CCCP-Л5821 (that is, SSSR-L5821 in Cyrillic characters) became the IL-18LL de-icing systems testbed. Here the test article installed on the centre fuselage is a section of the IL-62 airliner's wing.

The IL-18LL in a different configuration for testing an air intake de-icer. Judging by the shape of the test article, the intake assembly appears to be that of a supersonic tactical aircraft.

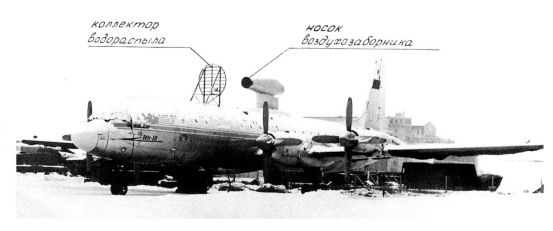

Yet another configuration of the IL-18LL with a test article resembling an F-104 style air intake – as used on the Tupolev Tu-128 heavy interceptor. Oddly, no sprinkler grid is fitted.

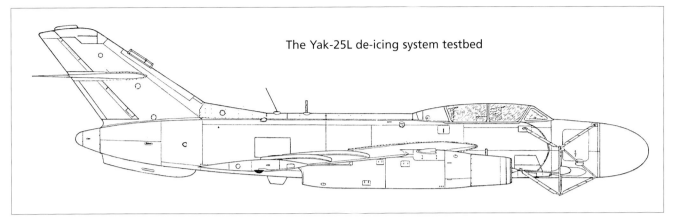

The Yak-25L de-icing system testbed

This view of Yak-25L '21 Red' shows the Yak-28 style air intake and the sprinkler grid carried on a system of struts.

Yakovlev Yak-25M de-icing system testbed (Yak-25L)

A Yak-25M coded '21 Red', was a de-icing system testbed for the Yak-28I *Brewer-C* and Yak-28L *Brewer-B* tactical bombers. The front ends of the engine nacelles were redesigned to replicate those of the Yak-28 powered by Tumanskiy R11AF2-300 afterburning turbojets, featuring extended sharp-lipped air intakes with movable shock cones. A circular grille with water sprinkler nozzles was mounted on struts in front of the starboard engine. Other non-standard features were an additional aerial on the fuselage spine just ahead of the fin fillet and a 'towel rail' aerial (possibly for data link) on the fin leading edge above the horizontal tail. Like the ejection system testbed, this aircraft has been referred to as the Yak-25L.

6. Boundary layer control system testbeds

Antonov An-12B BLC system testbed (?)

A Voronezh-built An-12B with the non-standard registration CCCP-29110 (c/n 402502)

had four cigar-shaped fairings on the wing upper surface in line with the engines. This, and the fact that it was used for soggy runway tests, suggests the aircraft was fitted experimentally with blown flaps. Later CCCP-29110 was reconverted to standard configuration.

Ilyushin DB-3 BLC system testbed

In 1942 a production DB-3 long-range bomber powered by Tumanskiy M-87B engines was converted by LII into the DB-3UPS testbed (*oopravleniye pogranichnym sloyem* – boundary layer control). The aircraft received modified wings with efficient high-lift devices (flaps and drooping ailerons) which were fitted with a BLC system. The boundary layer was sucked from the upper surface of the wings and high-lift devices through spanwise slits; for this purpose a 116-hp ZiS-101A car engine driving a suction fan was mounted in the bomb bay.

Mikoyan/Gurevich Ye-6V experimental fighter

In 1961 the Mikoyan OKB was tasked with developing a short take-off and landing (STOL) version of the MiG-21F-13 having a take-off and landing run of no more than 300-350 m

(990-1,150 ft); the fighter was to be submitted for state acceptance trials in the third quarter of the year. That year the OKB's experimental production facility modified two production MiG-21F-13s, which were designated Ye-6V/1 (c/n N74210301) and Ye-6V/2. These fighters served as testbeds for a number of features intended to give them STOL capability – namely SPRD-99 (*izdeliye* 314-2) JATO boosters, a BLC system (blown flaps) and a brake parachute relocated from the rear fuselage underside to a cigar-shaped housing at the base of the rudder.

The blown flaps underwent state acceptance trials with good results in 1961, whereupon the Mikoyan OKB decided to incorporate this feature on all subsequent versions of the MiG-21. The BLC system necessitated installation of a modified engine with air bleed valves, which entered production as the R11F2S-300; the S denoted *sdoov pogra**nich**novo **sloy**a* (boundary layer blowing). The air was bled from the third (final) high-pressure compressor stage and distributed via slots along the leading edges of the flaps; the latter were set at 45°. The BLC system was activated automatically when the flaps were deployed past 30°.

In order to conduct further research into BLC on the wing, aileron and stabilator leading edges and enhance the efficiency of the blown flaps, the Mikoyan OKB made the following modifications to the Ye-6V/1 in 1961. New hinged (that is, non-area-increasing) flaps with a maximum deflection in excess of 45° were fitted; bleed air was supplied to the flaps via manifolds with three slots along each flap's leading edge. The wing leading edges were fitted with boundary layer blowing manifolds and outlets. The maximum stabilator travel was increased to 30-35° and the leading edges were likewise fitted with boundary layer blowing manifolds and outlets.

During the Aviation Day air parade at Moscow-Tushino on 9th July 1961 one of the Ye-6Vs demonstrated a short take-off from Tushino's grass runway, assisted by SPRD-99 JATO boosters. This fact did not go unnoticed by Western observers and the Ye-6V received a separate NATO reporting name, *Fishbed-E*.

The test programme suffered a setback on 10th January 1962 when the Ye-6V/2's port JATO booster exploded on take-off, causing a massive fire that extensively damaged the aircraft. The tests continued with the first prototype.

In 1963 the Ye-6V/1 was retrofitted with an improved BLC system that reduced the approach speed and landing run dramatically and enhanced the ailerons' efficiency. After being verified on this aircraft the system

The Ye-6V/2 was a converted early-production MiG-21F-13, combining the old canopy (with lock fairings) and the new wide-chord vertical tail.

Close-ups of the Ye-6V/2's rear end with the brake parachute container doors closed and open. Note the tail bumper protruding from the ventral fin.

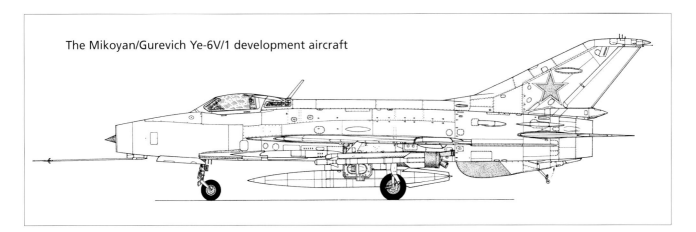

The Mikoyan/Gurevich Ye-6V/1 development aircraft

became a standard feature on production MiG-21s. In July 1963 the same aircraft was used in a brief test programme to explore the possibility of landing with angles of attack up to 16-18° (the MiG-21F's operational AoA limit was 10-11°) and determine the aircraft's landing performance with a specially fitted retractable tail bumper. The sprung tail bumper with a roller at the end was housed in the ventral fin. The BLC was deactivated for the duration of this programme. Mikoyan OKB test pilot Pyotr M. Ostapenko made four landings in this aircraft; a further two flights were performed by Aleksandr V. Fedotov.

Landings performed with an AoA of 14° (with 25° flap) and an AoA of 16°30' (with the flaps up) showed that the aircraft was fairly stable around all three axes and displayed adequate aileron and rudder authority. With the tail bumper extended, the Ye-6V could land smoothly with these high AoAs. In a flapless landing with an AoA of 16°30', the fighter touched down at 244-251 km/h (151-156 mph); with 25° flap and an AoA of 14°, the touchdown speed was 256 km/h (159 mph).

The pilots noted that stabilator authority was insufficient during these high-AoA landings. Also, the unusually strong nose-up attitude and high position of the cockpit above the ground in these conditions complicated the landing approach, demanding special skill and concentration on the part of the pilot. The report also said that the fighter's longitudinal stability was insufficient at certain CG positions (for example, 37.5% MAC), which also made it harder to fly the aircraft.

The report described the tail bumper's shock absorber as adequate; when the Ye-6V/1 touched down tail first, settling down on the mainwheels a few seconds later, no sharp jolts were experienced. Aileron authority remained adequate right down to 230-240 km/h (142-149 mph).

Of the six high-AoA landings mentioned above, only one involved the use of flaps. With 25° flap and the control stick pulled all the way back, the aircraft touched down with a smaller angle of attack at 240 km/h. Similar experiments involving high-AoA landings were later undertaken with the Ye-7/3 and Ye-7/4 (prototypes of the MiG-21PF *Fishbed-D* interceptor); interestingly, the Ye-7 touched down at the same speeds with no danger of tailstrike in these conditions. During landing approach the Ye-6V/1 glided at 320-350 km/h (198-217 mph); the holding-off was initiated at 3-4 m (10-13 ft) above the ground, which was higher than on the standard MiG-21F.

The pilots noted that the touchdown with the tail bumper was always unexpected, the pilot perceiving the mainwheels to be about 0.5 m (1 ft 7 in) above the ground at that moment. After scraping the runway with its tail bumper for a second or two the aircraft would settle down smoothly on the main-wheels and then drop the nosewheel.

The landing weight of the Ye-6V/1 during these tests varied from 5,450 to 5,850 kg (12,015-12,900 lb). The table on page 138 details the test results obtained during the high-AoA landings with the tail bumper.

The verdict of the test pilots and test engineers was unanimous and clear: the high-AoA landing technique making use of the tail bumper was complicated and harder to master for service pilots than the technique making use of blown flaps.

Sukhoi S-25 and S-25T BLC system testbeds

Work on a BLC system testbed known in-house as the S-25 began at OKB-51 in 1960. It was proposed to provide boundary layer blowing using a network of tiny apertures along the leading edges of the wings and flaps. Air

High-alpha landing tests of the Mikoyan Ye-6V/1 STOL development aircraft				
Date	Landing speed, km/h (mph)	Landing run, m (ft)	Stabilator deflection	Flap setting
4-7-63	256 (159)	1,387 (4,550)	29°	25°
4-7-63	n.a.	n.a.	n.a.	0°
9-7-63	258 (160)	716 (2,349)	29°	0°
9-7-63	244 (151)	578 (1,896)	28.6°	0°
19-8-63	260 (161)	1,045 (3,428)	24.4°	0°
21-8-63	251 (156)	925 (3,034)	29°	0°

was bled from the fifth compressor stage of the AL-7F engine, which had to be specially modified.

A production Su-7B (the aforementioned '02 Red', c/n 2502) was modified by removing all armament and weapons pylons; a system of tiny apertures was provided along the wing leading edges for boundary layer suction, and the standard flaps were replaced with blown flaps. Bearing the in-house designation S-25, the aircraft became a testbed for developing boundary layer control systems.

It was decided to gather preliminary data by placing the aircraft in TsAGI's T-101 wind tunnel. The tests took place in late 1960; in this configuration the aircraft was known as the S-25T. The suffix stood for **troob**nyy ('for wind tunnel tests'); this adjective was derived from **a**erodina**mich**eskaya troo**ba** – wind tunnel.

In 1961 the S-25 was returned to flight test status; it was flown by test pilot Yevgeniy S. Solov'yov. The boundary layer suction system turned out to be unreliable (the apertures could easily become clogged) and the programme was abandoned.

When testing of the BLCS had been completed, the S-25 found use in the laminar flow research programme described earlier.

Tupolev Tu-110L BLC system testbed (order No. 290)

On 18th July 1958 the Soviet Council of Ministers passed Directive No.786-378 whereby the Tupolev OKB was directed to design and build the Tu-124 *Cookpot* high-speed short-haul airliner powered by two 5,400-kgp (11,905-lbst) Solov'yov D-20P turbofans. The directive also envisaged the development of a version featuring blown flaps that would give the airliner STOL capability – the take-off and landing run was to be no more than 600-700 m (1,970-2,300 ft). To this end the STOL version was to be powered by D-20PO engines featuring an air bleed system to cater for the

blown flaps (hence the O suffix denoting ot**bor** [**voz**dukha] – air bleed). The directive specified that the engine was to be tested on a suitably modified Tu-110 *Cooker* airliner; the tests were to be completed in the second quarter of 1959.

In 1959 the Tupolev OKB undertook the necessary design work; all components of the BLC system were manufactured and ready for installation on the Tu-110 by mid-1960. Conversion of the Tu-110 prototype (serialled '5600 Blue') started in June; the air ducts were routed along the rear spar so that the air exited over the flaps through slits. The modified aircraft was designated Tu-110L for [*leta-yushchaya*] *labora**tori**ya* – 'flying laboratory'. In official documents the machine was referred to as 'order No. 290' for security reasons.

The conversion work turned out to be a lengthy process and was not completed until 1962. The Tu-110L entered flight test that year and the test programme continued until 1964; the aircraft was flown by pilots Vasiliy P. Borisov, Boris M. Timoshok, Aleksey P. Yakimov and Anatoliy S. Lipko, with engineer Nogtev as engineer in charge. The results were quite encouraging and the BLC system was recommended for installation on Tupolev aircraft.

Still, the system found no application on heavy aircraft at the time; the STOL version of the Tu-124 remained a paper project. The leaders of the Tupolev OKB justly considered that, despite all its advantages, the BLC system involved an unacceptably high risk on a commercial aircraft, as a BLC system failure on take-off or landing and the resulting sudden deterioration in field performance could lead to a crash – possibly with numerous fatalities as a result. It was deemed more expedient to use BLC on combat aircraft equipped with zero-altitude ejection seats that would give the crew a chance of survival if the system failed.

Yakovlev Yak-28P development aircraft with blown flaps

A production Yak-28P interceptor was experimentally fitted with blown flaps to give more lift. The modification was tested but did not find its way to the production line. It's hard to say if this was because the system worked unsatisfactorily or because the *Firebar* was about to be superseded by a brand-new interceptor, the Sukhoi Su-15 *Flagon*.

7. Aircraft carrier technology testbeds

MiG-23BN aircraft carrier technology testbed

In 1983-87, as part of the effort to develop a shipboard version of the MiG-23, a MiG-23BN *Flogger-H* fighter-bomber coded '03' and carrying the number 603 on the tail was fitted with an arrestor hook to execute aircraft carrier take-offs and landings on a simulated flight deck with arrestor wires. (Some sources refer to it as a MiG-27, calling it 'MiG-27LL'.) The modification required the *Flogger*'s characteristic folding ventral fin to be removed and the main gear units to be reinforced to permit no-flare landings typical of carrier operations.

The trials took place on the NIUTK 'unsinkable carrier' installation at Novofyodorovka AB near Saki. LII test pilots Aleksandr V. Krootov, Sergey N. Tresvyatskiy and Vladimir G. Gordiyenko teamed up to perform the successful trials but eventually the authorities decided against using the MiG-23 at sea. Instead, the test results were used in the development of the MiG-29K *Fulcrum* (the first aircraft thus designated, *izdeliye* 9.31), Su-27K (Su-33) *Flanker-D* and Su-25UTG *Frogfoot-B* shipboard aircraft.

The 'carrier-capable' MiG-23BN '03 Yellow' is seen during a weigh-in.

The modified MiG-23BN makes a ski jump take-off at Saki. Note the arrestor hook and the photo calibration markings.

Here the modified MiG-23BN's arrestor hook replacing the ventral fin catches the wire at the NIUTK 'unsinkable carrier'.

Mikoyan MiG-29KVP aircraft carrier technology testbed

In 1982, upon completion of the Phazotron N019 fire control radar's tests, the eighth prototype MiG-29 (*izdeliye* 9.12) *Fulcrum-A* coded '18 Blue' (and known as 'aircraft 918') was converted under the MiG-29K shipboard fighter programme. The objective was to test the type's compatibility with a conventional take-off and landing (CTOL) aircraft carrier and perfect carrier operations techniques. All unnecessary equipment was removed, reducing the gross weight to only 12 tons (26,455

lb), and an arrestor hook was fitted under the rear fuselage. This required the airframe to be reinforced to withstand the higher loads; the airbrake and brake parachute were deleted. In this guise the aircraft was unofficially designated MiG-29KVP (*korotkiy vzlyot i posahdka* – STOL).

Again, carrier operations trials proceeded on the NIUTK installation. On 21st August 1982 Mikoyan OKB test pilot Aviard G. Fastovets made the first take-off from the provisional T-1 ski jump (T = *tramplin*) in the MiG-29KVP. The aircraft became airborne at 240 km/h (133 kts) after a 250-m (820-ft)

Top: '18 Blue', the MiG-29KVP development aircraft used in the shipboard fighter programme. Note the photo calibration markings on the air intakes and the radome.

Above: The MiG-29KVP in front of a jet blast deflector.

The MiG-29KVP's arrestor hook. Note the absence of the usual split airbrake and brake parachute housing.

while the results were analysed. For example, GNIKI VVS test pilot Vladimir N. Kondaoorov made 200 flights at Novofyodorovka AB in 1988, including 65 arrestor wire engagements; some of these were made in the MiG-29KVP.

Initially the aircraft was repainted in a greenish-blue naval colour scheme. However, this turned out to be a bad idea – the naval camouflage was more of an 'anti-camouflage'; as Toktar Aubakirov put it, the aircraft was 'eye-catching as a butterfly'. Therefore, the MiG-29KVP soon reverted to its original grey colour scheme.

The aircraft was in the static park at MosAeroShow-92 in Zhukovskiy (11-16th August 1992). Later it was briefly used as an instructional airframe by the Moscow Energy Institute before being donated to the Central Russian Air Force Museum in Monino. For some obscure reason the MiG-29KVP had 'Mikron' written on both stabilators after being transferred to the museum.

Sukhoi T10-3 aircraft carrier technology testbed

Initially the third prototype *Flanker-A* (T10-3), '310 Blue', was used for powerplant testing, being the first Su-27 with the intended Lyul'ka AL-31F afterburning turbofans. After completing its flight test programme in 1982 the aircraft moved to the Soviet Naval Aviation's Flight Test Centre at Novofyodorovka AB near Saki. There on 28th August Sukhoi OKB test pilot Nikolay F. Sadovnikov made the first take-off from the provisional T-1 ski jump as the first step in adapting the T-10 to the shipboard fighter role. The take-off weight was 18,000 kg (39,680 lb) and the fighter became airborne after a take-off run of 200 m (660 ft). A few days later Sadovnikov took off in the T10-3 at the maximum permitted AoA, demonstrating that ski jump take-off was possible in various modes. Later, LII test pilot Vladimir G. Gordiyenko made a ski jump take-off with a 21-ton (46,300-lb) TOW.

In 1983 the T10-3 was fitted with an arrestor hook and recoded '03 Blue'. In this configuration the aircraft made numerous arrestor wire engagements on the NIUTK 'unsinkable carrier' with Sukhoi and LII test pilots. These were not really 'deck landings', as the aircraft was no longer airworthy – it simply accelerated and then caught the wire. Upon completion of the programme the T10-3 was retired at Novofyodorovka AB.

take-off run. The T-1 was unsatisfactory and was soon replaced; between 1st and 25th October 1984 the MiG-29KVP made a series of take-offs from the restyled T-2 ski jump.

Two other test pilots, Toktar O. Aubakirov (Mikoyan OKB) and Aleksandr V. Krootov (LII), also participated in the ski jump take-off and automatic carrier approach trials. Air Force test pilots had to wait their turn, using breaks in the manufacturer's flight test programme

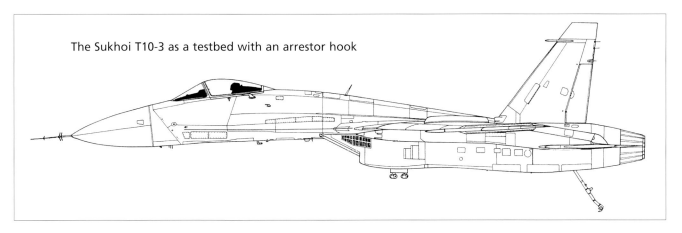

The Sukhoi T10-3 as a testbed with an arrestor hook

The as-yet hook-less T10-3 clears the edge of the T-1 ski jump. Note the video camera at the edge of the ski jump.

Two stills from a ciné film showing the T10-3 as it catches the wire of the S-2 arrestor wire system. The aircraft was not actually flown in this configuration.

The T10-3 ('03 Blue') in ultimate configuration with an arrestor hook

141

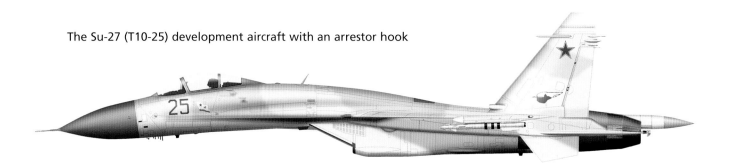

The Su-27 (T10-25) development aircraft with an arrestor hook

The T10-25 is catches the wire on the NIUTK 'unsinkable carrier' in September 1984. The fighter retained the early-type squared-off fin caps with anti-flutter booms.

Sukhoi T10-25 development aircraft

An early-production Su-27 *Flanker-B*, the T10-25 ('25 Blue', c/n 36911006…03, f/n 06-03), was also modified under the Su-27K programme. The aircraft was fitted with a reinforced landing gear, an arrestor hook and enlarged flaperons emulating the slotted flaps developed for the Su-27K. To underscore the naval nature of the programme it was involved in, the fighter sported the Soviet Navy's blue and white flag under the cockpit. The T10-25 still had the original straight-cropped fin tips and anti-flutter booms on the fins (which were replaced by raked fin tips on later *Flankers*).

This aircraft was much used to investigate and practice carrier operations, making ski jump take-offs and arrestor wire engagements at Novofyodorovka AB. Its first flight there probably took place on 3rd August 1984 at the hands of Nikolay F. Sadovnikov. Unfortunately the T10-25 crashed on 23rd November 1984 when a hydraulic pipeline burst, causing the rudder actuators to fail; Sadovnikov ejected safely from the uncontrollable aircraft.

8. Recovery system testbeds

Antonov An-12M LL multi-role testbed

In 1986 LII's An-12BK '43 Red' (c/n 8345902) – the An-12M LL – was converted for its first test programme undertaken jointly with the Parachute Systems Design Institute (*NII para**shoo**tostro**yen**iya*). Its functions was to test a recovery system intended for picking up film capsules ejected by surveillance satellites. Two long booms were hinged to the aft fuselage underside, their rear ends connected by a steel cable. The booms swung down so that the cable could snag the capsule's parachute as the capsule floated earthwards. A similar system had been used in the USA where a specially modified Lockheed JC-130 picked up film capsules ejected by Lockheed GTD-21 high-speed reconnaissance drones. The work continued until 1989 – that is, even after the An-12M LL had been further modified for ejection seat trials (see above).

Mil' Mi-8MT recovery system testbed

In 1984-89 an uncoded Mi-8MT (c/n 93115) was likewise used by LII and the Parachute Systems Design Institute to test the same recovery system. The intention was to use this system for recovering target drones, remotely piloted vehicles (RPVs) and cruise missiles undergoing trials and fitted with test equipment so that the flight data recorders could be deciphered. The clamshell cargo doors were removed and the system's hinged booms painted in black and white stripes were hinged to the rear fuselage near the aperture; these could be swung down to snag the RPV's parachute as it floated earthwards, its engine cut. The Yakovlev OKB used the results of this programme when developing the *Pchela* (Bee) subsonic tactical reconnaissance drone.

9. Other systems/equipment testbeds

Antonov An-12A trailing wire aerial testbed

In 1975 an uncoded Irkutsk-built An-12A *Cub* (c/n 9900902) operated by LII served as a testbed for the mighty BLT-5 winch driven by a ram air turbine; the winch was used to deploy a 2,500-m (8,200-ft) trailing wire aerial (TWA) with a stabilising drogue at the end. The unit had been specially developed for the Soviet Navy's Tupolev Tu-142MR *Bear-J* communications relay aircraft whose mission was to maintain very low frequency (VLF) communications between submerged nuclear missile submarines and land-based or airborne command posts in the event of a nuclear attack (R stands for *retranslyator* – communications relay installation). The aircraft also featured a whole mesh of wire aerials stretched from the tail unit to the wings, looking almost like a spider's web. The tests were filmed from a Tu-124 airliner acting as a chase plane.

Ilyushin DB-3 electric system testbed

A DB-3 bomber was used for flight-testing a powerful generator intended for use as an onboard electric power source.

An-12A c/n 9900902 was converted into a VLF communications equipment testbed with a mesh of wire aerials above the fuselage. Note the stabilising drogue of the trailing wire aerial near the cargo door threshold.

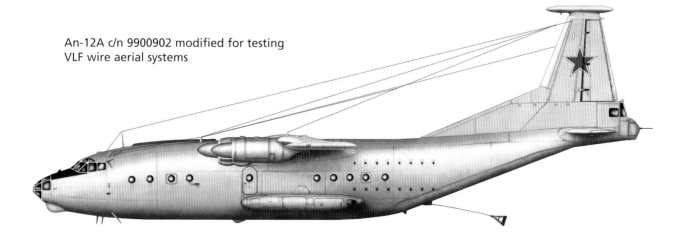

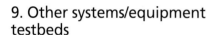

An-12A c/n 9900902 modified for testing VLF wire aerial systems

143

A still from a ciné film showing canopy jettisoning trials on MiG-19 '50 Red'; the shot was taken 0.048 seconds after jettison procedure commencement. Note the transverse metal strips on the canopy.

MiG-19 '18 Red' was used for drop tank separation tests; note the fairing housing ciné cameras immediately aft of the nose gear unit.

A still from a ciné film capturing the drop tank jettison sequence on the same aircraft.

Mikoyan/Gurevich MiG-17 testbeds/research aircraft at LII

LII had several late-production MiG-17s *sans suffixe* converted into systems testbeds. One of them, a Gor'kiy-built example serialled '611 Blue' (c/n N54210611), had a small teardrop fairing above the port airbrake and a long slender pitot on the intake upper lip. The type of equipment tested on this aircraft is unknown.

Another MiG-17 was used by LII for an aeromedical experiment to investigate the effect of altitude on the precision of the pilot's actions.

Mikoyan/Gurevich MiG-19 trials aircraft at LII (?)

An early MiG-19 *sans suffixe* coded '50 Red' was used to check the operation of the canopy jettison system. The aircraft had black photo-theodolite calibration markings painted on the nose, aft fuselage and fin. Interestingly, the canopy used in these trials was of a very early type, with transverse metal strips as on the first prototype (SM-9/1).

Another *Farmer-A* coded '18 Red' was used for drop tank tests aimed at ensuring reliable separation from the aircraft. To this end a rather large fairing housing two ciné cameras was installed immediately aft of the nose gear unit. Calibration markings were also applied – both to the aircraft and the drop tanks.

Mikoyan/Gurevich MiG-21US testbed

In 1970-75 LII used a modified MiG-21US *Mongol-B* trainer (quoted as 'No. 1214' – that is, c/n 12685114?) for 'researching the interaction between the pilot and the aircraft in abnormal flight situations and for testing onboard safety systems'. This flowery phrase could refer to tests of active flight safety systems, such as a 'panic button' bringing the air-

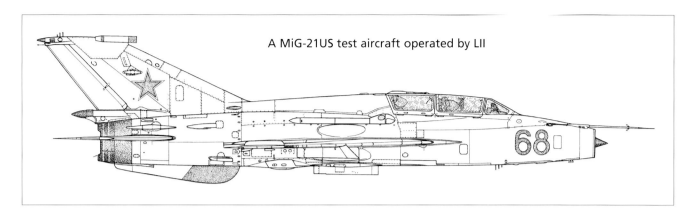

A MiG-21US test aircraft operated by LII

craft into straight and level flight from any attitude, should the pilot become disoriented; however, it could just as easily mean tests of ejection seats and canopy jettison systems.

No details are known. However, it is known that LII's extensive fleet of miscellaneous test and research aircraft included a MiG-21US coded '68'; this could be the aircraft in question. Another (and very probable) alternative is MiG-21US c/n 09685126, which was coded '68'. The aircraft had a cylindrical fairing with a hemispherical front end at the top of the fin housing a forward-looking ciné camera; another camera pointing obliquely aft was accommodated in the bulged front end of a semi-cylindrical fairing on the starboard side of the centre fuselage. This suggests that '68' was indeed used for canopy jettison/ejection seat trials.

After reaching the end of its service life MiG-21US c/n 09685126 was preserved as a gate guard in front of the Test Pilots School on the premises of the LII airfield. By then the camera fairing at the top of the fin had been removed and the normal fin tip reinstated; the lateral camera fairing remained, however. The tactical code was by then applied in low-visibility form as '68 White outline'; however, the truth may be that the coloured filling of the code was simply washed away completely by the elements.

Mikoyan MiG-23UB hydraulic system testbed

In 1987-89 the aforementioned 'MiG-23UB No. 919' served as a testbed for hydraulic system components.

Mil' Mi-24D foreign object damage protection systems testbed

A Mi-24D *Hind-D* assault helicopter coded '74 Red' was used to test an early model of the vortex-type intake filters which became standard on late *Hinds*. Unlike the production model resembling partly deflated footballs, these looked like large buckets. By 1998 the helicopter was derelict at the Mil' OKB's flight test facility in Panki, Moscow Region.

MiG-21US '68' was used by LII in an unidentified test programme. Note the camera fairings on the centre fuselage and at the top of the fin.

Sukhoi Su-9 cockpit lighting testbed

In 1961 a Moscow-built Su-9 (c/n 100000308) was used for testing a new red cockpit lighting system making the aircraft less observable to enemy aircraft at night.

Sukhoi Su-9 development aircraft with ram-air turbines

In 1963-64 two Su-9s were fitted experimentally with an ATG-2 ram-air turbine (RAT) providing electric power in an emergency. The RAT supplanted the ventral brake parachute bay and the parachute was relocated to a fairing at the base of the rudder. This feature was not incorporated on production aircraft.

Sukhoi T47-4 development aircraft

When the state acceptance trials of the T-3-8M aerial intercept weapons system had been completed, the fourth prototype Su-11 (T47-4, '35 Red') was fitted with an AP-28E-1 autopilot and an ESUV-1 electrohydraulic air intake control system (*elektroghidravlicheskaya sistema oopravleniya vozdookho-zabornikom*), serving as a testbed for these systems.

Sukhoi Su-15 multi-role testbed (c/n 1315340, 'aircraft 0009')

Tests of the Tumanskiy R13-300 afterburning turbojets on Su-15 c/n 0715311 revealed the engine's tendency to flame out at high Mach numbers because the air intakes were too small. Hence in 1970 the Sukhoi OKB modified a further production Su-15 (c/n 1315340) to feature wider air intakes with bulged outer faces which were later incorporated on the production Su-15TM. For security reasons this aircraft was referred to at the OKB as 'aircraft 0009' (possibly to fooling hypothetical spies into thinking it was the aircraft's c/n).

After lengthy manufacturer's tests the aircraft was turned over to GK NII VVS, making 44 test flights between mid-May and August 1971 as part of the Su-15TM's state acceptance trials. The aircraft showed encouraging performance in comparison with the Su-15TM prototype (c/n 0115305) which featured the early-type narrow intakes; acceleration at supersonic speeds was improved, the service ceiling increased from 17,600 to 18,500 m (from 57,740 to 60,695 ft) and effective range increased to 1,680 km (1,040 miles).

In June 1972 'aircraft 0009' was used to test the new ogival radome developed for the Su-15TM. This was meant to eliminate internal reflections of the radar pulse.

Sukhoi Su-15T multi-role testbed (c/n 0215306)

In 1972 the Sukhoi OKB and LII held a joint programme with the 16th production Su-15T (c/n 0215306) to explore the electromagnetic compatibility parameters of the avionics and test a trim mechanism in the roll control circuit. At the end of the year the aircraft was fitted with increased-area tailplanes, undergoing a special test programme in 1972-73 with a view to improving the handling in landing

The T47-4 development aircraft, seen here as a teaching aid at the Solntsevo tech school.

mode with the BLC activated. In 1973 Su-15T c/n 0215306 served as a testbed for the new Tester-U3 flight data recorder ('black box') which later became standard on many Soviet combat aircraft.

Tupolev Tu-134 miscellaneous testbeds and research aircraft

a) GosNII GA used Tu-134A CCCP-65047 (c/n (73)49600, f/n 3805) as a 'dogship' for studying aerodynamics, developing new piloting techniques and investigating flight safety issues. Among other things, it was used to study the propagation of smoke via the cabin ventilation system in the event of a fire in the rear baggage compartment or the No. 3 equipment bay when the crash of Tu-134AK CCCP-65120 on 2nd July 1986 was being investigated. To this end the aircraft was fitted with a GD-1M smoke generator (*ghenerahtor dyma*); two test flights were made on 12th and 20th August 1986.

b) In 1983 GosNII GA converted the second prototype Tu-134A, CCCP-65626 (c/n 9350704), into a testbed for an active static electricity neutralisation system designed to replace the traditional static discharge wicks. Two static dischargers were mounted at the wingtips, two at the tips of the main gear fairings and two more under the APU exhaust. Each discharger was a pointed rod 8 mm (0⁵⁄₁₆ in) thick with inner and outer contacts and a Teflon insulator in between. The system remained experimental.

c) In 1971 the first pre-production Tu-134 *sans suffixe*, CCCP-65600 (c/n 5350002), was converted into a testbed for some unspecified 'special equipment'.

Yakovlev Yak-25 'Burlaki' system testbed

The second prototype of the straight-wing Yak-25 fighter of 1947 ('15 Yellow', c/n 115002) was modified under the aforementioned Burlaki captive escort fighter programme by installing a pneumatically-operated telescopic probe (dubbed 'harpoon') atop the nose; this connected with a drogue at the end of a steel cable paid out by the bomber to be escorted.

Yakovlev Yak-25M testbed

A production Yak-25M coded '18 Red' was used to test the canopy jettison system.

Yakovlev Yak-28P testbed

A late-production (long-nosed) Yak-28P interceptor coded '57 Red' was modified by LII for some kind of test work, with two small projections (possibly cameras or aerials) on top of each wingtip fairing.

Yakovlev Yak-40 Fobos testbed/ research aircraft

This research aircraft converted from a production Yak-40 *Codling* short-haul airliner registered CCCP-87304 (c/n 9322028) was developed for the Lavochkin Science & Production Association which, after Chief Designer Semyon A. Lavochkin's demise, had

The second prototype Yak-25 fighter ('15 Yellow', c/n 115002) after being equipped with the probe of the Boorlaki captive fighter towing system.

The Yak-25 makes contact with the towing drogue.

Above: This Yak-25 coded '18 Red' was used for canopy jettison system trials. The photo was taken after 1955, as indicated by the absence of the star insignia on the fuselage.

switched to missile and space technology. The Yak-40 **Fob**os (Phobos) was to be used as a testbed for various equipment items developed by this enterprise for manned and unmanned space vehicles. In addition, it was intended to undertake a broad spectrum of research associated with the study of natural resources, environment, atmospheric phenomena, natural and man-made anomalies, solar radiation, air and water pollution and so on.

The results of this research provided the basis for economic analysis of the environment situation with a view to evolving the necessary practical measures. The Yak-40 Fobos filled this role in the late 1980s/early 1990s, whereupon it was reconverted to passenger configuration and was in airline service since 1993 as RA-87304 – first with Aeroflot and then with Belgorod Avia.

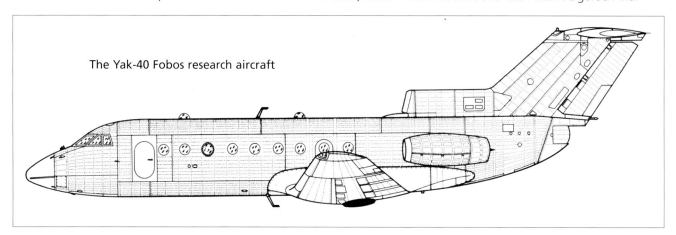

The Yak-40 Fobos research aircraft

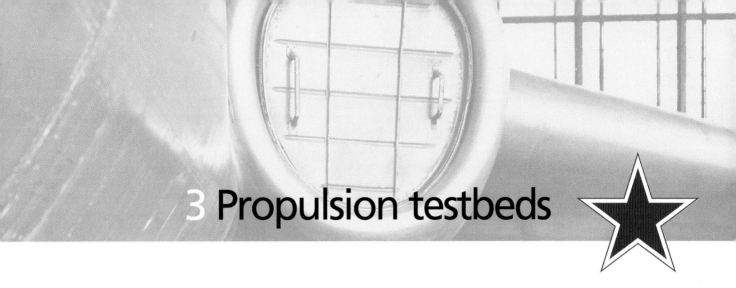

3 Propulsion testbeds

A n important class of testbeds encompasses aircraft used for testing new powerplants – engines, propulsors and propulsion technologies, such as the use of cryogenic fuel.

Antonov An-12 engine testbed

From the outset the An-12 was extensively used for test and research purposes, since its large payload and capacious freight hold accommodating a lot of test equipment made it eminently suitable for this role. Thus, as early as 1959 an An-12 (identity unknown) was used as a testbed for the Ivchenko AI-24 turboprop (initially rated at 2,500 ehp). The engine, and the associated AV-72 four-blade reversible-pitch propeller of 3.9 m (12 ft 9½ in) diameter developed by the Stoopino Machinery Design Bureau (now NPP Aerosila), had been created for the An-24 *Coke* twin-turboprop regional airliner, which entered flight test on 20th October 1959. The An-24 turned out to be highly successful and became the progenitor of a whole family of twin-turboprop aircraft; most of them were powered by the same engine which, in its ultimate form (AI-24VT) was uprated to 2,820 ehp.

Antonov An-12M LL multi-role testbed

As an alternative to the ejection seat test module, LII's An-12M LL multi-role testbed (An-12BK '43 Red', c/n 8345902) could be fitted with a module housing an auxiliary power unit aft of the tail gunner's station. Thus by May 1994 it was fitted with the tail-cone of the Ilyushin IL-114 twin-turboprop regional airliner housing a VD-100M APU. In this guise the aircraft was displayed statically at the MAKS-95 airshow (22nd-27th August 1995).

A drawing from LII's documents shows that the development APU could also be installed in a conformal pod low on the forward fuselage port side (ahead of the entry door) to replicate the APU installation on a typical high-wing military transport. An additional cooler for the APU was located at the rear of the cargo hold to simulate low ambient temperatures and was apparently extended in flight. The greater part of the test equipment was likewise in the unpressurised cargo hold.

According to LII's advertising materials, testing was possible at altitudes up to 10,000 m

The rear end of the An-12M LL at the MAKS-95 airshow; an IL-114 tailcone has been fitted for the purpose of testing the airliner's VD-100M APU. Note the video camera above the tail gunner's station window.

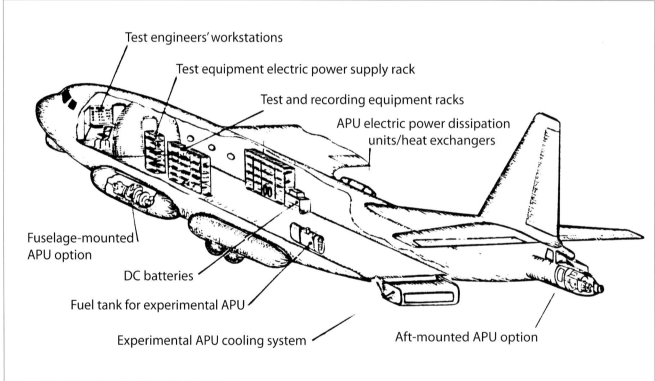

Test engineers' workstations

Test equipment electric power supply rack

Test and recording equipment racks

APU electric power dissipation units/heat exchangers

Fuselage-mounted APU option

DC batteries

Fuel tank for experimental APU

Experimental APU cooling system

Aft-mounted APU option

The location of test articles and test equipment on the An-12M LL in APU testbed configuration.

(32,810 ft), indicated airspeeds of 250-550 km/h (155-341 mph), ambient air temperatures of +60°C to –60°C (140°F to –76°F) and air humidity up to 100% (!). High and low ambient temperatures could be simulated by means of special ground support equipment, eliminating the need to travel to hot or cold climatic zones and saving certification time; three patents were issued for this technique. The tests were held in 1993-98.

Antonov An-24 engine testbed

An An-24 airliner enigmatically referred to in LII's records as 'An-24 No. 807' was used in 1962-74 for perfecting the powerplant and improving the reliability and service life of the Ivchenko AI-24 turboprop. 807 cannot be the last three digits of the c/n, as Batch 8 contained only five aircraft; nor can it mean An-24B CCCP-46807, as this aircraft was built in 1965.

This An-24RV (CCCP-46271?) was used by the Stoopino Machinery Design Bureau for testing experimental eight-bladed low-noise propellers. Here, maintenance work is under way on the aircraft, with a special vehicle providing hot air.

Antonov An-24RV propeller testbed

In 1979 a production An-24RV in 1973-standard Aeroflot livery was fitted with experimental eight-blade propellers of an unknown model replacing the standard AV-72s. The new propellers were intended to reduce external noise levels by reducing propeller speed. The registration has been quoted as CCCP-46271 but this is doubtful, as this was a fairly early-production An-24B (c/n 77303604?); much later-production 'Bs were usually upgraded to An-24RV standard. The aircraft was eventually preserved at the NPP Aerosila facility in Stoopino.

Antonov An-32LL propulsion testbed

Upon completion of the trials the first prototype of the An-32 *Cline* tactical transport (CCCP-83966 No. 1, c/n 1006) was converted into a testbed for an advanced eight-blade propeller, receiving the non-standard An-24-style registration CCCP-46961 (Soviet/CIS An-32s

were normally registered in the 48xxx block). The blades of the development propeller featured straight trailing edges and scimitar-shaped leading edges. Initially only one such unit was fitted, replacing the port AV-68DM four-blade propeller. Later in the course of testing, an identical eight-blade propeller was fit-

Close-up of the starboard propeller, showing the cropped blade tips. Note the single ventral fin.

An-32LL CCCP-46961 with an experimental eight-blade propeller on the port engine.

ted to the starboard engine as well. The aircraft has been referred to as the An-32LL.

This propeller was presumably a technology demonstrator for the Stoopino Machinery Design Bureau SV-36 contra-rotating propeller used with the Lotarev (ZMKB) D-236T experimental propfan engine. An eight-blade propeller of similar design formed the front stage of the SV-36, the aft stage having six blades. The D-236T/SV-36 combination was envisaged for the An-70 military transport. On the actual aircraft, however, Muravchenko (ZMKB) D-27 propfans driving SV-27 contraprops of similar design had to be used because of the much-increased gross weight (see IL-76LL below).

North American B-25J engine testbeds

One of the B-25J-20-NC Mitchell bombers delivered to the USSR under the Lend-Lease agreement (44-29347, the c/n is quoted variously as 108-32622 or 108-33672) became a testbed for the RD-10F turbojet – the afterburning version of the RD-10 which was a Soviet copy of the Junkers Jumo 004B. The afterburning version had been developed locally. Unusually, the development engine was mounted on a pylon above the bomber's fuselage rather than beneath it.

B-25J-25-NC 44-30041 (c/n 108-33316 or 108-34366) with a modified rear fuselage was used as a testbed for the D-5 pulse-jet engine delivering a thrust of 420-440 kgp (930-970 lb). The engine was intended for the 14Kh cruise missile (see weapons testbed chapter).

North American B-25 rocket booster testbed

Another B-25 stripped of armament was used for testing Soviet-designed JATO boosters; these included the Type 93-1 solid-fuel booster and the SU-1500 liquid-fuel booster. (One source states that a B-25B was used at LII for testing the Type 93-1 boosters.)

B-25J-20-NC 44-29347 was a testbed for the RD-10F afterburning turbojet.

This early-model Mitchell (probably a B-25B) served as a testbed for the Type 93-1 JATO boosters.

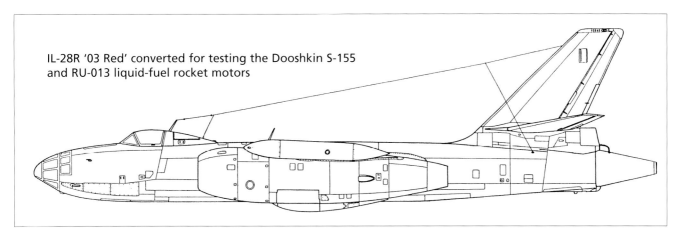

IL-28R '03 Red' converted for testing the Dooshkin S-155 and RU-013 liquid-fuel rocket motors

Boeing B-29 engine testbed

As part of the effort to reverse-engineer the Boeing B-29 Superfortress heavy bomber and put it into production as the B-4 (Tu-4), one of the Superfortresses captured by the Soviets after force-landing in the Soviet Far East – specifically, Wichita-built B-29-5-BW '256 Black' (ex-USAAF 42-6256 'Ramp Tramp', c/n 3390) – served as a testbed for the 2,400-hp Shvetsov ASh-73TK 18-cylinder radial selected to power the Tu-4. The development engine was fitted instead of the starboard inner Wright R-3350 Twin Cyclone.

After passing a rigorous test programme the ASh-73TK was used successfully for many years on the Tu-4 and the Beriyev Be-6 *Madge* flying boat. Later B-29 '256 Black' was ferried to the Kazan' aircraft factory No. 22 and converted into a 'mother ship' for rocket-powered experimental aircraft in accordance with MAP order No. 210 of 16th April 1948.

Ilyushin IL-18 engine testbed

An IL-18 referred to as 'No. 604' (presumably IL-18B CCCP-75660, c/n 188000604) was used in 1960-75 for perfecting the powerplant of the production IL-18 and improving the reliability and service life of the Ivchenko AI-20 turboprop.

Ilyushin IL-28R engine testbed

A single IL-28R reconnaissance aircraft coded '03 Red' (c/n unknown) was modified to test two liquid-propellant rocket motors developed by Leonid S. Dooshkin – first the 3,800-kgp (8,380-lbst) S-155 designed for the Mikoyan Ye-50 mixed-power high-altitude interceptor and then the RU-013. Both models ran on TG-02 hypergolic kerosene (***top**livo ghipergo**lich**eskoye* – hypergolic, that is, self-igniting fuel) and AK-20 oxidiser (AK stood for *a**zot**-naya kislo**ta*** – nitric acid); grade T hydrogen peroxide was used to work the turbo pump supplying the fuel and oxidiser to the rocket

IL-28R '03 Red' with a liquid-fuel rocket motor developed by L. S. Dooshkin instead of the gun barbette.

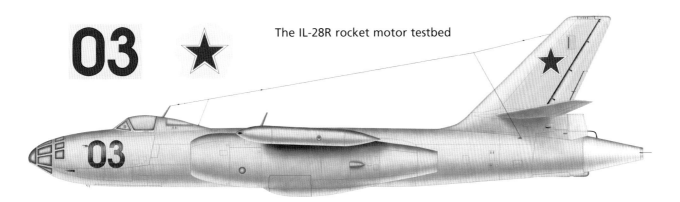

The IL-28R rocket motor testbed

The Lotarev D-18T turbofan is fitted to IL-76LL CCCP-86891 in one of LII's hangars.

booster. The experimental powerplant was installed at the rear in a conical fairing supplanting the IL-K6 cannon barbette. The tests took place in 1953-57; the testbed was flown by LII test pilot Pyotr I. Kaz'min.

Ilyushin IL-76LL engine testbeds

The IL-76LL engine testbed is perhaps the best-known of the R&D versions of the *Candid* – probably because it was used to test mainly 'civilian' engines and thus could be demonstrated publicly without causing a security breach. In the case of the IL-76, the LL suffix applies only to the engine testbeds.

Development of the IL-76LL began in the late 1970s when LII was faced with the need to test new powerful jet and turboprop engines. The converted bombers (Tu-4LL, Tu-16LL and Tu-142LL) that had filled this role previously had a major deficiency, namely limited space for test equipment. By then real-time on-board processing of test data had become a priority task. This meant the aircraft

had to carry a sizeable crew of test engineers which could decide if a particular test mode needed to be repeated, and there was simply no room for them in a bomber.

The experimental engine is fitted on a special pylon instead of the No. 2 (port inboard) Solov'yov D-30KP turbofan. The pylon has special fittings enabling different engines to be installed quickly; hence the aircraft has also been referred to as ULL-76 or ULL-76-02 (*ooniversahl'naya letayushchaya laboratoriya* – versatile testbed). The main shortcoming of this installation is that the experimental engine can create a thrust asymmetry which has to be countered by differential thrust of the other engines and/or control input.

The IL-76LL can be used to test engines rated at up to 25,000 kgp (55,110 lbst) and having a nacelle diameter up to 3.56 m (11 ft 8⁵⁄₃₂ in). Since the experimental engine can be both heavier and more powerful than the D-30KP, the wing centre section and No. 2 pylon attachment points have been reinforced.

The freight hold houses five test engineer workstations and two equipment modules for recording and monitoring engine parameters. The modules can be changed to suit the mission; part of the equipment (for example, video recorders) is of Western origin. Test equipment heat exchangers, a characteristic feature of the IL-76LL, are installed on the fuselage sides immediately aft of the wings; their quantity differs on different examples. The electric system has been modified to supply 208 V/115 V/36 V (400 Hz) AC, 220 V/127 V (50 Hz) AC and 27 V/6 V DC for the test equipment.

Test flights can be performed at altitudes up to 12,000 m (39,370 ft) and speeds up to 600 km/h (372 mph) or Mach 0.77; the minimum flight speed is 280 km/h (174 mph). Maximum take-off and landing weights are

This view of CCCP-86891 accentuates how much larger the D-18T is in comparison with the IL-76's D-30KP engines. Note the shape of the No. 2 pylon.

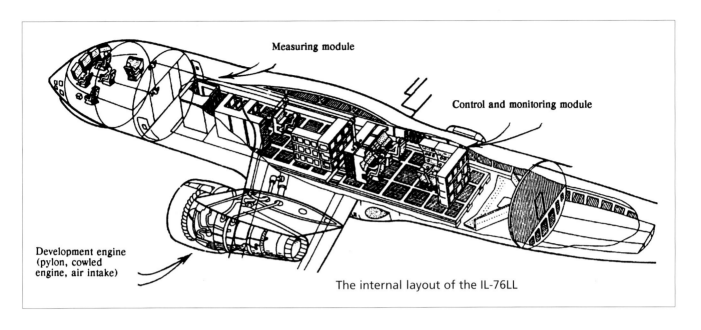

The internal layout of the IL-76LL

MEASUREMENT SCHEME

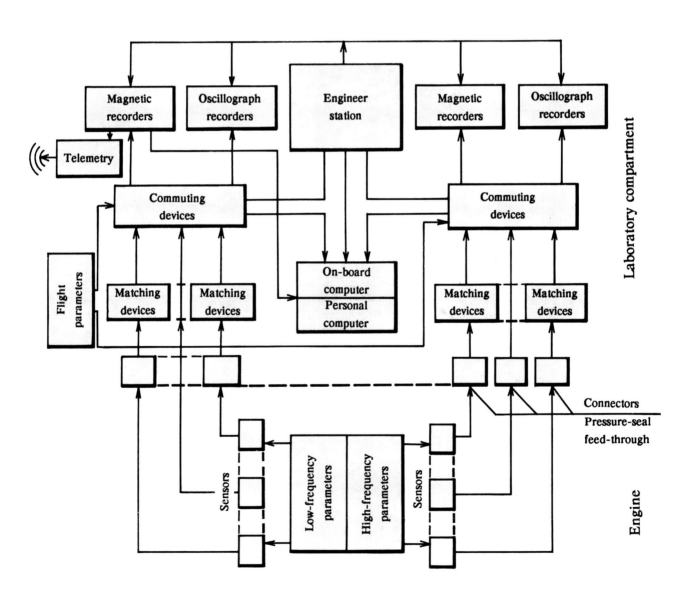

CCCP-86891 parked in front of a jet blast deflector; the cowlings of the D-18T development engine are open.

Another view of the same aircraft in a winter setting. The 1950s-vintage GAZ-51A lorry is an MZ-51 oil replenishment vehicle.

170,000 and 140,000 kg (374,785 and 308,650 lb) respectively; the G limits in flight are +1.0/–0.3, the maximum bank angle is 30°, and the maximum angle of attack and side-slip are 15°. The IL-76LL can operate from runways at least 3,000 m (9,840 ft) long.

Seven *Candids* were converted to IL-76LLs; one more aircraft, IL-76T 'Falsie' CCCP-76528 (c/n 073410293, f/n 0804), was in 'hot reserve' but never converted. The first testbed was the first prototype IL-76 *sans suffixe*, CCCP-86712 (c/n 0101), which ended its days as a testbed for the 13,000-kgp (28,660-lbst) Kuznetsov NK-86 turbofan developed for the Ilyushin IL-86 *Camber* wide-body airliner. Oddly, the development engine lacked the cascade-type thrust reverser. Tests began in 1975. By then CCCP-86712 had reached the limit of its useful life, and the aircraft was scrapped once the trials programme had been completed in the late 1970s.

The second aircraft, IL-76M 'Falsie' CCCP-86891 (c/n 093421628, f/n 1607A),

was used to test the 23,400-kgp (51,590-lbst) Lotarev D-18T high-bypass turbofan created by ZMKB *Progress* (*Zaporozhskoye motorno-konstrooktorskoye byuro* – Zaporozh'ye Engine Design Bureau, ex-OKB-478) in the Ukraine. This engine developed for the Antonov An-124 Ruslan (*Condor*) heavy transport had nearly twice the diameter of the D-30KP. One test equipment heat exchanger was fitted to port and two to starboard. A small cigar-shaped fairing of unknown purpose was mounted above the flight deck.

Tests began in 1982 and the engine logged 1,285 hours in 418 test flights. Black and white stripes were applied to the intake de-icer for icing visualisation and a video camera was fitted in place of the wing/air intake inspection light at a late stage of the trials. Reregistered RA-86891, the aircraft was in the static park at the MAKS-95 airshow. By August 1999 the aircraft was retired and the 'commercial' tail-cone had been scavenged from it – presumably for an IL-76MD-to-IL-76TD conversion.

As for the D-18T, ZMKB Progress has developed growth versions – the 25,000-kgp D-18TM and 27,500-kgp (60,630-lbst) D-18TP/TR intended for the stillborn An-218 wide-body airliner and the 27,500-kgp D-18T Srs 5 intended for the new-production An-124-100M-150. However, these are too large for the IL-76LL and will have to be tested on a suitably modified An-124; besides, ZMKB Progress will surely source a testbed in the Ukraine, not in Russia.

The third IL-76LL (sometimes called IL-76LL3) was an IL-76MD 'Falsie' initially registered CCCP-76492 (c/n 0043452549, f/n 3908) – the first aircraft to have this registration. (The registration was reused for IL-76T 'Falsie' RA-76492 (ex-Iraqi Air Force YI-AKT, c/n 093418548, f/n 1407) in 1994.) Initially CCCP-76492 was fitted with the 13,500-kgp (29,770-lbst) Solov'yov D-90 (PS-90A) turbofan and had two heat exchangers per side – a late addition. Testing in this configuration began on 26th December 1986 and the prototype engine logged about 400 hours in 188 flights. It transpired that cruise specific fuel consumption (SFC) was better than anticipated but the engine was delivering less power than it should. This was proved by the Tu-204's poor single-engine performance and led the Solov'yov OKB (Aviadvigatel' JSC) to uprate the PS-90A to 16,000 kgp (35,280 lbst), whereupon reliability problems began.

Other versions of the PS-90 family were developed in the early 1990s, including the PS-90P (a joint effort with Pratt & Whitney, MTU and MAN) and the derated 12,000-kgp (26,455-lbst) PS-90A-12 for the Yakovlev Yak-242 airliner project. These would probably have been tested on the same IL-76LL but the collapse of the Soviet Union and ensuing economic problems foiled these plans.

CCCP-76492 was on display at MosAeroShow '92 and visitors were even allowed inside the aircraft. By 2001 it had been withdrawn from use.

In 2006 the aircraft was reactivated for testing the Kuznetsov NK-93 contra-rotating integrated shrouded propfan (CRISP) rated at 18,000 kgp (39,680 lbst). The engine, which has eight scimitar-shaped blades on the front row and ten on the rear row, was envisaged for several projects, all of which remain 'paper aeroplanes' – the IL-90-200 long-haul wide-body airliner (a twin-engined spinoff of the IL-96-300), the IL-106 heavy transport, the Myasishchev MGS-6 *Gherakl* (Hercules; MGS = *mnogotselevoy groozovoy samolyot* – multi-role cargo aircraft) super-heavy transport and the Tu-214 (the first aircraft to have this designation – a re-engined Tu-204; the actual Tu-214 is a high gross weight version of the Tu-204 with the same PS-90A engines). Five prototype engines had been completed by 1995 but the trials were postponed indefinitely because the Kuznetsov OKB (now known as

IL-76LL RA-86891 in the static park of the MAKS-95 airshow.

Close-up of the D-18T on the IL-76LL. Note the icing visualisation markings on the air intake de-icer.

Front view of IL-76LL CCCP-76492 carrying a Solov'yov D-90 development engine.

SNTK 'Trood', the 'Labour' Samara Scientific & Technical Complex named after Nikolay D. Kuznetsov) could not afford to pay for them.

Eventually SNTK 'Trood' managed to secure the necessary funding (RUR 80 million) and a prototype engine (c/n 10) was installed on IL-76LL CCCP-76492 in December 2006. In the meantime, the manufacturer had not been sitting idle: measures had been taken to cut the engine's dry weight and improve the service life of the propfan's reduction gearbox. No problems were experienced with the engine's core, which was in series production as the NK-38 – a drive unit for natural gas pumping stations. The first flight with the NK-93 shut down and the propfan feathered took place on 26th December but a failure of an oil pump supplying lubricant in autorotation mode necessitated a premature removal of the engine. The NK-93 was reinstalled in April 2007 and the aircraft flew again with the registration applied as 76492 (with no prefix, as per current Russian practice for experimental aircraft) in May. In August of that year it was

displayed statically at the MAKS-2007 airshow. By then the aircraft had been repainted in an all-white colour scheme. The aircraft carried 'IL-76LL' nose titles, the LII logo and 'M. M. Gromov Flight Research Institute – Testbed' titles in Russian, as well as the logo of Gazbank (which had provided funding for the trials).

Soon, however, the money ran out and the aircraft stayed firmly on the ground for more than a year. SNTK Trood and the governor of the Samara Region went to great lengths to secure additional funds, even approaching the then President of Russia Vladimir V. Putin. In the autumn of 2008 the engine was run on the ground again, and on 7th October 2008 IL-76LL 76492 resumed flights with the NK-93 propfan. However, in April 2009 the development engine was removed and, sadly enough, the NK-93 programme appears to have been terminated – 'thanks' to the efforts of competing engine makers from Perm' (NPO Aviadvgigatel') and the foreign aircraft manufacturer's lobby.

A rare shot of CCCP-86891 flying near Zhukovskiy. The 'civil' tailcone is clearly visible here.

ИВЧЕНКО
ПРОГРЕСС

The logo of ZMKB Progress (Ivchenko-Progress) whose engines were tested on the IL-76LL.

In October 2010 IL-76LL 76492 was fitted with a test engine pod taken straight from one of the decommissioned Tu-16LLs (see below). The pod, which came complete with lateral test equipment heat exchangers (whose inlets had been blanked off, as the IL-76LL has its own heat exchangers), was empty then. There are rumours that 76492 is to serve as a testbed for the GTX-35VS Kaveri afterburning turbofan developed by India's Gas Turbine Research Establishment (GTRE) for the Hindustan Aeronautics Ltd. (HAL) Tejas light fighter. Plans to test this engine on an IL-76LL had been in hand since 2007. (The Tejas prototypes are powered by General Electric F414-GE-INS6 afterburning turbofans.) The first flight in this configuration took place on 3rd November.

The fourth and perhaps best-known example (IL-76LL4) was a demilitarised IL-76 *sans suffixe*, CCCP-76529 (ex-Iraqi Air Force YI-AIP, c/n 073410308, f/n 0807). Initially this aircraft was fitted with the first Soviet propfan engine – the 10,900-ehp (8,128-kW) Lotarev D-236T driving SV-36 contraprops developed by the Stoopino Machinery Design Bureau. The engine was a derivative of the D-136 turboshaft powering the Mil' Mi-26 *Halo* heavy-lift helicopter. This, in turn, was based on the core of the 6,500-kgp (14,330-lbst) D-36 turbofan powering the Antonov An-72/An-74

IL-76LL CCCP-76529 in its early configuration with a Lotarev D-236T propfan parked on the taxiway near runway 12 at Zhukovskiy.

Close-up of the
D-236T engine
and SV-36 con-
traprops on
CCCP-76529.

The same aircraft
in later guise as
RA-76529, with a
Muravchenko
D-27 propfan
engine and SV-27
contraprops.
Note the test
equipment heat
exchangers and
the video camera
further aft; the
cable connected
to the propeller
hub serves a
vibration sensor.

Coaler STOL transport and Yakovlev Yak-42 *Clobber* short/medium-haul airliner.

Propfan airliner projects, which were plentiful in the late 1980s, invariably had engines in pusher configuration. The D-236T, however, had tractor propellers, being intended for the Antonov An-70 transport developed since 1981 as an An-12 replacement, and could be readily installed on the IL-76LL. The SV-36 had glassfibre blades with a hollow composite spar and integrated electric de-icing threads. The front and rear rows had eight and six blades respectively, running at 1,100 and 1,000 rpm respectively; the 100-rpm difference was meant to reduce noise and vibration.

Tests began in 1987. A model of the IL-76LL with the D-236T propfan and the non-existent registration CCCP-86786 (?!) was displayed at the 38th Paris Aerospace Salon in

June 1989. In August 1990 the real aircraft appeared at the ILA'90 airshow in Hannover, creating a veritable sensation. It was immediately seen that the abovementioned model had been inaccurate: the SV-36 had straight blades with slightly raked tips, not the scimi-

The fifth IL-76LL,
CCCP-06188, with
a Klimov
TV7-117A devel-
opment engine.

Above right:
Close-up of the
TV7-117A engine
and feathered
SV-34 propeller;
the many small
air intakes were
featured on this
trials installation
only. Note the
vibration sensor
cable.

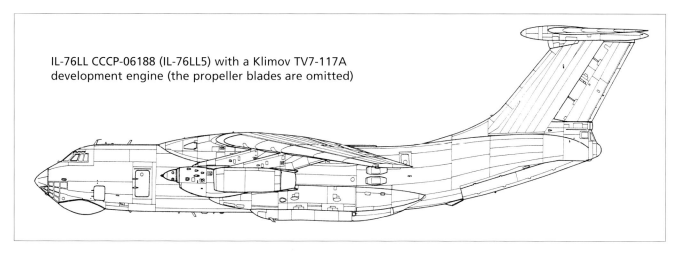

IL-76LL CCCP-06188 (IL-76LL5) with a Klimov TV7-117A
development engine (the propeller blades are omitted)

tar-shaped blades of the model. CCCP-76529 took part in the flying display and the engine demonstrated remarkably low noise levels, both on the ground and in flight, thanks to the low propeller speed.

As originally flown the propeller blades were grey with red and yellow calibration markings on the front row, and a vibration sensor cable ran from the propeller hub to the fuselage. By the time of the aircraft's Hannover appearance the cable had been removed and two test equipment heat exchangers installed on each side of the fuselage. The propeller blades were painted bright blue with yellow tips; black stripes were applied to the engine pylon and wing leading edge for icing visualisation, and large orange Cyrillic 'LII' titles added to the forward fuselage.

Because of the tractor configuration, straight blades and the fairly large propeller diameter (4.2 m; 13 ft 9⅜ in) Western observers tended to regard the D-236T as an 'advanced turboprop' rather than a pure propfan. In contrast, experimental US propfan engines (the General Electric GE36 and Allison Model

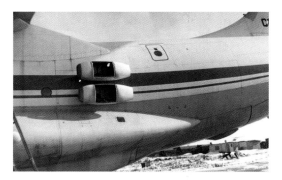

This is how the IL-76LL's heat exchanger fairings look with the heat exchangers removed.

A beautiful shot of IL-76LL CCCP-06188 with the TV7-117A development engine running. The aircraft has a tail gunner's station, despite the 'IL-76T' nose titles.

IL-76LL 76454 was used for testing the PowerJet SaM146 turbofan.

Two close-ups of the SaM146 development engine.

A fine landing study of 76454 returning after a test flight.

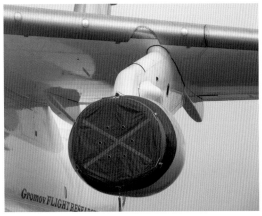

378-DX) had pusher propellers of 3-3.5 m (9 ft 10 in to 11 ft 5⁵⁄₆₄ in) diameter with sharply curved blades.

This was probably because Western prop-fan airliner projects (Boeing 7J7 and the like) had a T-tail, rear-engine layout with the engines mounted as close to the fuselage as possible to reduce thrust asymmetry in the event of an engine failure. This inevitably resulted in small-diameter props which had to turn at high speed to generate adequate thrust. At 1,300 rpm the blade tips reached almost supersonic speeds, producing a deaf-ening roar, which of course was totally unacceptable; hence scimitar-shaped blades were used to cure the problem.

(Incidentally, the model had misled Western observers in more ways than one. Several reference books, including *JP Airline-Fleets International*, listed the non-existent IL-76LL CCCP-86786 with an Izotov TV7-117 turboprop!)

The D-236T logged 70 hours in 36 flights on the IL-76LL. Further trials were made on the Yak-42LL CCCP-42525, which is described later in this chapter. Meanwhile, however, the

IL-76LL 76492, with appropriate nose titles, at the MAKS-2007 air-show with a Kuznetsov NK-93 shrouded prop-fan as a development engine. The NK-93's large diameter is apparent in these views.

An-70's MTOW had increased from the original 93,100 kg (205,250 lb) to 123,000 kg (271,160 lb). The Antonov OKB apparently considered the D-236T too small and by 1990 the project was altered to feature 14,000-ehp (10,290-kW) Muravchenko (ZMKB) D-27 propfans driving Stoopino SV-27 contraprops of 4.49 m (14 ft 8⁴⁹⁄₆₄ in) diameter.

In late 1990 a prototype D-27 engine was fitted to IL-76LL CCCP-76529. Outwardly the engine installation was very similar to the D-236T, except that the blades were scimitar-shaped. The engine nacelle was somewhat more streamlined and the oil cooler was recontoured (the result was vaguely reminiscent of the 'Andy Gump' nacelles of the

Close-up of the NK-93 on 76492. The engine bears the crest of the Samara Region.

IL-76TD RA-76792 was a testbed for the D-30KP-3 Burlak carried on the No. 2 pylon, which makes an interesting comparison with the standard D-30KP No. 1.

Overall view of RA-76792 at the MAKS-2007.

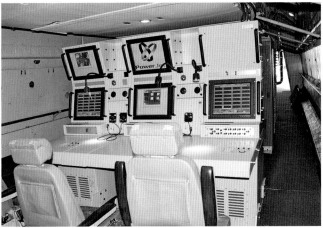

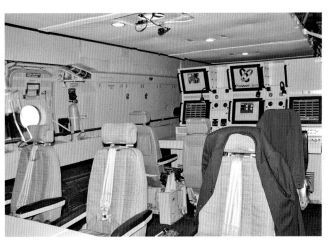

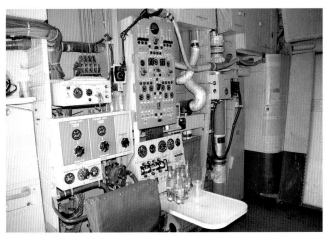

Top and centre: The test engineers' workstations in the cargo cabin of IL-76LL 76454, with PowerJet logos on some of the displays.

Above and left: The cabin of IL-76LL 76492, looking towards the nose. Part of the test and recording equipment is carried on the detachable upper deck (normally used for carrying troops on the IL-76MD). Note the power supply cables leading to the test equipment on the 'first floor'.

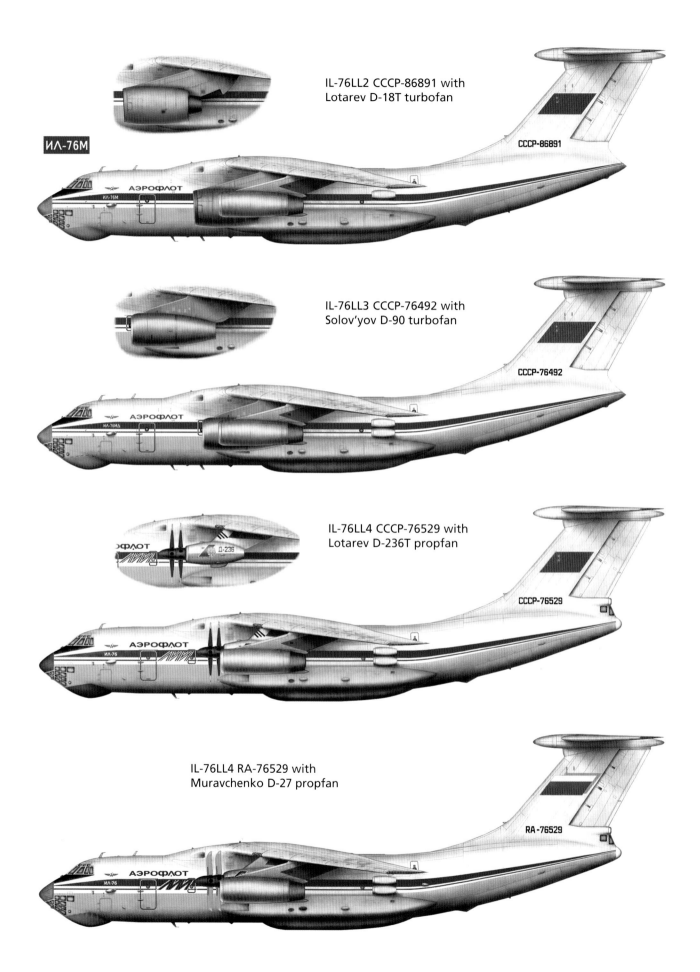

IL-76LL2 CCCP-86891 with
Lotarev D-18T turbofan

IL-76LL3 CCCP-76492 with
Solov'yov D-90 turbofan

IL-76LL4 CCCP-76529 with
Lotarev D-236T propfan

IL-76LL4 RA-76529 with
Muravchenko D-27 propfan

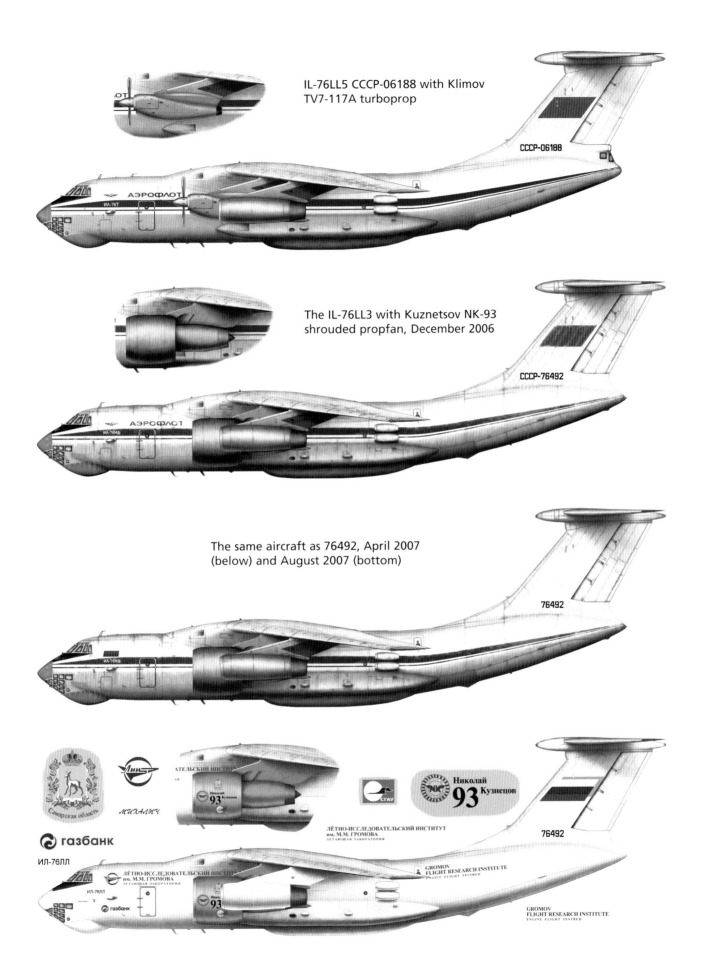

IL-76LL5 CCCP-06188 with Klimov TV7-117A turboprop

The IL-76LL3 with Kuznetsov NK-93 shrouded propfan, December 2006

The same aircraft as 76492, April 2007 (below) and August 2007 (bottom)

IL-76LL 76454 with PowerJet SaM146 turbofan

The logo of NPO Saturn featuring the stylised Cyrillic letters AL (АЛ) to denote Arkhip M. Lyul'ka – an engine designer closely associated with the enterprise.

Boeing B-50). Again a vibration sensor cable was fitted to the propeller hub, and a video camera was added aft of the port side heat exchangers to monitor engine operation.

In this guise CCCP-76529 was displayed statically at the MAKS-93 airshow. Next year it participated in the FI'94 and ILA'94 airshows, appropriately reregistered RA-76529; the sensor cable was removed and a normal spinner installed for the occasion. Meanwhile, the first prototype An-70 was rolled out in Kiev on 20th January 1994 and entered flight test on 16th December 1994, becoming the world's first aircraft to fly solely on propfan power. RA-76529 was on show again at MAKS-97.

The fifth IL-76LL (IL-76LL5) was a demilitarised IL-76T 'Falsie' with the non-standard registration CCCP-06188 (ex-Iraqi Air Force IL-76M YI-AKQ, c/n 093421635, f/n 1609). This aircraft was used to test the Izotov (Klimov) TV7-117A turboprop driving a Stoopino Machinery Design Bureau SV-34 six-bladed propeller. The engine was developed for the Ilyushin IL-114 feederliner and rated at 2,350 ehp (1,760 kW); the certified TV7-117S

production version delivered 2,500 ehp (1,840 kW). It was also selected for several transport aircraft projects, including the IL-112, Mikoyan MiG-101M, MiG-110, MiG SVB and Sukhoi S-80. (The actual S-80 (Su-80GP) prototypes were powered by General Electric CT7-9B turboprops with Hamilton Standard propellers.) The original plan was to test the TV7-117A on a modified IL-18, but this was rejected in favour of the *Candid*.

The aircraft had two heat exchangers to port and one to starboard. Unlike all other IL-76LLs, CCCP-06188 retained the standard No. 2 engine pylon (the development engine is attached via an adapter). The nacelle of the turboprop has an unusual banana-like shape and numerous small cooling air intakes, most of which were omitted on the IL-114. Once again the propeller was rigged with a vibration sensor cable and had red and yellow calibration markings on the blades.

CCCP-06188 was the first IL-76LL to be demonstrated publicly, taking part in the Aviation Day flypast in Zhukovskiy on 16th August 1990. It also made a single demo flight

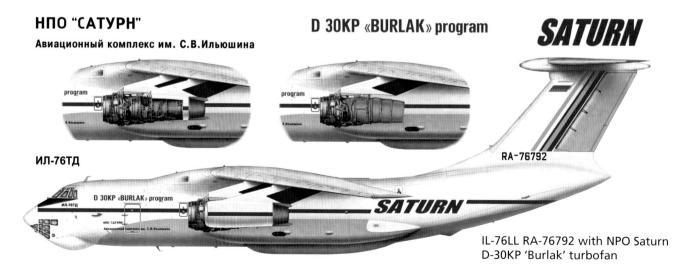

IL-76LL RA-76792 with NPO Saturn D-30KP 'Burlak' turbofan

at MosAeroShow '92 on 12th August, one of the press days. The basic idea behind many-bladed props is that lower propeller speed is needed to produce the required thrust, thereby reducing noise. Sure enough, the IL-114 is an extremely quiet aircraft – in flight. Contrary to all logic, however, the TV7-117A produced an almighty roar at ground idling rpm, earning the IL-114 the disparaging nickname of *lesopilka* (sawmill). And when the IL-76LL lined up for take-off at MosAeroShow '92, the unmistakable turboprop sound could be heard even through the whine of the three D-30KP turbofans.

Obviously the engineers were unhappy about it, and apparently changes were made to the propellers or to the engine control software (the IL-114 has FADEC). Anyway, production IL-114s are somewhat quieter on the ground than the prototypes (CCCP-54000, CCCP-54001 and RA-54002).

Starting in 1989, the TV7-117A logged 210 hours in 70 flights on the IL-76LL. By 1995 CCCP-06188 was withdrawn from use and

Here, IL-76LL 76492 is seen in October 2010 being fitted with a development engine nacelle taken from a retired Tu-16LL. The nacelle is reportedly to house the Indian GTRE Kaveri turbofan.

Close-ups of the new test engine pod fitted to 76492. Note the cowling doors and the lateral test equipment heat exchanger housings; their inlets are blanked off because the IL-76LL's own heat exchangers remain.

NPO Saturn and the French aero engine maker SNECMA (part of the Safran Group), for the Sukhoi SuperJet regional airliner (formerly RRJ – Russian Regional Jet). SNECMA was responsible for the core and the FADEC, while NPO Saturn designed the fan and bypass duct and is responsible for final assembly and airframe integration; the nacelle for the SaM146 was designed by Aircelle. Depending on the normal or high gross weight version of the aircraft, the engine delivers 6,125 or 6,895 kgp (13,500 or 15,400 lbst) for take-off, with a 6,895- or 7,940-kgp (15,400- or 17,500-lbst) contingency rating; it has a bypass ratio of 4.43, a dry weight of 1,708 kg (3,770 lb) and an SFC of 0.629.

Painted white overall with IL-76LL nose titles and LII, Safran Group, SNECMA and PowerJet logos, the aircraft was displayed statically at the MAKS-2007 airshow. The first flight with the SaM146 took place on 6th December 2007; in the course of manufacturer's flight tests the aircraft made 28 flights. The trials were completed on 27th May 2010, and on 23rd June the engine received its EASA type certificate.

One more IL-76LL testbed operated by NPO Saturn was intended for testing the D-30KP Burlak (alias D-30KP-3) – an upgraded version of the D-30KP-2 powering the standard IL-76TD/IL-76MD. The tightening of European noise and environmental regulations for aircraft in 2002 had barred the IL-76 from flying to Western Europe (except for humanitarian and relief missions, such as firefighting, for which special permits would be granted). This undermined the trade of many CIS cargo operators, and the D-30KP Burlak programme was meant to overcome this hurdle.

The D-30KP Burlak (D-30KP-3) featured a new single-stage fan of greater diameter with broad-chord blades replacing the standard three-stage low-pressure compressor. This increased the take-off thrust from 12,000 to 13,000 kgp (from 26,455 to 28,660 lbst) and reduced the fuel burn by 9%. IL-76s powered by remanufactured D-30KP Burlak engines would comply with ICAO Chapter 16/Stage IV noise regulations and CAEP 6 environmental regulations. According to NPO Saturn, the re-engining would extend the IL-76TD's range by 450 km (280 miles) with any payload; field performance remained virtually unchanged. Besides, the D-30KP Burlak was marketed as a cheaper alternative to the PS-90A-76 produced by the competing Aviadvigatel' JSC and hence to the IL-76TD-90/IL-76MD-90.

was still sitting at LII minus the propeller and Nos. 1 and 3 engines in August 2003.

Unusually, the sixth IL-76LL was converted from one of the five Ilyushin/Beriyev 'aircraft 976' *Mainstay-C* (a radar picket aircraft for cruise missile tests based on the IL-76MD), not from a regular *Candid*. The aircraft in question was the fourth example, CCCP-76454 (c/n 0073469074, f/n 5209). The characteristic revolving saucer radome (rotodome) was removed together with the twin supporting pylons – only the root portions of the pylons remained, since they were integrated into the rear fuselage structure. The four probe aerials on the nose and the six L-shaped aerials on the fin were also removed, but the distinctive cylindrical antenna pods at the wingtips remained, as did the dorsal satellite navigation system fairing ahead of the wings and the bulbous radome supplanting the UKU-9K-502-I cannon barbette.

Now wearing the registration as 76454, the aircraft served for testing the SaM146 two-spool turbofan developed by PowerJet International, a 50/50 joint venture between

This LaGG-3 was used to investigate the possibility of increasing the fighter's speed by means of Bondaryuk VRD-1 ramjet boosters.

In mid-2007 NPO Saturn outfitted a former Samara Airlines IL-76TD, RA-76792 (c/n 0093497942, f/n 7406) as a testbed for the D-30KP Burlak. Painted in a red/white/grey colour scheme with Saturn and Burlak titles, the aircraft was demonstrated statically at the MAKS-2007 with the standard No. 1 engine and the Burlak in the No. 2 position uncowled for comparison. Tests were scheduled to begin in the second quarter of 2010 and end in the fourth quarter of 2011, with production commencing in 2012.

It may be noted that the fifth 'aircraft 976', CCCP-76456 (c/n 0073474208, f/n 5602), was converted into an engine testbed in similar manner to 76454. This aircraft, however, was sold to the China Flight Test Establishment after initial flight tests with the temporary registration 76456 and delivered to the CFTE at Yanliang, gaining the military serial '760 Black'.

Lavochkin/Gorboonov/Goodkov LaGG-3 development aircraft with ramjet boosters

In July-August 1942 a LaGG-3 fighter was fitted experimentally with a pair of wing-mounted VRD-1 ramjet boosters (*vozdooshno-reaktivnyy dvigatel'*) designed by Mikhail M. Bondaryuk jointly with G. A. Varshavskiy. Some sources maintain the work began as early as 1941. Tests showed that, with the boosters running, the fighter was 30 km/h (18.6 mph) faster than a standard LaGG-3; however, with the boosters inoperative it was 50 km/h (31 mph) slower due to the high drag of the booster pods.

Lavochkin La-7 development aircraft with ramjet boosters

Lavochkin's last wartime fighter, the La-7, also served as a testbed for experiments with jet propulsion. An example fitted with two Bondaryuk PVRD-430 ramjet boosters under the wings was expected to attain a speed of 800 km/h (497 mph) at 6,000 m (19,685 ft). However, the high drag of the underslung ramjet units prevented it from exceeding 670 km/h (416 mph).

Lavochkin La-7R development aircraft

The La-7R (*reaktivnyy* – literally, jet-powered) was more promising. This experimental aircraft which emerged in late 1944 was fitted with an additional RD-1KhZ liquid-propellant rocket motor designed by Valentin P. Glushko, giving 300 kgp (660 lbst) of additional thrust. (KhZ stands for *khimicheskoye zazhigahniye* – chemical ignition.) The rocket booster was

A La-7 fitted experimentally with PVRD-430 ramjet boosters.

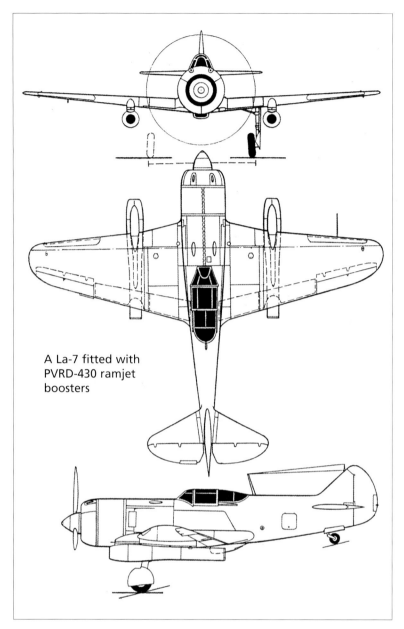

A La-7 fitted with PVRD-430 ramjet boosters

made 15 test flights, but not without accident. On 12th May 1945 the rocket motor exploded while being run on the ground; luckily there were no fatalities. Later, another explosion occurred in flight, but fortunately Shiyanov managed to land the crippled aircraft. Despite all the difficulties, the La-7R took part in the Aviation Day flypast at Moscow-Tushino on 18th August 1946 with the rocket motor running.

Lavochkin '120R' fighter prototype

The third Lavochkin machine to be fitted with a liquid-fuel rocket booster (after the two La-7R prototypes) was the 'aircraft 120' fighter prototype which received the manufacturer's designation '120R'; once again the R referred to the rocket booster. The installation of a rocket motor led to a complete revision of the rear fuselage and vertical tail. The DC battery previously accommodated in the rear fuselage was transferred to the gun mount; the central fuel tank was replaced by a nitric acid tank, the oil tank was moved from the gun mount (where a kerosene tank and an air bottle were accommodated instead) to a position atop the engine's reduction gearbox and its capacity was reduced by 5 kg (11 lb). The armament comprised one 23-mm (.90 calibre) Nudel'man/Suranov NS-23 cannon and one 20-mm (.78 calibre) Berezin B-20 cannon instead of two NS-23s. The main engine – an ASh-83 radial – was moved forward by 70 mm (2 ¾ in). The fuel feed system of the RD-1KhZ rocket motor installed on the '120R' was identical to that installed on the second example of the La-7R. The '120R' aircraft featuring improved aerodynamics as compared to its predecessors held a promise of making the most effective use of the booster engine.

Flight tests of the '120R' commenced on 2nd July 1945. The first two flights performed on 12th July and 25th September quickly revealed an overheating of the engine oil caused by inefficient design of the oil system, and the aircraft was sent to the workshop for modifications. Meanwhile, the Lavochkin OKB moved to new premises at Plant No. 301, this process entailing further delays. As a result, the '120R' did not resume its flight testing until 12th April 1946. In all, 23 live rocket motor runs were made, five of them in flight. The combustion chamber logged only 11 minutes and 55 seconds of running, whereupon cracks were discovered along the burner ring and in the nozzle throat. It was replaced again

housed in the rear fuselage, the rudder being cropped at the base to mate room for the nozzle and the tailplane being slightly raised. The fuel supply for the rocket motor was 90 litres (19.8 gallons) of kerosene, with 180 litres (39.6 gallons) of nitric acid as an oxidiser. To save weight the fuel load for the main engine (a Shvetsov ASh-82FN 14-cylinder radial) was reduced to 215 kg (474 lb), but still the La-7R weighed some 3,500 kg (7,720 lb) versus 3,370 kg (7,430 lb) for the standard La-7.

Two La-7s were thus converted. Compared with the standard fighter, top speed increased by 80 km/h (49.7 mph) with the rocket motor operating for 3-3.5 minutes, but the La-7R's handling and manoeuvrability had deteriorated. During the first three months of 1945 pilots Gheorgiy M. Shiyanov and A. V. Davydov

The La-7R development aircraft. The nozzle of the rocket booster can be seen beneath the rudder.

Three-quarters front view of the La-7R. Note the dorsal engine inlet scoop, rather like that of the La-5FN.

The aftermath of a rocket booster explosion on the La-7R. The starboard elevator and half the rudder have been wiped out.

The Lavochkin '120R' with the rocket booster uncowled.

The '120R' mixed-power fighter outwardly differed from the original '120' in having a new, larger vertical tail. Note the slit at the base of the rudder above the housing of the rocket booster cowling.

The rear end of the '120R' with the booster's cowling in place.

Here the cowling is removed, exposing the RD-1KhZ rocket motor and its mount.

and on 27th August another live test run of the engine fitted with a new combustion chamber was made. Barely two days later, cracks appeared again during the second test run. On 31st August the work resumed, using the fourth combustion chamber; this time cracks appeared after the third run. Finally, on 1st October a combustion chamber made of stainless steel was installed; it demonstrated stable functioning during four ignition trials and three live runs totalling 50 minutes.

In all, the '120R' aircraft performed 16 flights in the course of its test programme, including seven with the rocket motor running. In four flights the starting and functioning of the RD-1KhZ at altitudes of 3,000; 2,000; 800 and 70 m (9,840; 6,560; 2,625 and 230 ft respectively) was checked out. In 1946, during the annual Aviation Day air display, the '120R' with test pilot A. V. Davydov at the controls made a pass over Tushino airfield with the rocket motor running.

The maximum speed of the '120R' aircraft with the RD-1KhZ running could be measured only in one flight; the measurements made at the altitude of 2,150 m (7,050 ft) showed that top speed without the use of the rocket motor was 622 km/h (386 mph), while with the booster switched on it rose to 725 km/h (450 mph). That was equivalent to a speed increase of 103 km/h (64 mph). In all, five combustion chambers were used up during ground and flight testing of the RD-1KhZ and 63 start-ups of the booster were effected, five of them with the aircraft airborne. The combustion chambers logged in all 28 minutes and 19 seconds of operation. The testing revealed that operating an aircraft fitted with the RD-1KhZ rocket motor was an arduous task which required the development of special ground facilities for filling the aircraft's tanks with nitric acid and kerosene under pressure.

Lavochkin '164' fighter prototype

When manufacturer's testing of the Lavochkin '126' fighter was completed, a decision was taken to equip it with two Bondaryuk PVRD-430 ramjet boosters. The aircraft thus modified received the manufacturer's designation '164'. Both outer wing panels were provided with four easily detachable attachment points for the boosters, metal fairings were installed between the lower wing surface and the booster engine casing, and a pitot tube was mounted on the left wing outer panel above the booster engine. In addition to this,

the stabiliser attachment unit and the elevator spar were reinforced. Equipment items associated with the work of the booster engines were installed in the cockpit. The space previously occupied by the port and outer starboard cannons and their ammunition boxes was used to accommodate control and feed units of the booster engine system.

In the period between 26th June and 4th September 1946 joint factory tests of the '164' aircraft were held at the flight test facility of Plant No. 301. Their purpose was to obtain basic data for determining the efficiency

Close-up of the RD-1KhZ, the engine mount and accessories. Note the spherical combustion chamber.

Above: Ground-running the RD-1KhZ booster of the '120R' with the cowling removed.

Left: Though of poor quality, this air-to-air shot is remarkable, being one of very few pictures of the '120R' prototype in flight.

Head-on view of the '164' fighter prototype, showing how close the ramjet boosters were located to the main gear units.

Side view of the '164'; the fighter was painted medium grey overall with red trim. As these views show, the PVRD-430 boosters were relatively compact units.

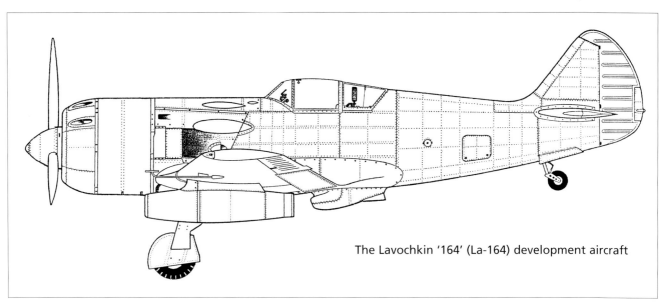

The Lavochkin '164' (La-164) development aircraft

Two more views of the '164' during trials. Note the soot stains on the cowling above the engine exhaust ports, apparently caused by firing the cannons.

and operational properties of the booster engines. Factory test pilots A. V. Davydov and A. A. Popov made 34 flights (29 and 5 respectively), logging a total of 12.5 hours at altitudes of 200; 500; 1,000; 2,000; 3,000 and 4,800 m (660; 1,640; 3,280; 6,560; 9,840 and 15,750 ft). The ramjet booster engines were turned on in 30 flights; their total time of operation in the air amounted to 46 minutes. The boosters were ignited 110 times; of these, in 20 cases there were ignition failures caused by malfunctions in the ignition and fuel feed systems. The maximum speed attained with the booster engines running was 694 km/h (431 mph) at the altitude of 2,340 m (7,680 ft) and 663 km/h (412 mph) at 1,235 m (4,050 ft). The speed increase was 109 and 104 km/h (67 and 65 mph) respectively, which tallied

well with estimated data and bore witness to the good functioning of the booster engines. However, the boosters created a lot of drag which had a negative effect on the fighter's performance. As a result, the speed increase as compared to the aircraft with no boosters amounted to a mere 64 km/h and 62 km/h (39.8mph and 38.5 mph) at the indicated altitudes.

Lavochkin La-9RD experimental fighter

A production La-9 fighter coded '09 White' (c/n 48210509) was fitted with two RD-13 pulse-jet engines designed by Vladimir N. Chelomey at Plant No. 51. Known as the La-9RD, the aircraft passed state acceptance

The La-9RD ('09 White'), seen here without pulse-jet boosters. This picture shows well the large pylons for the boosters protruding far beyond the wing leading edge.

trials between 21st November 1947 and 13th January 1948. The aircraft's project test pilot was Ivan M. Dziuba. Modifications to the aircraft included a revised fuel system, deletion of the armoured backrest and of two NS-23 cannons; additionally, the airframe was suitably restressed. The speed increase with the pulse-jets running amounted to 70 km/h (43.5 mph). The pilot noted strong vibrations and noise when the pulse-jets were in operation; also, the draggy booster engines under the wings impaired the aircraft's manoeuvrability and field performance. Start-up of the pulse-jets was unreliable, endurance dropped sharply, and maintenance became much more complicated. The work conducted on this aircraft proved useful only with regard to perfecting pulse-jets intended for installation on the first Soviet cruise missiles.

The La-9RD took part in air displays and invariably awed the public with its thunderous flypast. According to eyewitness reports, groups numbering from three to nine pulse-jet

equipped fighters took part in different air displays. Tests with the pulse-jets reached their culmination in the shape of a flypast of nine La-9RDs during the Tushino air display in in the summer of 1947. The aircraft were flown by GK NII VVS test pilots Vasiliy I. Alekseyenko, Aleksey G. Koobyshkin, Leonid M. Koovshinov, Andrey A. Manucharov, Viktor G. Masich, Gheorgiy A. Sedov, Pyotr M. Stefanovskiy, Andrey G. Terent'yev and Viktor P. Trofimov.

Lavochkin '130R' fighter prototype

In the spring of 1946 OKB-301 began construction of the '130R' experimental aircraft fitted with a liquid-fuel rocket booster. In those years this type of engine, despite its enormous fuel consumption and highly toxic oxidants, was mounted on many fighters. Fitting the RD-1KhZ liquid-fuel rocket motor to this particular machine necessitated considerable changes in its configuration. The central fuel tank was replaced by a nitric acid tank.

The La-9RD development aircraft (c/n 48210509) with RD-13 pulse-jet boosters

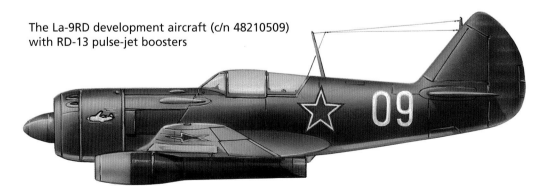

178

The La-9RD with the boosters attached. The RD-13 pulse-jets were quite bulky units, hence the large size of the pylons.

The actuators of the pumps feeding fuel components to the rocket motor were geared to the ASh-83FN engine. The kerosene tank was placed between the engine and the cockpit under the gun mount on which only two cannons were retained. To ensure the proper CG position the main engine was moved forward by 170.5 mm (6⁴⁵⁄₆₄ in), which necessitated the manufacture of new engines and cowlings. However, the longitudinal stability margin of the '130R' proved insufficient and the empennage area had to be increased; the fabric covering of the rudder and elevator was replaced by metal skinning. The additional weight of the tail unit caused by these modifications made it necessary to reinforce the tailwheel fork.

In 1946, when airframe assembly was already under way, all work on the '130R' was stopped. The reason for this may have been the successful testing of the first Soviet turbojet-powered fighters and the generally unsuccessful attempts to install the still immature and dangerous liquid-fuel rocket motors in aircraft of other types.

Lavochkin '138' (La-138) and '138D' (La-138D) fighter prototypes

Building on the results of the '164' (La-164) prototype's tests, GK NII VVS recommended that, after rectifying the defects revealed, the '130' aircraft possessing greater fuel capacity should be equipped with PVRD-430 ramjet boosters; after completion of factory tests the aircraft was to be submitted for state acceptance trials. This is how the prototype construction plan for 1947 came to include the '138' (La-138) prototype aircraft. This was a derivative of the production La-9 fighter; it featured two Bondaryuk PVRD-430 ramjet boosters, each of which was attached to the wing undersurface by three adjustable fittings. The installation of the boosters required the removal of the outer starboard cannon together with its ammunition box; the fighter was fitted with the necessary equipment for the maintenance and control of the ramjets.

Two aircraft with ASh-82FN engines and PVRD-430 boosters were built, undergoing manufacturer's flight tests. In all, the first prototype designated '138' performed 20 flights,

The '138' (La-138) development aircraft under construction at plant No. 301, showing the compact PVRD-430 ramjet boosters.

Close-up of the starboard PVRD-430 fitted to the '138' prototype.

the second one designated '138D' (*dooblyor* – second prototype) making another 38. However, the test missions were successfully fulfilled only in ten flights. In 19 other cases, failure to fulfil the mission was due to malfunction of the booster engines. Their reliable operation could only be ensured at altitudes up to 3,000 m (9,840 ft) where the use of the ramjet boosters gave a speed increase of 107-112 km/h (66.5 to 69.6 mph). Compared to the production La-9 at different altitudes the increase ranged from 45 km/h (28 mph) at 3,000 m (9,840 ft) to nearly 60 km/h (37 mph), although it had been expected that the increase would amount some 70-100 km/h (43-62 mph). However, with the ramjets

The same aircraft in completed form during trials.

turned off the speed proved to be 60-80 km/h (37.3 to 44.7 mph) lower than that of the 'aircraft 130' (La-9 prototype) without boosters. With all engines running the range did not exceed 112 km (70 miles) and endurance was limited to ten minutes.

The '138' fighter could wage combat against the Boeing B-29 and B-50 Superfortress bombers, also when chasing them, but in a duel with enemy fighters, piston-engined and jet-powered ones alike, it had few chances of success. As a positive fact, it might be noted that a considerable improvement of high-altitude performance of the mixed powerplant was revealed. However, due to low reliability of the PVRD-430 engines the work on the '138' aircraft was suspended. It was presumed that the work would be restarted as soon as improved engines running reliably within the whole range of operational altitudes and speeds were made available. However, such engines never appeared.

Mikoyan/Gurevich MiG-3 with six-bladed propeller

Despite its impressive high-altitude performance compared to other Soviet fighters of the day, the standard MiG-3 could not reach the German reconnaissance aircraft flying high over Moscow. Therefore in September 1941 a single MiG-3 was fitted with an experimental six-bladed propeller to increase thrust at high altitudes. The propeller retained the same diameter as the standard VISh-61, featuring a new hub to which six stock blades were attached. Unfortunately no further details are known.

Mikoyan/Gurevich MiG-15 (*izdeliye* SL-1, SL-2, SL-3, SL-4) engine testbeds

In order to refine the Klimov VK-1 turbojet and bolster the trials programme of *izdeliye* SD (the future MiG-15*bis Fagot-B*) the Mikoyan OKB built four powerplant testbeds designated SL-1 through SL-4. The L stood for [*letayushchaya*] *laboratoriya* – flying laboratory (in this instance, engine testbed).

The SL-1 made its first post-conversion flight with the new engine on 29th August 1949, piloted by Aleksandr N. Chernoboorov. The SL-2 followed on 26th December with Aleksandr M. Yershov at the controls; the SL-3 first flew three days later at the hands of Sergey F. Mashkovskiy, while the SL-4 took to the air on 9th May 1950 – again piloted by Chernoboorov.

This picture taken on 18th September 1941 shows a MiG-3 equipped with an experimental six-bladed propeller. Note the different colouring of the blades and the stripes on the black-painted ones.

MiG-15*bis* (*izdeliye* SL-5) engine testbed

In June-July 1951 the Mikoyan OKB converted a production MiG-15*bis* under an agreement with OKB-45 (the Moscow/Tushino-based engine design bureau headed by Vladimir Ya. Klimov) to take the new Klimov VK-5 centrifugal-flow turbojet – a derivative of the VK-1 rated at 3,000 kg (6,613 lb). To this end the main engine bearers were changed, the fuselage tail fairing was modified and a new extension jetpipe installed. The manufacturer's designation was *izdeliye* SL-5.

The converted aircraft was delivered to Zhukovskiy on 20th July 1951 for flight tests which lasted from 15th August to 31st October 1951. The idea was not pursued further because axial-flow engines were clearly superior to centrifugal-flow engines.

Mikoyan/Gurevich MiG-17 engine test rig

Additionally, LII used a MiG-17 erected vertically to investigate the peculiarities of jet engine operation in the vertical position (that is, in a vertical climb) and the impact of jet exhaust on dirt and concrete landing strips as part of vertical take-off and landing (VTOL) technology research.

Mikoyan/Gurevich MiG-21DF (MiG-21D) development aircraft

Each successive modification led to an increase in the MiG-21's take-off weight and hence to a reduction in manoeuvrability. This led some of the Soviet Air Force's top commanders to believe that the fire control radar was not really necessary for close-in dogfighting and, by replacing it with a simpler but lighter radar rangefinder, a more nimble dogfight version could be obtained. Hence in 1969 a production MiG-21 (probably a MiG-21S/MiG-21SM *Fishbed-J*) was refitted with the experimental Tumanskiy R13F2-300 turbojet uprated to 6,600 kgp (14,550 lbst) in full afterburner and the Kvant (Quantum) radar rangefinder. The aircraft was armed with a 23-mm Gryazev/ Shipunov GSh-23L built-in cannon and could carry AAMs, rocket pods and bombs; the maximum ordnance load was increased to 2,000 kg (4,410 lb).

The modified aircraft was designated MiG-21DF, although the military used the simplified designation MiG-21D (for *dahl'nomer* – rangefinder). The machine underwent trials at GNIKI VVS in January-April 1971, being flown by several Air Force pilots, including the well-known test pilot Vladimir N. Kandaurov. The trials revealed that manoeuvrability was indeed improved; yet the MiG-21DF was not put into production, nor were any operational aircraft updated to this standard.

Mikoyan/Gurevich MiG-21PD STOL technology demonstrator (Ye-7PD, *izdeliye* 23-31, *izdeliye* 92)

In the mid-1960s two of the Soviet 'fighter makers' – Pavel O. Sukhoi's OKB-51 and Artyom I. Mikoyan's OKB-155 – began experimenting with STOL aircraft. At the time the Sukhoi OKB was working on the T-58 (Su-15 *Flagon*) interceptor and the T-6 tactical bomber (the future Su-24 *Fencer*), while the Mikoyan OKB was developing the MiG-23 *Flogger* tactical fighter. Both design bureaux opted for the same solution of the STOL capability task, choosing a mixed powerplant with one or two cruise engines and several lift engines buried in the fuselage. By then a suitable engine – the 2,350-kgp (5,180-lbst) RD36-35 turbojet – had been developed by Pyotr A. Kolesov's OKB-36 in Rybinsk.

OKB-155 envisaged two alternative layouts for the MiG-23 STOL fighter. One of the versions, known in house as *izdeliye* 23-01, was a fairly conventional aircraft featuring a tailed-delta layout with mid-set wings and semi-circular lateral air intakes with centrebodies. It was powered by one 8,500-kgp (18,740-lbst) Khachatoorov R27F-300 afterburning turbojet and a pair of RD36-35 lift-jets installed almost vertically near the aircraft's CG between the cruise engine's inlet ducts. This aircraft was code-named *Faithless* by NATO and erro-

The Ye-7PD (*izdeliye* 23-31) STOL technology demonstrator created as part of the MiG-23 programme was a heavily modified MiG-21PFM. The photos on these pages show the open air intake scoop for the lift engines, the ventral bulge incorporating lift engine exhaust control vanes, the revised vertical tail, the abbreviated fuselage spine and the 'towel rail' data link aerial under the nose. Note the photo calibration markings.

neously referred to as the MiG-23PD (*pod-yomnyye **dvig**ateli* – lift engines). The other version, *izdeliye* 23-11 (the prototype of the *Flogger-A*), lacked lift engines; instead, STOL capability was obtained by using shoulder-mounted variable-geometry wings. In order to speed up development of the *izdeliye* 23-01

the Mikoyan OKB decided to build a propulsion testbed/STOL technology demonstrator based on the MiG-21 for verifying the RD36-35 lift engines.

The demonstrator was built by extensively modifying a production MiG-21PFM *Fishbed-F* interceptor (*izdeliye* 94). Since the aircraft was

The Ye-7PD takes off at Moscow-Domodedovo during an air-show on 9th July 1967.

developed under the MiG-23 programme, it sported two product codes at once – *izdeliye* 23-31 and *izdeliye* 92 – and the designation Ye-7PD (*pod**yom**nyye **dvig**ateli*). In conversation and in some publications the demonstrator was referred to unofficially as the MiG-21PD.

The MiG-21PFM airframe was cut up and a 900-mm (2 ft 11⁷⁄₁₆ in) 'plug' of greater diameter was inserted into the fuselage. As in the case of the *izdeliye* 23-01, the two lift engines were mounted in tandem and tilted slightly forward (their axes were at 85° to the fuselage waterline). During take-off and landing they breathed through a hydraulically actuated aft-hinged intake door incorporating 16 spring-loaded suction relief doors located four-abreast. Hence the fuselage spine was replaced by a short fairing aft of the cockpit terminating immediately ahead of the lift engine bay. The lift engines exhausted through a rectangular aperture in a flat-bottomed bulge on the centre fuselage underside. This aperture had a series of swivelling deflector vanes enabling the pilot to vector the thrust of the lift-jets 10° aft for take-off and 5° forward for landing to reduce the landing speed. The cruise engine was a Tumanskiy R13F-300 rated at 6,400 kgp (14,110 lbst) in full afterburner.

Since the cruise engine's inlet ducts positioned around the lift engine bay left no room for the mainwheel wells, the Ye-7PD had a

The Ye-7PD flies at the Moscow-Domodedovo air-show. Note the faired-over mainwheel wells (the landing gear was non-retractable).

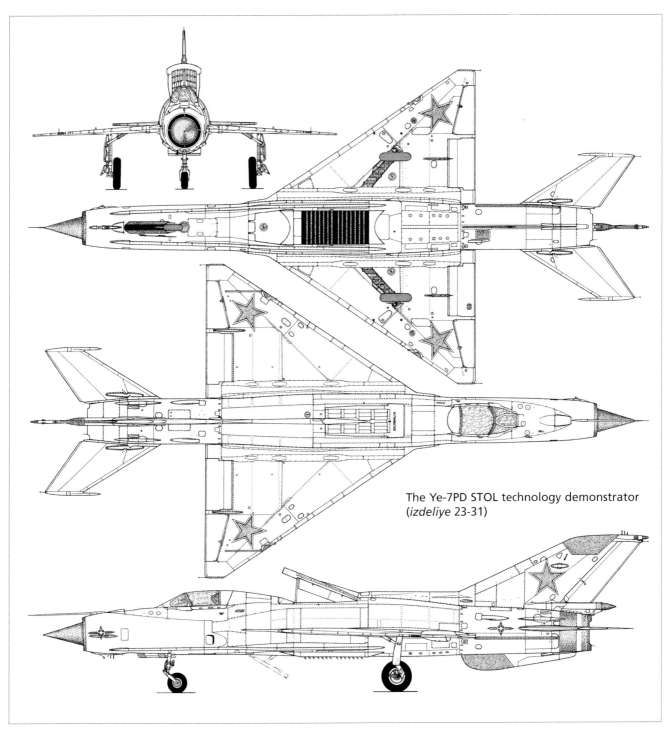

The Ye-7PD STOL technology demonstrator
(*izdeliye* 23-31)

fixed landing gear. However, the demonstrator was not intended for high-speed flight and there was no need to retract the gear anyway. Because of the wider-than-usual centre fuselage to which the wings were attached, the wheel track was also wider.

The vertical tail featured a long fin fillet whose front end was integral with the dorsal intake door. Two piping conduits ran along the lower forward/centre fuselage sides from the front of the former nosewheel well to the rear of the lift engine bay. A 'towel rail' aerial for

the data link system was located ahead of the nose gear unit.

Despite the fuselage stretch, the lift engines ate up a lot of internal space, causing a reduction in fuel capacity, not to mention the greatly increased fuel consumption. Therefore the Ye-7PD's endurance was no more than 15-17 minutes.

The Ye-7PD (*izdeliye* 23-31) made its first flight on 16th June 1966 with Pyotr M. Ostapenko at the controls. Another Mikoyan OKB test pilot, Boris A. Orlov, also took part in the demonstrator's flight tests. The test programme went ahead successfully, and it was even deemed advisable to show the aircraft publicly. On 9th July 1967 the Ye-7PD demonstrated its short-field capability at the impressive airshow at Moscow-Domodedovo airport.

Boris Orlov had the following recollections of the Ye-7PD. *'Taking off and landing in this aircraft was not easy. At ground level the efflux of the lift engines spread out in all directions, creating a suction effect that reduced the wing lift. The influence of the lift jets both on overall lift and on the aircraft's stability and handling varied, depending on the speed and altitude. While there were few problems on take-off (only a slight change in longitudinal trim which was easily counteracted after becoming airborne), on landing this suction effect caused the aircraft to lose altitude fast and decelerate sharply. Therefore it was necessary to accelerate the main engine to full military power or even engage the afterburner immediately before touchdown.'*

Having mastered the piloting of a fighter with lift engines on the Ye-7PD, Pyotr M. Ostapenko was able to commence the flight tests of the *izdeliye* 23-01 prototype in April 1967. However, the disadvantages of this type of powerplant outweighed its strong points and the *izdeliye* 23-01 was quickly abandoned in favour of the 'swing-wing' *izdeliye* 23-11, which became the MiG-23S in production form.

Upon completion of the tests the Ye-7PD was donated to the Moscow Aviation Institute, serving as an instructional airframe. Sadly, this unique aircraft did not survive.

Mikoyan MiG-25M (Ye-155M) experimental interceptor

The 13th April 1972 directive ordering the MiG-25RB, MiG-25RBK and MiG-25RBS into service also elaborated on the upgrade possibilities of the basic design. The military wanted an increase in range at low and medium altitude and an increase in service ceiling and maximum speed.

The Mach 2.83 speed limit imposed on the MiG-25 was purely theoretical, since the aircraft had the potential to go faster from the very start. High speeds reduced lateral stability and service life, but there were cases of pilots exceeding the speed limit without harming the aircraft. Therefore, the designers intended to reach a Mach 3.0-3.2 top speed so that the MiG-25 could outperform its arch-rival, the SR-71A – the world's fastest reconnaissance aircraft. This could be achieved by fitting the MiG-25 with more powerful and fuel-efficient engines.

As far back as the early 1960 a group of engine designers at OKB-300 led by Shookhov and Rotmistrov proposed a comprehensive upgrade of the Tumanskiy R15B-300 afterburning turbojet. The idea materialised as the uprated R15BF2-300 (*izdeliye* 65M). The improvement in performance was achieved by adding a compressor stage and increasing the combustion chamber and turbine temperatures. As compared to the basic R15B-300, the R15BF2-300 had a lower SFC, a higher thrust (10,000 kgp/22,045 lbst dry and 13,230-14,500 kgp/29,170-31,970 lbst reheat) and a higher engine pressure ratio (4.95 versus 4.75).

'710 Blue', one of the two MiG-25M development aircraft, in the museum at Moscow-Khodynka. The 'solid' nozzle shrouds of the R15BF2-300 engines are clearly visible from this angle.

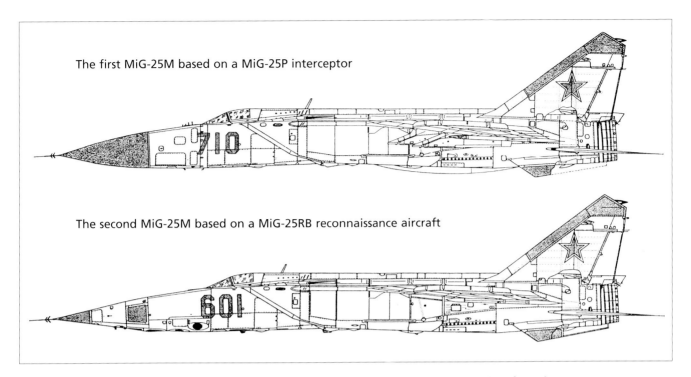

The first MiG-25M based on a MiG-25P interceptor

The second MiG-25M based on a MiG-25RB reconnaissance aircraft

The two engines were perfectly interchangeable, having identical dimensions and mountings. Providing the airframe was made more heat-resistant because of the new engine's higher turbine temperature, the R15BF2-300 offered a substantial increase in rate of climb, service ceiling, range and speed (up to 3,500 km/h, or 2,173 mph).

Teaming up with other aircraft industry enterprises and establishments, the Mikoyan OKB started a massive research effort with a view to increasing the MiG-25's top speed, concentrating mainly on aerodynamic stability and airframe/engine thermal limits. The aircraft's principal structure was made of steel and thus was heat-resistant enough. Some airframe components, however (such as the radome and forward fuselage, wingtips, flaps and ailerons), were made of duralumin and composites. They were not subjected to significant structural loads but experienced high temperatures and had to be replaced with steel or titanium honeycomb structures. This, in turn, called for new technologies. Therefore the Mikoyan OKB suggested to split the work into two stages – that is, test and refine the engine on a structurally standard MiG-25 first and come back to the speed issue later.

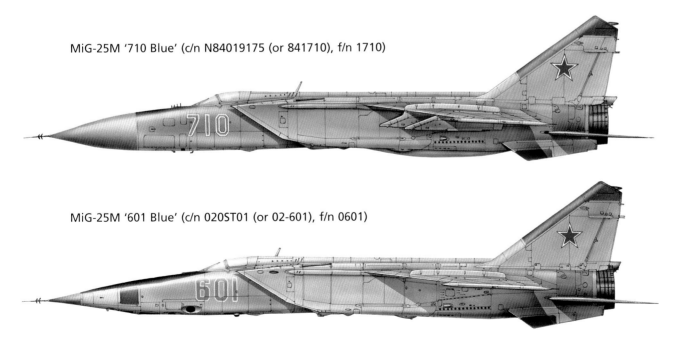

MiG-25M '710 Blue' (c/n N84019175 (or 841710), f/n 1710)

MiG-25M '601 Blue' (c/n 020ST01 (or 02-601), f/n 0601)

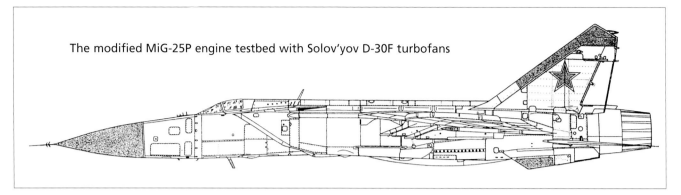

The modified MiG-25P engine testbed with Solov'yov D-30F turbofans

Both the Air Force and MAP went along with this approach and gave the go-ahead for Stage 1. In September 1964 the Ministry of Aircraft Industry issued an order detailing the test programme of the re-engined MiG-25. Yet the theoretical part, as well as the manufacturing and bench testing of the R15BF2-300 prototypes, took longer than predicted; development work was completed in 1971-72 but the flight tests did not begin until 1973.

The development aircraft was converted from a standard MiG-25P manufactured in the summer of 1973 (c/n N84019175, f/n 1710); the modification was performed by the manufacturer in accordance with OKB documents. Still powered by standard R15B-300 engines and equipped with a Smerch-A2 radar, the machine made its first flight on 12th June 1973 with Pyotr M. Ostapenko at the controls. Later it was flown by Aleksandr V. Fedotov, Aviard G. Fastovets, Boris A. Orlov and other Mikoyan OKB test pilots. On 30th August 1973 the aircraft was refitted with the intended R15BF2-300 engines, receiving a new non-standard c/n (841710) – albeit on paper only – and the appropriate tactical code '710 Blue'. From then on the aircraft bore the unofficial designation MiG-25M (*modifitseerovannyy* – modified); in official paperwork, however, it was referred to as 'MiG-25P No. 710'.

Flight tests of the re-engined interceptor began in 1974. In 1975 a tentative conclusion was issued, recommending the aircraft for production. That year '710 Blue' was subject-

ed to further modifications, receiving a set of wings taken from a modified MiG-25RB coded '601 Blue' (c/n 020ST01, f/n 0601; see below) and the horizontal tail of an initial-production MiG-25P coded '502 Blue' (c/n 840SE02, f/n 0502). More updates followed in 1976, this time to the aircraft's control system and electric system.

MiG-25M '710 Blue' underwent state acceptance trials in 1976. Meanwhile, in keeping with a ruling issued by the Council of Ministers' Commission on Defence Industry Matters (VPK – *Voyenno-promyshlennaya komissiya*) on 17th April 1974 the ministries of aircraft, electronics and defence industries were working on the MiG-25-40M aerial intercept weapons system comprising the Ye-155M interceptor powered by R15BF2-300 engines, the upgraded Smerch-A4 radar and the improved K-40M AAM (R-40M; NATO AA-6 *Acrid*). The longer detection/tracking range of the radar (which had 'look-down/shoot-down' capability) and the longer launch range of the missiles should allow the interceptor to destroy targets within a wider range of speeds and altitudes, including the destruction of fast low-flying targets. For close-in engagements the Ye-155M interceptor was to be armed with K-60 or K-60M short-range IR-homing AAMs (the future R-60/R-60M; NATO AA-8 *Aphid*) and a 23-mm (.90 calibre) Gryazev/Shipunov GSh-23L twin-barrel fast-firing cannon.

Yet eventually the Powers That Be decided against launching production of the Ye-155M.

The modified MiG-25P engine testbed with Solov'yov D-30F turbofans

The reason was that the more advanced Ye-155MP interceptor (*izdeliye* 83/1) – the prototype of the future MiG-31 – and the MiG-25P/*izdeliye* 99 engine testbed (see below) had been actively undergoing tests since 1975. Both of these aircraft were powered by the advanced and fuel-efficient Solov'yov D-30F6 (PS-30Г) afterburning turbofans with a similar rating but a lower SFC. Moreover, at the time the Soviet aero engine plants were shouldering a considerable workload, and putting the R15BF2-300 into production could have been a problem. Finally, the PVO top command was more interested in the MiG-31 than in upgrading the existing MiG-25. Hence the R15BF2-300 was not built in quantity, and '710 Blue' remained a one-off.

Upon completion of the state acceptance trials the aircraft was used as a testbed until withdrawn from use in April 1977. After sitting idle at the OKB's flight test facility at Zhukovskiy for a while the interceptor was transferred to one of the Air Force's Junior Aviation Specialists Schools as a ground instructional airframe. Later, '710 Blue' ended up in the open-air aviation museum at Moscow-Khodynka established in 1994 (incidentally, wearing its original c/n). Regrettably the museum is now closed and the aircraft, like all the other exhibits, is in serious danger.

Mikoyan MiG-25M (Ye-155M) experimental reconnaissance/strike aircraft

The Ye-155M development programme also involved three MiG-25RB reconnaissance/strike aircraft refitted with more powerful and fuel-efficient R15BF2-300 engines and the upgraded Peleng-2 navigation suite. The maximum bomb load of these aircraft was increased to 5,000 kg (11,020 lb).

The first of the three (f/n 0703) was coded '703 Blue', entering flight test in February 1976. The R15BF2-300 engines were also verified on MiG-25RB '401 Blue' (c/n 020SL01, f/n 0401), which had previously served as a testbed for some of the MiG-25BM's equipment items, and MiG-25RB '601 Blue' (c/n 020ST01, f/n 0601), which had been one of the four *Foxbats* temporarily deployed to Egypt in 1971. When the latter aircraft was re-engined, its c/n was changed to 02-601 – probably again in the record card only.

The conversion work was completed very quickly but refining the engine took longer than expected. Still, it was worth the sweat:

the engine did produce the claimed performance. The service ceiling exceeded 24,200 m (79,396 ft) and supersonic cruise range was 1,920 km (1,200 miles) in clean condition or 2,530 km (1,581 miles) with a 5,300-litre (1,177 Imp gal) drop tank.

The modified MiG-25RB was used to set a number of world time-to-height and altitude records. In the documents submitted to the FAI (*Fédération d'Aviation Internationale*) for registration purposes the aircraft was referred to under the fake designation Ye-266M. On a single day (17th May 1975) Aleksandr V. Fedotov and Pyotr M. Ostapenko set three time-to-height records, reaching 25,000 m (82,020 ft), 30,000 m (98,425 ft) and 35,000 m (114,830 ft) in 2 min 23.2 sec, 3 min 9.85 sec and 4 min 11.7 sec respectively. For these record flights all non-essential equipment was removed to reduce weight.

In the summer of 1977 Fedotov bettered his own altitude world records. On 22nd June he took the aircraft to 37,800 m (121,653 ft) with a 2,000-kg (4,410-lb) payload, and reached 37,650 m (123,523 ft) on 31st August with no payload. However, soon after the record flights, a pressure valve in the fuel system failed in a regular flight, causing one of the fuselage fuel tanks to become overpressurised and burst. A good-sized chunk of the upper fuselage skin came off in mid-air; test pilot Aviard G. Fastovets displayed no mean skill and bravery, managing to land safely. The aircraft was repaired but tests did not resume.

The test flights of the re-engined MiG-25Ms confirmed the possibility of improving the aircraft's performance considerably. In lightened form for the record-breaking flights the aircraft had a thrust/weight ratio better than 1.0 – for the first time in Mikoyan OKB history. As a result, the brakes couldn't hold the aircraft in full afterburner, and a special mobile detent had to be developed (a heavy vehicle with a jet blast shield to which the aircraft was connected by a strong steel cable and lock).

Mikoyan MiG-25 (*izdeliye* 99) development aircraft

The production MiG-25P interceptor met all design requirements except range. To increase range one MiG-25P was experimentally re-engined with Solov'yov D-30F (aka PS-30F) afterburning turbofans rated at 15,500 kgp (34,170 lbst) – both for the purpose of verifying the engine itself and as part of the new Ye-155MP's development programme. This engine was a derivative of the 'pure' D-30 commercial turbofan rated at 6,800 kgp (14,990 lbst) powering the Tu-134 short-haul airliner, not the later and much larger D-30KU/D-30KP turbofan which is a totally different engine. The testbed was designated *izdeliye* 99 and appropriately coded '991 Blue', receiving the new c/n 990001 (*izdeliye* 99, batch 00, 01st aircraft in the batch). The conversion was completed in April 1976 and the aircraft was delivered to the Zhukovskiy flight test facility in June. Later, a MiG-25R was similarly converted and coded '992 Blue' (c/n 990002).

As distinct from the MiG-25M described above, the new engines required major modifications to the airframe. Still, outwardly the aircraft was little different from standard MiG-25s and the internal fuel capacity remained unchanged at 19,700 litres (4,377 Imp gal). The new turbofan was expected to improve rate of climb and especially range (particularly at subsonic speed) thanks to its lower SFC.

A short while earlier, the abovementioned MiG-25M (Ye-155M) testbeds powered by Tumanskiy R15BF2-300 turbojets had been tested, but there was no knowing if and when this engine would enter production. The new and fairly complex MiG-31 fighter weapons system could also take a long time testing. Thus, a MiG-25 fitted with the new fuel-efficient engines could supplant the standard MiG-25PD on the Gor'kiy production line for a while until the MiG-31 would be ready.

The scope of the *izdeliye* 99 programme was much larger than with the MiG-25M. Yet, with assistance from the Gor'kiy aircraft factory and due largely to the insistence of lead engineer M. Proshin the technical problems were solved quickly enough. Shortly after test flights commenced a subsonic cruise range of 3,000 km (1,863 miles) on internal fuel was achieved. Supersonic flight, though, caused more problems.

Normal take-off weight during tests was 37,750 kg (83,220 lb), including 15,270 kg (33,660 lb) of internal fuel; MTOW with drop tank was 42,520 kg (93,740 lb). Range was increased to 2,135 km (1,326 miles) in supersonic cruise or 3,310 km (2,068 miles) at transonic speed, and service ceiling was boosted to 21,900 m (71,850 ft); rate of climb was also improved.

Tests of the *izdeliye* 99 continued until 1978. However, the MiG-31 was designed around the D-30F engine from the outset. Thus, when the Ye-155MP-1 (the first prototype MiG-31) entered flight test in the autumn of 1975, interest in the MiG-25/D-30F re-engining project waned. In fact, no one took the trouble to study the aircraft's performance completely. The two modified aircraft were relegated to the role of engine testbeds under the MiG-31 development programme. Moreover, '992 Blue', which was transferred to LII and was to be flown by LII test pilot Aleksandr A. Shcherbakov, never flew, being used only for ground engine runs (right on the runway) in 1978. After being struck off charge this aircraft became a cutaway instructional airframe at the Moscow Aviation Institute.

MiG-25PD engine testbed (*izdeliye* 84-20)

In 1991-92 the second prototype MiG-25PD interceptor, '306 Blue' (c/n N84042680, f/n ...306), was used by the Mikoyan OKB as a testbed for the experimental Lyul'ka-Saturn AL-41F afterburning turbofan (*izdeliye* 20). This engine was intended for Russia's fifth-generation multi-role fighters, including the projected Mikoyan '1.42' and its demonstrator version, the '1.44'. The development engine was fitted instead of the port R15BD-300 turbojet. The modified MiG-25 was designated *izdeliye* 84-20 (that is, *izdeliye* 84 modified for testing *izdeliye* 20) and used for testing the AL-41F in various flight modes, including supersonic flight.

The test programme was a rather complex one, since the new engine had a totally different control system – full authority digital engine control (FADEC) – and much higher thrust than the standard R15BD-300 (about 17,700 kgp/39,020 lbst versus 11,200 kgp/24,690 lbst in full afterburner), which had its adverse effect on the aircraft's handling. Therefore, much

Airframe components of the second *izdeliye* 99 engine testbed ('992 Blue') serve as a teaching aid at the Moscow Aviation Institute. Note the nose section on the right clearly indicating the machine's MiG-25R lineage.

At the end of its flying career the MiG-25PD prototype '306 Blue' was converted into the *izdeliye* 84-20 engine testbed. The longer nozzle of the AL-41F development engine on the port side is visible.

thought was given to safety measures in case the development engine or its air intake control system should fail, especially on take-off and in supersonic flight up to Mach 2.

Stage A of the flight test programme (including supersonic flight) was completed without major difficulties. Aircraft 84-20 made more than 30 test flights, some of which proceeded at high supersonic speeds and high altitudes; some of the missions involved inflight engine shutdown and restarting. The tests yielded invaluable data, allowing Lyul'ka-Saturn to make the necessary changes to the design of the AL-41F. *Izdeliye* 84-20 was post-Soviet Russia's first supersonic engine testbed.

Mikoyan MiG-29 testbeds for uprated RD-33 (*izdeliye* 21) engine

After completing its state acceptance trials the tenth MiG-29 prototype, '21 Blue' ('aircraft 921'), was converted into a testbed for a version of the Izotov RD-33K turbofan (*izdeliye* 21) uprated to 8,800 kgp (19,400 lbst) in full afterburner. This version was intended for the MiG-29M (*izdeliye* 9.15). The prototype engine supplanted the fighter's port RD-33 (the standard engine in the starboard nacelle was retained). The air intakes were converted to MiG-29M standard, with downward-hinging FOD protection grilles instead of solid blocker doors and no dorsal auxiliary

The *izdeliye* 84-20 testbed parked at the RSK MiG flight test facility in Zhukovskiy.

Upon completion of the tests of the RD-33K engine MiG-29 '21 Blue' ('aircraft 921') became an instructional airframe at the Dolgoye Ledovo tech school.

intakes in the LERXes. The testbed first flew in this configuration on 27th September 1985. Later it became an instructional airframe at the Dolgoye Ledovo tech school near Moscow.

Similarly, the first production MiG-29, '24 Blue' ('aircraft 924', c/n 0390501625, f/n 0101), was used for testing an improved version of the RD-33 engine upon completing its basic test programme.

'24 Blue' ('aircraft 924'), another MiG-29 used as a testbed for modified RD-33 engines. It is seen here in storage at Zhukovskiy.

MiG-29OVT development aircraft

The highlights of the MAKS-2005 airshow included a breathtaking flying display by a development aircraft presented by RSK MiG – the MiG-29OVT experimental super-agile multi-role fighter (OVT = *otklonyayemyy **vektor tyagi*** – thrust vector control) flown by the corporation's CTP Pavel N. Vlasov. Actually the fighter ('156 White'; c/n 2960905556?) had been unveiled at the MAKS-2001 airshow… but more on that later. The aircraft, which had been converted from the final prototype of the MiG-29M (*izdeliye* 9.15) – in effect, a pre-production example – was really a propulsion testbed meant to verify a version of the RD-33 engine fitted with an all-aspect vectoring nozzle. In other words, '156 White' was a TVC technology demonstrator – though one might also regard it as a control-configured vehicle.

The story began in the mid-1990s when NPO Klimov commenced development of its own TVC concept as an answer to the rival Lyul'ka-Saturn design bureau's AL-31FP thrust-vectoring afterburning turbofan created for the Su-27 family. NPO Klimov's vectoring nozzle was intended for light fighters and branded KLIVT (***Klim**ovskiy **vek**tor **tya**gi* or Klimov's Vectoring Thrust). The same enterprise had developed the RD-33 afterburning turbofan in the mid-1970; this engine had been progressively modified over the years to increase its service life, improve reliability, increase the thrust and cut the fuel consumption. However, while addressing these burning issues (no pun intended), the designers had been working on long-term development projects; one of these was a thrust-vectoring version of the RD-33.

After analysing the TVC research undertaken in the Soviet Union and abroad by then, the engineers of NPO Klimov concluded that tilting the entire axisymmetrical nozzle (as was the case with the AL-31FP) was inexpedient; it made more sense to deflect the nozzle's supersonic section (the petals). This option made the design not only simpler and lighter but also reduced the deflection time – and, most importantly, allowed the nozzle petals to move in any direction, not just in the vertical plane. All the petals of the supersonic section were deflected at the required angle simultaneously, the motion being imparted via push-pull rods from a single control ring; this, in turn, was powered by three hydraulic actuators located at equal intervals on a fixed ring running around the afterburner. The position of the actuator pistons rigidly determined the spatial orientation of the control ring and hence the position of the nozzle petals and the thrust vector. The actuators were enclosed by relatively compact fairings. Because of the additional forces applied to the engine casing and the nozzle during thrust vectoring, some elements of the afterburner had to be reinforced.

A number of other changes were to be introduced into the design of the RD-33 concurrently with the vectoring nozzle. The engine was to receive a new FADEC system and be uprated from 5,040 to 5,600 kgp (from 11,110 to 12,345 lbst) at full military power and from 8,300 to 9,000 kgp (from 18,300 to 19,840 lbst) in full afterburner. In this guise the engine was initially known as the RD-133, making its public debut at the Dvigateli-98 (Engines '98) exhibition in Moscow; the following year the engine was on display again at the MAKS-99 airshow in Zhukovskiy. In the late 1990s NPO Klimov had plans to develop modernised versions of the RD-33 delivering up to 10,000-12,000 kgp (22,045-26,455 lbst) in full afterburner. At various airshows and industry fairs these projects were announced as the RD-333, RD-33-10M and VKS-10M. Later, however, the company reverted to displaying the RD-133, which was nothing but a production RD-33 mated to a KLIVT nozzle. Recently the RD-133 designation was dropped altogether, the engine being henceforth referred to simply as the 'thrust-vectoring RD-33'.

The first prototype of the KLIVT nozzle had been completed by early 1997. After installation on an RD-33 the nozzle underwent 50-hour bench tests which involved close to 1,000 changes of thrust vector in various operating modes, including full afterburner. The thrust vectoring angles were ±15° in all directions; the maximum angle speed was eventually increased from an initial 30°/sec to 60°/sec.

Flight tests of the thrust-vectoring engine on a suitably modified MiG-29 were due to commence in late 1997, but even then, when the Russian bank crisis of 1998 had not yet struck, the prospective customer (the Russian Air Force) was unable to fund the project and the work was halted. In 2001 two non-functional RD-133s were installed on MiG-29M '301 Blue' (ex-'156 Blue'). Built in July 1991, this aircraft had made 86 flights when the programme was suspended in September 1993. After that the aircraft was mothballed, staying on the ground for nearly ten years (although in August 1995 it had an 'outing' in the static park of the MAKS-95 airshow as '301 Blue').

The logo of RSK MiG whose fighters served as engine testbeds.

The logo of NPP Klimov whose RD-33 engines were tested on MiG-29 fighters.

The MiG-29OVT superagility demonstrator received this smart colour scheme in 2003. The KLIVT vectoring nozzles are clearly visible. Note the missile-shaped smoke generator pods.

Eventually it was reactivated in 2001 for testing the RD-133 engine, making its debut at the MAKS-2001 airshow. The fighter was repainted for the occasion in a two-tone grey camouflage strikingly similar to the USAF's 'Egypt One' scheme, with 'МиГ-29ОВТ' (MiG-29OVT) titles on the fins, and became '156 White' (the original code remained on the fin caps). Concurrently, launch rails for R-73E AAMs were added to the wingtips in the manner of the Su-27 or the F-16.

Two years later '156 White' finally received a shipset of flight-cleared RD-133s and a smart new red/white colour with blue stylised 'MiG' titles and a huge red star on the upper fuselage; the normal wingtips were reinstated in so doing. In August 2003 test pilot Pavel N. Vlasov took the re-engined fighter into the air for the first time. When the MAKS-2003 airshow opened a few days later, the MiG-29OVT had not yet logged enough flight hours to be admitted to the flying display and thus stayed

firmly on the ground. At the show it was announced that future production MiG-29Ms and MiG-29M2s would be powered by similar thrust-vectoring engines, while the accompanying data placard said the MiG-29M/M2 would have the 9,000-kgp thrust-vectoring version of the RD-33MK.

By early August 2005 RSK MiG test pilots Pavel Vlasov and Mikhail Belyayev had made more than 50 flights in '156 White' in which the TVC system was put through its paces and its integration with the FBW control system was verified. This prompted the decision to demonstrate the aircraft at the MAKS-2005 airshow. Those who witnessed the aircraft's demo flights at the show said the MiG-29OVT was at least equal, or maybe even superior, in manoeuvrability to the super-agile Su-30MKI which had by then become a regular airshow performer; the aircraft could do loops and spirals literally around its own nose. However, there's more to it than just spectacular and unique aerobatics at an airshow. RSK MiG Chief Designer Nikolay N. Boontin, who is in charge of the MiG-29K/KUB, MiG-29M/M2 and MiG-29OVT programmes, says that all-aspect TVC endows the new MiG with entirely new capabilities both in normal flight modes and in superagility mode.

The airframer and the engine maker alike were entirely happy with the results obtained with the MiG-29OVT. The specialists involved in the test programme believe that production of the thrust-vectoring RD-33 could begin right away for the future new-build MiG-29M

(*izdeliye* 9.61) and MiG-29M2 (*izdeliye* 9.67). Five such engines had been manufactured by August 2005; two of them were in use for bench tests, two more had been fitted to the MiG-29OVT. The fifth engine (manufactured as a spare) found itself unwanted – the two examples fitted to '156 White' behaved faultlessly.

In the course of its research into thrust vectoring control NPO Klimov came to the conclusion that the design of the KLIVT nozzle can be adapted to other engine types, including Western ones. The MiG-29M/M2 and its derivatives with TVC may find a market both in Russia and abroad; among other things, the thrust-vectoring MiG-35 has been entered for the Indian Air Force's MRCA tender.

Close-up of the MiG-29OVT's distinctive KLIVT nozzles with three actuator fairings installed at 120° intervals. Note that the legends on the inner faces of the fins read 'OVT' (in Russian) to port and 'VTC' (vectoring thrust control) to starboard.

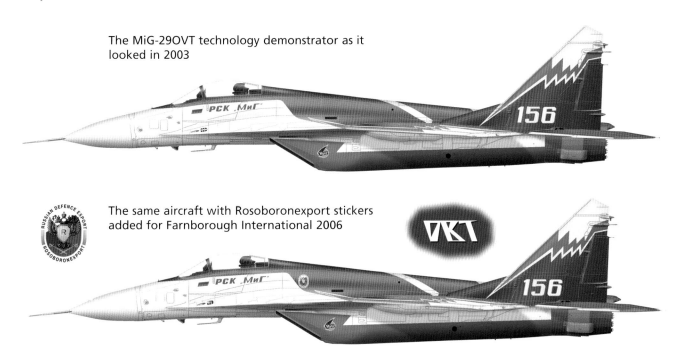

The MiG-29OVT technology demonstrator as it looked in 2003

The same aircraft with Rosoboronexport stickers added for Farnborough International 2006

SOVIET AND RUSSIAN TESTBED AIRCRAFT

The MiG-29OVT's demo flights at the MAKS-2005 overshadowed those of its Sukhoi competitors; the machine could literally manoeuvre around its own nose. Note the 'MiG' titles styled as lightning bolts (in Russian on the outer face of the port fin and in English on the inner face of the starboard fin).

In addition to the usual aerobatics and high-alpha/low-speed passes, at the MAKS-2005 the MiG-29OVT performed 360° turns with 90° bank and literally hung in the air on the thrust of its engines, barely moving forward. Note the wing vortices and the simple 'lightning bolts' on the starboard side of the tails.

Mil' Mi-6 engine testbed

In 1969 a Mi-6 heavy-lift helicopter became a testbed for the Solov'yov D-25VF turboshaft engine with a take-off rating of 6,500 shp. This was an uprated version of the Mi-6's 5,500-shp D-25V engine (hence the F for *forseerovannyy*) developed for the Mil' V-12 *Homer* twin-rotor heavy-lift helicopter – the world's largest rotorcraft. The Mi-6 testbed was also used for studying ways and means of improving the performance of the basic Mi-6.

Mil' Mi-24 engine testbed

On 30th May 2000 the Mil' OKB began trials of a Mi-24 converted into an engine testbed. The helicopter is fitted with Klimov VK-2500 turboshafts – the then-latest version of the proven TV3-117 (originally the engine was known as TV3-117VMA-SB3 but then someone decided the designation was much too long!). The VK-2500 is rated at 2,400 shp for take-off, with a 2,700-shp contingency rating; engine life is increased to 7,500 hours and time to first overhaul is 3,000 hours (twice that of the basic TV3-117V).

Petlyakov Pe-2LL engine testbed

In the 1940s a Pe-2 dive-bomber was converted into a testbed for a pulse-jet engine intended for the cruise missiles of the 10Kh/14Kh/16Kh family (see weapons testbed chapter). The Pe-2's twin-fin tail unit made it a convenient vehicle for such tests: the pulse-jet was installed above the rear fuselage so that the jetpipe was located between the fins, extending beyond the tailcone.

The Pe-2LL testbed with a pulse-jet on top of the rear fuselage.

Petlyakov Pe-8LL engine testbeds

After the Great Patriotic War a single Pe-8 (TB-7) four-engine heavy bomber (c/n 4218) – the sole surviving example powered by Mikulin AM-35A liquid-cooled Vee-12 engines – was converted into an engine testbed designated Pe-8LL. The development engine, a 1,850-hp Shvetsov ASh-82FN 14-cylinder radial, was installed in the extreme nose, supplanting the navigator's station glazing and turning the Pe-8LL into a five-engine aircraft. A test engineer's station was located immediately aft of the experimental engine.

A different Pe-8LL served as a testbed for the 4,600-kgp (10,140-lbst) Lyul'ka TR-3 axial-flow turbojet developed intended for the Sukhoi Su-17 supersonic interceptor of 1949 (the first aircraft to bear the designation, aka *izdeliye* R). Pavel O. Sukhoi's OKB-134 was closed in 1949 and reborn in 1953 as OKB-51.

Above and left: The Pe-8LL engine testbed featuring an ASh-82FN development engine in the specially modified nose. Note the ventral oil cooler for the radial engine.

Polikarpov I-15*bis*DM (I-152DM?) development aircraft

An I-15*bis* biplane fighter (c/n 5942) made a strong claim to fame in December 1939, when, with P. Ye. Loginov at the controls, it made the first-ever flight of a ramjet-equipped piloted aeroplane. Under each lower wing it carried a DM-2 ramjet booster (*dopolnitel'nyy motor* – supplementary motor) developed by Igor' A. Merkoolov from the smaller DM-1. Hence the modified fighter was known as the I-15*bis*DM; some sources, though, call this aircraft I-152DM. The ramjet had a diameter of 400 mm (1 ft 3¾ in), a length of 1.5 m (4 ft⁵⁹⁄₆₄ in) and weighed only 19 kg (62 lb), including fasteners; it ran on the same fuel as the main engine (a Shvetsov M-25 nine-cylinder radial), delivering a thrust of 0.187 kN (42 lbst).

The first five flights were used to test and perfect the boosters' starting procedure. Subsequent flights recorded the increase in speed with ramjets ignited and were used to devise methods of reducing the start-up time and to test improvements made to the combustion process. An increase in maximum

The I-15*bis*DM development aircraft (also known as the I-152DM) with DM-2 ramjet boosters under the lower wings.

Side view of the I-15*bis*DM, showing the fittings on which the compact ramjets are attached.

The I-153DM (c/n 6024) in early configuration with DM-2 ramjet boosters, September 1940.

Here, the same aircraft is seen a month later with DM-4 boosters; note how much larger the new model is.

Side view of the
I-153DM with
DM-4 boosters.
The boosters'
extremely small
ground clearance
is noteworthy.

speed of approximately 22 km/h (14 mph)
was recorded and tests continued until May
1940 by which time 54 flights had been successfully completed.

Polikarpov I-153DM (I-153PVRD) development aircraft

In September 1940 a production I-153 **Chai**ka
(Seagull) fighter (c/n 6034) – the retractable-
gear derivative of the I-15 – was fitted with
two Merkoolov DM-2 ramjet boosters under
the lower wings, becoming the I-153DM or
I-153PVRD (pryamo**toch**nyye voz**doosh**no-
reak**tiv**nyye **dvig**ateli – ramjets). A 30-km/h
(19-mph) increase in top speed with the
boosters running was recorded by test pilots
P. Ye. Loginov, A. V. Davydov and A. I. Zhookov.
Further trials were carried out in October 1940
with 500-mm (1 ft 7¹¹⁄₁₆ in) diameter DM-4
boosters when a maximum speed increase of
51 km/h (32 mph) was achieved. In total, 74
test flights were completed with both types of
ramjet.

Sukhoi Su-6 M-82FN/RD-1 (Su-7R) high-altitude fighter prototype

In 1942, building on his previous experience
with the Su-6 ('aircraft A') single-seat attack
aircraft, Pavel O. Sukhoi developed a high-altitude tactical fighter around the new 2,200-hp
Shvetsov M-71F 14-cylinder radial with two
TK-3 turbosuperchargers. According to the
design predictions, the aircraft would have a
maximum speed of 520 km/h (323 mph) at sea
level and 635 km/h (394 mph) at altitude, a
service ceiling of 12,500 m (41,010 ft), a range

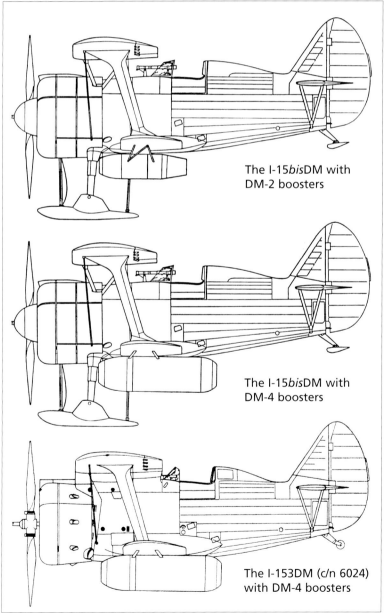

The I-15*bis*DM with
DM-2 boosters

The I-15*bis*DM with
DM-4 boosters

The I-153DM (c/n 6024)
with DM-4 boosters

Two aspects of the Su-6 M-82FN/RD-1 fighter (aka Su-7R), showing the heat shield protecting the fuselage from the supercharger flame. The RD-1 rocket booster is enclosed by a long fairing.

The rocket booster and its engine bearer, with enough room for tail-wheel retraction inside the bearer.

of 1,240 km (770 miles) and a gross take-off weight of 4,340 kg (9,570 lb).

Designated Su-7 (the first aircraft to bear the designation), the fighter was based on the structure of the Su-6 with its wooden fuselage, mixed-construction wings and aft-sliding cockpit canopy. The armour protecting the powerplant and its accessories was omitted and the bomb bay was eliminated. The outer wings featured leading-edge slats, as introduced on the Su-6 ('aircraft SA'). The landing gear was unchanged, with 750 x 250 mm (29.5 x 9.8 in) mainwheels and a 300 x 125 mm (11.8 x 4.9 in) tailwheel.

The engine drove a four-blade propeller; steel heat shields were attached to the fuse-

The rocket booster is refuelled with kerosene and nitric acid.

lage sides aft of the turbosuperchargers to protect the wooden structure. The armament was limited to two 20-mm ShVAK cannons installed in the wings.

Although the Su-7 was successfully undergoing manufacturer's flight tests, in 1943 it suddenly found itself without an engine when the original M-71F ran out of service life. Since the M-71F was not produced in series, a specially modified Shvetsov ASh-82FN derated to 1,800 hp was installed; it had several additional systems, including an alcohol/water cylinder head cooling system. Unfortunately, the power of this replacement engine was not sufficient to provide the predicted performance; the heavy Su-7 could manage only 491 km/h (305 mph) at 2,300 m (7,500 ft) and 506 km/h (314 mph) at 6,300 m (20,600 ft).

In order to provide short-term increases in the maximum speed, the designers decided to install a liquid-propellant rocket booster designed by Valentin P. Glushko. Suitable rocket motors from the Glushko OKB were the RD-1, RD-1KhZ, RD-2 and RD-3; all of them ran on kerosene with nitric acid as an oxidiser.

The RD-1, delivering 300 kgp (660 lbst) of thrust, was considered the best choice for the Su-7. Concurrently the heat shields protecting the fuselage from turbosupercharger torching were extended beyond the wing trailing edge.

The RD-1 was installed in the suitably modified tailcone. During the flight tests, which began in late 1944, 84 rocket motor start-ups were performed on the ground and in flight. From 31st January to 15th February 1945 eighteen rocket booster runs were made on the ground, using an ether/air starting system. Test flights with a new RD-1KhZ rocket motor featuring a chemical ignition system took place from 28th August to 19th December 1945.

According to manufacturer's estimates, the fighter would have a top speed of 590 km/h (367 mph) at 7,500 m (24,600 ft) with the rocket booster inoperative and 680 km/h (423 mph) with the booster running. The appropriate speeds at 12,000 m (39,370 ft) were predicted as 510 km/h (317 mph) and 705 km/h (438 mph). The estimated service ceiling was 12,750 m (41,830 ft).

Another view of the Su-7R with the RD-1 booster exposed.

Here the Su-7R is seen with the rocket booster running.

The Su-6 M-82FN/RD-1 (Su-7R)

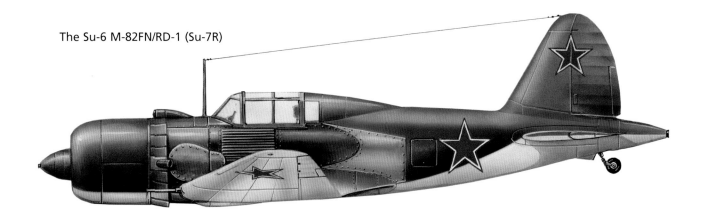

Frequent failures of the RD-1KhZ caused the tests to drag on longer that expected. In late 1945 test pilot G. Komarov established that the rocket booster increased the top speed from 491 to 566 km/h (from 305 to 351 mph) at 2,300 m (7,380 ft) and from 506 to 597 km/h (from 314 to 371 mph) at 6,300 m (20,660 ft) – that is, a mere 75-91 km/h (47-57 mph). The RD-1KhZ was unreliable, and after five rocket motors had been replaced, the designers decided to abandon the booster altogether. Nevertheless, the development and flight testing of prototypes with mixed powerplants was an important stage in Soviet aircraft technology development. Through these experiments, the Sukhoi OKB acquired experience in the design of high-speed aircraft.

Sukhoi T-5 experimental interceptor (*izdeliye* 81-1)

To meet an Air Force requirement the Sukhoi OKB developed a much-modified version of the T-3 (*izdeliye* 81) interceptor – the immediate precursor of the Su-9 *Fishpot-A* – designated T-5 (*izdeliye* 81-1). Instead of a single Lyul'ka AL-7F-1 afterburning turbojet rated at 7,500 kgp (16,530 lbst) dry and 10,000 kgp (22,045 lbst) reheat the aircraft was powered by two R11F-300 axial-flow afterburning turbojets developed by Sergey K. Tumanskiy's OKB-300, with a take-off thrust of 4,200 kgp (9,260 lbst) dry and 6,120 kgp (13,490 lbst) reheat, and was effectively a propulsion testbed designed to verify this twin-engine powerplant.

Being lighter and more compact than the AL-7F-1, the Tumanskiy engines were installed side by side in a completely redesigned and much wider rear fuselage; the result was a pronounced 'waist' at the wing trailing edge. The fuselage break point was moved aft from frame 28 to frame 34. The section between frames 28 and 34 was a sort of adapter between the existing forward fuselage structure and the new rear fuselage, incorporating a bifurcated inlet duct for the two engines; it also housed an enlarged No. 3 fuselage tank increasing the total internal fuel capacity from the T-3's 3,130 litres (688.6 Imp gal) to 3,480 litres (765.6 Imp gal). (In reality, however, not more than 3,330 litres (732.6 Imp gal) was filled during trials so as not to exceed the maximum take-off weight, the missing 150 litres (33 Imp gal) being distributed between the wing tanks.)

Other structural changes included a new fuselage nose with an axisymmetrical air intake identical to the one tested on the T43-1. The inboard ends of the flaps were slightly cropped to cater for the increased width on the fuselage. Unlike the AL-7, which was started by a jet fuel starter (a small gas turbine engine), the smaller and lighter R11-300 had electric starting; hence the single 12-kilowatt GS-12T generator was replaced by two 9-kilowatt GSR-ST-9000A starter-generators. Finally, the BU-30 and BU-34 hydraulic control surface actuators were replaced with identical BU-49 actuators in all three control circuits, and the control cables in the rear fuselage gave place to push-pull rods. The rest of the equipment remained unaltered.

The T-5 was converted from the T-3 prototype (c/n 01) at the Sukhoi OKB's experimental plant (MMZ No. 51). Due to the extent of the changes the conversion job took eight months (from October 1957 to June 1958). In early July the aircraft was delivered to the flight test facility in Zhukovskiy to commence ground checks, with M. I. Zooyev as engineer in charge of the tests. On 18th July the T-5 made its first flight at the hands of Vladimir S. Ilyushin. The manufacturer's flight test pro-

The T-5 development aircraft had a sharp-lipped axisymmetrical air intake with a shock cone as used on the T-43 (Su-9). Note the gun blast plates near the wing roots.

A side view of the T-5, showing the forked fairing between the engine nozzles.

This rear view illustrates the T-5's wide rear fuselage housing two Tumanskiy R11F-300 engines side by side.

203

One more aspect of the T-5; note the engine bay cooling air scoops.

much the same story with the MiG-19 where the two engines breathed through a single air intake divided by a splitter into individual inlet ducts; the downwind engine had to be started first, otherwise the other engine would literally take all the air away from it!)

Furthermore, the T-5's longitudinal stability proved to be unacceptably low (to be precise, the aircraft had virtually zero longitudinal stability due to the CG being positioned well aft). In May 1959 the aircraft's chief project engineer Yevgeniy G. Fel'sner called a halt to the test programme.

Sukhoi PT-95 testbed

The second pre-production T-3 interceptor (c/n 0015302) – that is, the second PT-8 (the PT8-2) – never flew as such but was immediately converted into a propulsion testbed designated PT-95. Receiving a specially instrumented engine with test equipment sensors and a new forward fuselage patterned on the T43-1, the aircraft was delivered to its new owner, LII, in 1958. In 1958-59 the PT-95 served as a testbed for the new AL-7F-1, helping the engine's teething troubles to be overcome, and was used to explore the interaction between the air intake and the engine. The programme was performed by LII test pilot Valentin P. Vasin; on one occasion he had to make a dead-stick landing in the PT-95 when the engine quit and refused to restart.

gramme was completed in full, involving 26 flights. The tests showed that, with the engines in afterburner mode, the aircraft was overpowered and, in spite of the decidedly higher drag (primarily due to the wider rear fuselage), could reach much higher speeds than the T-3; the engines' structural strength was the limiting factor, the R11F-300 being designed to withstand speeds below Mach 2.

On the down side, the engines' automatic fuel control units still had a few bugs to be eliminated; as a result, the engines ran unstably, the afterburners shutting down frequently of their own accord during climb. Engine starting proved problematic as well, since both engines used a common air intake and, figuratively speaking, were short of breath. (It was

Sukhoi Su-15/R11F3-300 engine testbed

The fourth production Su-15 *Flagon-A* interceptor (c/n 0115304) was used by LII as a testbed for the Tumanskiy R11F3-300 engine featuring a contingency rating. Actually the term 'contingency rating' was used in a non-standard way, as it referred to increasing thrust at low altitudes to improve performance, not to an automatic power reserve activated in the event of an engine failure.

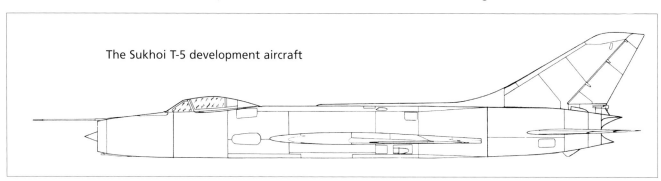

The Sukhoi T-5 development aircraft

This view of the T-58VD experimental STOL aircraft shows the closed dorsal air intake scoops for the lift engines and the modified double-delta wings. Note the photo calibration markings on the nose.

Sukhoi T-58-95 engine testbed

In keeping with MAP and Air Force orders issued in October 1967 and May 1968 respectively several production Su-15s were modified for testing the new Gavrilov R13-300 afterburning turbojets. The R13-300 (*izdeliye* 95) was a derivative of the R11F-300 developed by the Ufa-based Soyuz (Union) engine design bureau led by Sergey A. Gavrilov. At that time the young OKB had little design experience of its own and thus ran into major problems with the R13-300, which differed significantly from the precursor, the R11F-300 (among other things, the number of high-pressure compressor stages was increased from three to five and

a second afterburner stage was added). Thanks to these changes the engine delivered 4,100 kgp (9,040 lbst) at full military power and 6,600 kgp (14,550 lbst) in full afterburner versus 3,900 kgp (8,600 lbst) and 6,175 kgp (13,610 lbst) respectively for the R11F2S-300.

Upon completion of the bench tests a flight-cleared R13-300 was installed on Su-15 c/n 0415302 by August 1967, replacing the starboard R11F2S-300. The resulting 'lopsided' aircraft was known at the Sukhoi OKB as the T-58-95, the last two digits referring to the development engine's product code. Suitably fitted with data recording equipment, the T-58-95 made 11 flights under the initial

The T-58VD as it appeared at the 1967 Domodedovo air show. The tactical code '58 Red' is gone and the nose is painted differently.

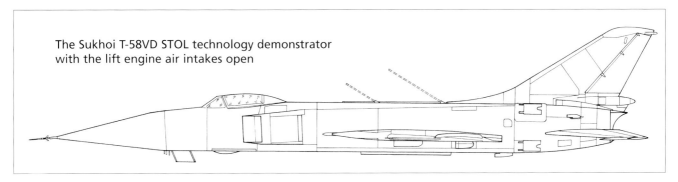

The Sukhoi T-58VD STOL technology demonstrator with the lift engine air intakes open

The T-58VD as it appeared at the 1967 Domodedovo air show

flight test programme before being transferred to LII where manufacturer's tests of the R13-300 engine were held in 1967-68.

Sukhoi Su-15/R13-300 engine testbeds

By mid-December 1968 the Sukhoi OKB had installed a complete shipset of R13-300 engines in a production Su-15 coded '11' (c/n 0715311). After passing manufacturer's flight tests in March 1969 the aircraft was flown to the GK NII VVS facility in Akhtoobinsk for state acceptance trials of the new engine and the re-engined interceptor as a whole. The trials programme involving 53 flights showed that the service ceiling, acceleration time, effective range, combat radius and field performance had improved thanks to the new engines.

Between November 1969 and February 1970 Su-15 c/n 0715311 underwent addition-

al testing to explore the R13-300's resistance to surging; the test programme was performed by Vladimir S. Ilyushin. Another Su-15 coded '37 Red' (c/n 1115337), an aircraft owned by LII and one of the first examples to be powered by R13-300 engines, joined the test programme in December 1969. In the course of the tests engine flameouts at high Mach numbers were experienced on both aircraft when the engines were running in full afterburner. It turned out that the engines were simply 'short of breath' because the air intakes were too small. To eliminate this dangerous phenomenon the air intakes had to be widened but this was impracticable in mass production at the time, since it required major structural changes and, as a consequence, changes to the tooling. Hence it was decided simply not to engage the second afterburner stage on Su-15s *sans suffixe* equipped with R13-300 engines, since the first afterburner stage provided an adequate increase of flight performance.

Nevertheless, a third production Su-15 (identity unknown) was re-engined with R13-300s, undergoing tests in 1970-71. It featured wider air intakes to cater for the new engines' greater mass flow; this intake design, together with the R13-300 engines, was later introduced on the ultimate Su-15TM.

Sukhoi T-58VD STOL technology demonstrator

In early 1965, when the T58D-1 (the first prototype Su-15 *sans suffixe*) had completed a brief flight test programme with the new double-delta wings intended for the Su-15T *Flagon-E*, the OKB decided to use this aircraft as a propulsion systems testbed and a STOL technology demonstrator in conjunction with the development of the T-58M low-altitude attack aircraft. (The latter designation proved to be short-lived; the T-58M, which later became the T-6, was a totally unrelated design and a much larger aircraft which evolved into the Su-24 tactical bomber and lies outside the

Two views of the T-58VD parked on the grass at Zhukovskiy. Note the bulges into which the upper airbrakes are incorporated – a non-standard feature.

This close-up of the T-58VD's centre fuselage with all engines running shows the open dorsal intake scoops and the auxiliary blow-in doors of the cruise engines.

scope of this book.) This involved installing small turbojet engines vertically inside the fuselage to generate lift. The lift-jet concept was quite popular then both in and outside the Soviet Union. Western development aircraft making use of lift engines included the Dassault Balzac supersonic fighter and the Short SC.1 technology demonstrator, although both of them were vertical take-off and landing (VTOL), not STOL, aircraft.

On 6th May 1965 MAP issued an order requiring the Sukhoi OKB to build and test a proof-of-concept vehicle in order to verify the STOL technology using lift-jets. The engineers set to work; by mid-year they had completed the project documents for the conversion of the T58D-1 into such a vehicle. The extensive conversion involved remanufacturing the centre fuselage to accommodate three 2,350-kgp (5,180-lbst) Kolesov RD36-35 turbojets. The lift engines were installed in a bay between the cruise engines' inlet ducts with the axes inclined forward 10°, breathing through two large scoop-type intakes on the fuselage's upper surface (the forward intake served the

foremost engine and the rear intake the other two). The exhaust aperture was closed by louvres which had two operational settings, directing the jet exhaust aft on take-off to add a measure of forward thrust or forward on landing to slow the aircraft down. In cruise flight (in 'clean' configuration) the lift-jets' intakes and exhaust louvres closed flush with the fuselage skin. The conversion also included modifications to the wings, relocation of various equipment and piping and so on.

All fuel was now carried in the wing tanks – and that means less fuel and three more engines guzzling away at it. But then, range and endurance were not crucial for a pure technology demonstrator that was not meant to operate far away from its base.

Designated T-58VD (for *verti**kahl'nyye dvigateli*** – 'vertical engines', that is, lift-jets), the rebuilt aircraft was completed at the end of 1965, commencing tethered tests on a purpose-built ground rig on the Sukhoi OKB's premises; Roman G. Yarmarkov retained his assignment as this aircraft's engineer in charge of the tests. The rig featured an 'open-air wind

The T-58VD makes a short take-off from an unpaved airstrip, the lift jets creating a minor dust storm.

The T-58VD climbs away with the landing gear already retracted and the lift engines running.

The LL-UV (KS) TVC testbed seen flying near Zhukovskiy.

tunnel' – a Kuznetsov NK-12 turboprop engine driving ducted propellers emulated the slipstream at simulated speeds up to 400 km/h (248 mph), creating proper operating conditions for the lift engines. The pad on which the aircraft sat was rigged with pressure sensors to capture the off-loading of the landing gear and thus assess the efficiency of the lift engines.

The tethered tests allowed the T-58VD's aerodynamics with the lift engines running to be explored and the operation of all principal systems to be checked. Unfortunately they were marred by a tragic incident in February 1966. A mechanic from the OKB's propulsion laboratory was careless enough to approach the aircraft's forward fuselage when the blower was running; the powerful stream of air immediately swept him off his feet and hurled him savagely against the aircraft, killing him.

On 26th April 1966 upon completion of the ground test phase the aircraft was trucked to LII, making its first flight on 6th June at the hands of Yevgeniy S. Solov'yov. Later, the T-58VD was flown by both Solov'yov and Sukhoi OKB chief test pilot Vladimir S. Ilyushin; by the end of the year it had made 37 real flights and 19 taxi runs and short hops, including high-speed runs on a dirt strip. The manufacturer's flight tests showed that the lift created by the auxiliary engines reduced the unstick speed from 390 to 285 km/h (from 242 to 177 mph) and the landing speed from 315 to 225 km/h (from 195 to

139 mph). The take-off run was shortened from 1,170 to 500 m (from 3,840 to 1,640 ft) and the landing run from 1,000 to 560 m (from 3,280 to 1,840 ft) – an impressive result. On the other hand, it became apparent that the chosen location of the lift engines was not the optimum one, as the thrust of the forward engine caused a strong tendency to pitch up during landing approach; the problem was solved by using only the centre and rear lift-jets for landing.

On 9th July 1967 the T-58VD participated in the air show at Moscow-Domodedovo, giving a short take-off and landing demonstration with Yevgeniy S. Solov'yov at the controls. After that, the STOL version received the reporting name *Flagon-B*.

The results of the T-58VD's flight tests gave the Sukhoi OKB valuable experience in designing, building and testing STOL aircraft and allowed the test pilots to master the technique of flying such aircraft. This knowledge was incorporated into the design of the delta-wing T6-1 strike aircraft prototype. However, the trade-off for the good short-field performance turned out to be too high; the lift engines reduced the space available for fuel dramatically while significantly increasing fuel consumption on take-off and landing. Also, the operation of the lift-jets worsened longitudinal stability somewhat, and in cruise mode the lift engines were just a lot of useless weight which reduced the payload. Hence the second prototype, the T6-2I, was radically reworked to feature variable-geometry wings which gave the desired results.

Upon completion of the test programme the T-58VD was donated to the Moscow Aviation Institute where it served as an instructional airframe for a while. Sadly, later this unique aircraft was apparently scrapped to free up space for new teaching aids in the MAI hangar; no trace of it could be found in the autumn of 2002.

Sukhoi LL-UV (KS) TVC technology testbed

As part of the efforts to enhance the agility of future fighters, in the mid-1980s the Sukhoi OKB began experimenting with thrust vector control (TVC). To this end in 1986 NPO Saturn (as the Lyul'ka OKB was known in the 1980s; now Lyul'ka-Saturn) headed by Viktor M. Chepkin started work on a movable axisymmetric convergent-divergent nozzle. The AL-31F's project chief A. V. Andreyev was

in charge of the programme. An AL-31F engine fitted with this nozzle underwent a lengthy bench testing programme in 1987-88; the nozzle was powered by the aircraft's hydraulic system and could be deflected up and down within ±15°.

In 1989 the Sukhoi OKB and NPO Saturn converted one of the early-production Su-27 *Flanker-B* 'dogships', the T10-26 ('26 Blue', c/n 36911007...02, f/n 07-02), for TVC research. Only one engine on this aircraft had a vectoring nozzle, the other one was standard. In this guise the aircraft was designated LL-UV (KS). The acronym stood for *le**ta**yushchaya lab-ora**tori**ya s oopravl**lya**yemym **vek**torom [**tya**gi], **kroog**loye **sop**lo* – TVC research air-craft, 'round' (= axisymmetrical) nozzle.

The modified aircraft entered flight test on 21st March 1989 with Sukhoi OKB test pilot Oleg G. Tsoi at the controls. Flight tests gave encouraging results, confirming that the mov-able nozzle improved the fighter's agility, espe-cially at low speeds. Hence it was decided to develop a production version of the thrust-vec-

toring engine whose nozzle controls would be included into the Su-27's control system.

Sukhoi LL-UV (PS) TVC technology testbed

In parallel, the Sukhoi OKB began experiment-ing with two-dimensional vectoring nozzles. In 1990 an initial-production Su-27UB *Flanker-C* trainer built by the Komsomol'sk-on-Amur Aircraft Production Association (KnAAPO) and known at the OKB as the T10U-8 ('08 Blue', c/n 49021002502, f/n 02-02) was converted into a propulsion/control technology testbed called LL-UV (PS). The acronym stood for *le**ta**yushchaya labora**tori**ya s oopravl**lya**-yemym **vek**torom [**tya**gi], **plos**koye **sop**lo* – TVC research aircraft, 'flat' (= two-dimension-al) nozzle. Some sources referred to it as the Su-27UB-PS, but this designation is probably fictional. Deputy Chief Designer Mikhail A. Pogosyan, who subsequently became the OKB's Chief Designer, was placed in charge of the programme.

Two views of the LL-UV (KS) devel-opment aircraft ('26 Blue'). Note the instrumented nose probe, as well as the short tail 'stinger' an anti-flutter booms on the fins characteristic of very early pro-duction Su-27s.

209

Three views of the two-dimensional vectoring nozzle fitted to the LL-UV (PS) testbed based on the Su-27UB.

The LL-UV (PS) featured a long boxy structure terminating in a two-dimensional movable nozzle (reminiscent of that used on the McDonnell Douglas F-15S/MTD testbed) on the port engine; this fairing protruded far beyond the tip of the tail 'stinger'. The experimental nozzle was manufactured by the Ufa-based NPO Motor and the conversion took place at KnAAPO, whereupon the aircraft was airlifted to Zhukovskiy by

Here, the nozzle doors are seen in the fully open position.

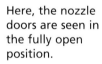

More views of the 2-D vectoring nozzle, showing the narrow aperture.

Far left: The LL-UV (PS) on take-off and in a steep turn. Note the asymmetrical carriage of a dummy R-27 missile as a counterweight.

A series of stills from a ciné film showing the 2-D nozzle in action.

an Antonov An-22 *Cock* transport to undergo trials.

The LL-UV (PS) made a total of 20 flights. The tests showed a marked reduction of the fighter's heat signature, which reduced its vulnerability to IR-homing missiles; also, the nozzle could be easily adapted to operate as a clamshell thrust reverser. Unfortunately funding shortfalls prevented further research with this aircraft. Besides, two major problems surfaced. Firstly, the transition from circular to rectangular cross-section caused unacceptably high thrust losses amounting to 14-17%; secondly, the 2-D nozzle was much heavier than an axisymmetrical one because of the need to

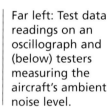

Far left: Test data readings on an oscillograph and (below) testers measuring the aircraft's ambient noise level.

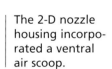

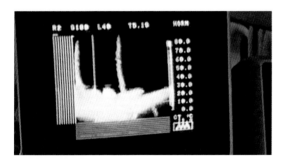

The 2-D nozzle housing incorporated a ventral air scoop.

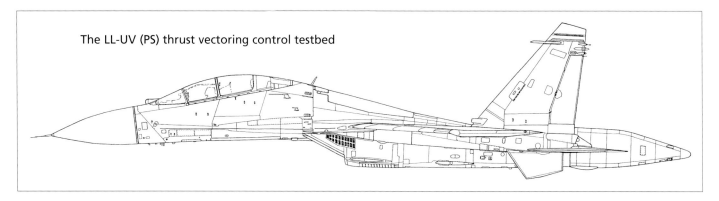
The LL-UV (PS) thrust vectoring control testbed

The logo of the MMPP Salyut engine plant applied to the nose of Su-27LL '595 Blue outline'.

absorb the bending loads. The weight penalty could be negated by using carbon-carbon materials and ceramic heat insulation; however, these materials were still at an early development stage. All things considered, the idea was shelved; instead, the Sukhoi OKB concentrated on axisymmetrical vectoring nozzles, which found use on the production Su-30MKI.

Sukhoi Su-27LL engine testbeds

At the turn of the century the 'Salyut' Moscow Engine Production Enterprise (MMPP Salyut), one of the two factories producing the Lyul'ka AL-31F afterburning turbofan for the Su-27 family, began a multi-stage programme to refine the engine. Unlike the efforts of Lyul'ka-Saturn, this was primarily with mid-life updates in mind. The first stage of the effort resulted in the AL-31F-M1 engine incorporating a new KND-924 low-pressure compressor (*kompressor nizkovo davleniya*) with a diameter increased from 905 to 924 mm (from 2 ft 11⅝ in to 3 ft 0⅜ in). This modification increased the mass flow by 6% and the turbine temperature by 25°C, with a resulting thrust increase to 8,300 kgp (18,300 lbst) at full military power and 13,300 kgp (29,320 lbst) in full afterburner. The engine also featured an improved automatic control system.

After completing its bench test cycle a prototype AL-31F-M1 engine was installed in the starboard nacelle of LII's Su-27P '595 White'

(c/n 36911037511, f/n 37-11) – one of the two *Flanker-Bs* operated by Anatoliy N. Kvochur's aerobatic team. The machine first flew in this configuration on 25th January 2002 with LII test pilot Aleksandr V. Pavlov at the controls. Repainted in the same violet/grey LII house colours that the LMK-2405 now wears and recoded '595 Blue outline', the machine was displayed statically at the MAKS-2003 and MAKS-2005 airshows.

In 2004 the aircraft was refitted with a more advanced version of the development engine – the AL-31F-M2. This engine features a new combustion chamber, a modified turbine utilising new materials and a new cooling system, FADEC and a three-axis vectoring nozzle controlled by moving the jointed nozzle petals in the required direction; the nozzle was created jointly by MMPP Salyut and the St. Petersburg-based NPO Klimov engine design bureau. The new version has a 100°C (212°F) higher turbine temperature as compared to the baseline AL-31F and a maximum afterburner thrust of 14,100 kgp (31,080 lbst); it also has a longer service life and time between overhauls.

The next stage of the programme will see the aircraft equipped with the 14,600-kgp (32,190-lbst) AL-31F-M3. This engine has an all-new three-stage LP compressor with blisks (discs manufactured integrally with wide-chord blades). Like the preceding two versions, it will be completely interchangeable with the

The Su-27LL engine testbed (c/n 36911037511) with AL-31F-M2 engines, Zhukovskiy, 1985

standard AL-31F; the latter may be upgraded to M1 or M2 (and possibly M3) standard in the course of an overhaul.

The static park at the MAKS-2005 airshow also featured the IFR-capable Su-27PD '598 White' (c/n 36911037820, f/n 37-20). Billed as a Su-27LL with AL-31F-M1 thrust-vectoring engines, the fighter featured an AL-31F-M1 engine with a vectoring nozzle in the port nacelle (the starboard engine appeared to be standard). The aircraft was fitted with a test instrumentation suite allowing information to be downloaded to the ground command post in real time via data link.

The Su-27LL engine testbed ('595 Blue outline') in the static park of the MAKS-2003 airshow. Note the Russian flag on the tails in lieu of a red star and the missile rails painted red for better conspicuity to avoid damage by ground vehicles.

Three views of the Su-27LL's rear end at the MAKS-2003. The aircraft is fitted with AL-31F-M2 engines with vectoring nozzles; note the starboard nozzle at maximum downward deflection.

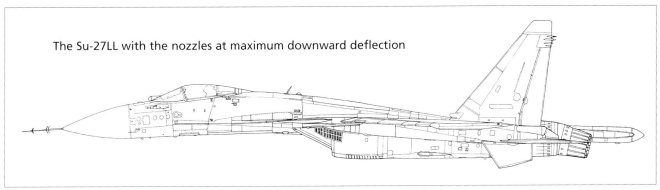

The Su-27LL with the nozzles at maximum downward deflection

213

The tenth Su-27M (Su-35) prototype, T10M-10 ('711 Blue'), became a testbed for the AL-41F engine. Here the aircraft is seen on final approach to Zhukovskiy with an AL-41F-1A in the starboard nacelle.

The T10M-10 passes overhead, showing the differing engine nozzles.

Here the T10M-10 is seen with a pair of Lyul'ka-Saturn *izdeliye* 117S afterburning turbofans intended for the T-50 fifth-generation fighter. Note the ceramic tiles in the nozzles.

Sukhoi Su-35 fifth-generation fighter engine testbed

In 2004 the tenth Su-27M/Su-35 prototype, T10M-10 ('710 Blue', c/n 79871011002, f/n 11-02), which was retained by the Sukhoi OKB, became a testbed for the Lyul'ka-Saturn *izdeliye* 117S afterburning turbofan – a growth version of the AL-31F intended for the Sukhoi T-50 fifth-generation fighter currently undergoing trials. (The designation Su-35 has since been reused for a newer version of the *Flanker*, the T-10BM Generation 4++ fighter.) The engine is an advanced derivative of the AL-31F and delivers 15,000 kgp (33,070 lbst), with potential for further uprating. It may also be fitted to aircraft of the Su-27/Su-30 family as a mid-life update.

Originally a single development engine was fitted instead of the starboard AL-31F; the first flight in this configuration took place on 5th March 2004. Later the T10M-10 was seen flying with two *izdeliye* 117S engines.

Tupolev I-4 rocket booster testbed

Between 1932 and 1936 a Tupolev I-4 (ANT-5) sesquiplane fighter was fitted with both liquid-fuel and solid-fuel booster rockets in rocket-assisted take-off (RATO) experiments.

Tupolev Tu-2LL engine testbed

A total of five Tu-2 twin-engined bombers were converted into Tu-2LL testbeds for testing early Soviet turbojet engines. Four of these (including c/n 3/48) served with LII in 1947-54; the fifth was operated by GK NII VVS. In all cases the development engine was housed in a prominent ventral pod.

Most of the engines tested on the Tu-2LL were reverse-engineered versions of foreign designs: the 900-kgp (1,980-lbst) RD-10 axial-flow turbojet (Junkers Jumo 004 Orkan copy), the 800-kgp (1,760-lbst) RD-20 axial-flow turbojet (BMW 003 Sturm copy), the 2,040-kgp (4,500-lbst) RD-45 centrifugal-flow turbojet (Rolls-Royce Nene I copy), the 2,270-kgp (5,000-lbst) RD-45F (Rolls-Royce Nene II copy), the 1,590-kgp (3,505-lbst) RD-500 centrifugal-flow turbojet (Rolls-Royce Derwent V copy), the 2,700-kgp (5,950-lbst) Klimov VK-1 (a further refined and uprated version of the RD-45F). These engines found use on such Soviet combat jets as the Ilyushin IL-28 (RD-45F and VK-1), Lavochkin La-15 (RD-500), Mikoyan/Gurevich MiG-9 (RD-20), MiG-15 (RD-45F), MiG-15*bis* (VK-1), Tupolev Tu-14 (VK-1), Yakovlev Yak-15

(RD-10) and various experimental aircraft (the La-200, Mikoyan I-320, Yak-25 single-engined fighter and the like).

Indigenous engines tested on the Tu-2LL were the Klimov VK-5 centrifugal-flow turbojet and the 2,000-kgp (4,410-lbst) Mikulin AM-5 axial-flow turbojet which powered the twin-engined Yak-25 (Yak-120).

Tupolev Tu-4LL engine testbeds (including DR-1/DR-2, '94/1' & '94/2')

The Tu-4 strategic bomber was also used for testing new piston, turboprop and turbojet engines. As mentioned earlier, the first testbed of this kind was not a Soviet-built Tu-4 but a genuine Boeing B-29 used for testing the Tu-4's ASh-73 engine.

In 1950 the ninth Kazan'-built Tu-4 (c/n 220204), aptly serialled '9 Black', was modified for testing the 2,000-hp Dobrynin VD-3TK four-row radial; this supercharged liquid-cooled engine was intended for the projected Alekseyev I-218 attack aircraft, which never materialised. Two VD-3TKs driving AV-28 four-bladed contra-rotating propellers were fitted in place of the Nos. 1 and 4 ASh-73TKs. Like most Tu-4s filling the engine testbed role, the aircraft was designated Tu-4LL.

In the second half of the 1950s the Tupolev OKB converted two other *Bulls* into engine testbeds. Initially these aircraft were used to test the engines intended for the Tupolev 'aircraft 80' and 'aircraft 85' long-range bombers – the 4,000-hp Shvetsov ASh-2TK and 4,700-hp ASh-2K 28-cylinder four-row air-cooled radials and the 4,300-hp Dobrynin VD-4K 24-cylinder water-cooled radial driving an AV-44 four-bladed propeller. (The ASh-2TK never flew on anything except a testbed, but the other two engines subsequently powered the two prototypes of the '85'.) In all cases the development engine was installed in the No. 3 (starboard inner) position.

When turboprop engines came on the scene these two aircraft were further modified for testing early Soviet turboprops – the 5,163-ehp Kuznetsov TV-2, the 6,250-ehp TV-2F (*forseerovannyy* – uprated), the 7,650-ehp TV-2M (*modifitseerovannyy* – modified) and the 12,500-ehp 2TV-2F coupled engine (the Tu-4LL with the latter engine was also known as the 'aircraft 94/2'). The other machine, 'aircraft 94/1', had the entire forward fuselage of the 'aircraft 91' (Tu-91) *Boot* naval strike aircraft, complete with a single

One of LII's Tu-2LL engine testbeds with a RR Nene I in the ventral pod. Note the air intake splitter matching that of the future MiG-15.

This Tu-2LL wears LII's old logo above the c/n (3/48) on the fins.

This view shows well the shape of the Tu-2LL's test engine pod.

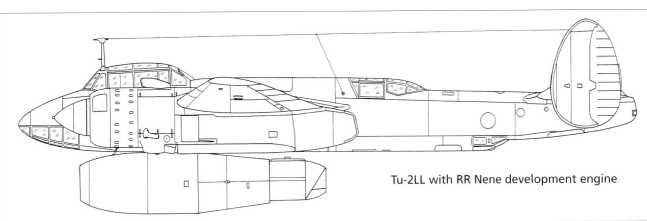

Tu-2LL with RR Nene development engine

Tu-4LL c/n 230113 in its early configuration with a Lyul'ka AL-5 turbojet (known as the DR-1). The aircraft is parked over a concrete-lined trench for ground-running the engine; the test engine pod is in the lowered position.

More views of the DR-1. The aircraft is parked over the trench in a different way to create a nose-down attitude; the mainwheels rest on built-up ramps. Note the safety grille ahead of the turbojet's air intake to protect the ground personnel. The cloth covers on the wings and main engines are also noteworthy.

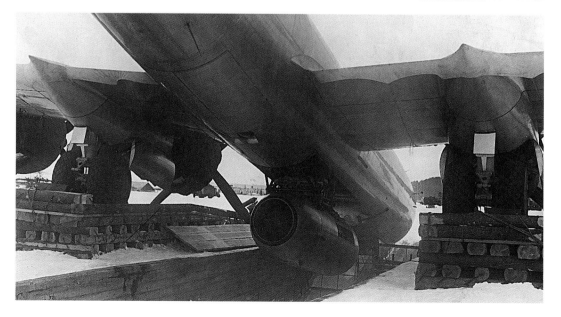

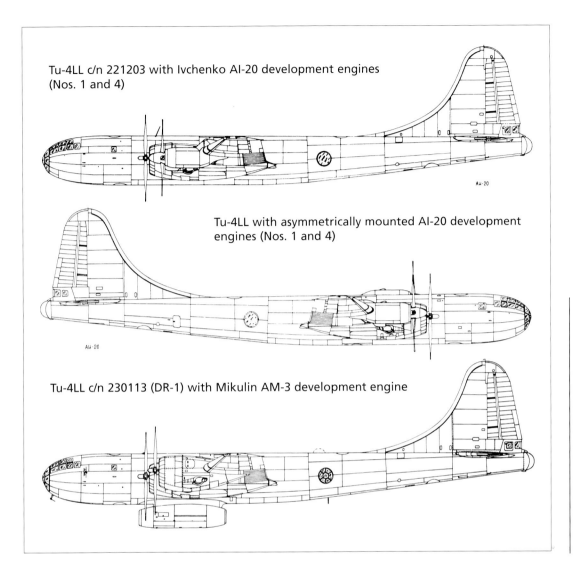

Tu-4LL c/n 221203 with Ivchenko AI-20 development engines (Nos. 1 and 4)

Tu-4LL with asymmetrically mounted AI-20 development engines (Nos. 1 and 4)

Tu-4LL c/n 230113 (DR-1) with Mikulin AM-3 development engine

One of the Tu-4LL testbeds known as 'aircraft 94/2'. The No. 3 nacelle houses a Kuznetsov 2TV-2F coupled turboprop with contrarotating propellers as fitted to the first prototype of the Tu-95 ('95-1'). Note the air data boom mounted on the flight deck glazing.

Tu-4LL '22 Red' (c/n 221203) was used to test the Ivchenko AI-20 and Kuznetsov NK-4 turboprops with AV-68 propellers. The port side overwing installation is an AI-20 as fitted to the Ilyushin IL-18 airliner while the underwing installation on the starboard wing is an NK-4 as fitted to the Antonov An-10 airliner!

One more view of Tu-4LL c/n 221203. Here, two AI-20 turboprops in the Nos. 1 and 4 nacelles are both installed in the underwing position.

TV-2F turboprop and three-bladed contraprops, supplanting one of the ASh-73TK radials. An air data boom looking like a tripod was installed on the fuselage nose.

Tu-4LL c/n 225402 (serial unknown) had a pair of TV-2 turboprops (c/ns 16 and 17) replacing the Nos. 1 and 4 ASh-73TKs. In this configuration it made 27 flights totalling 72 hours 51 minutes. Unfortunately the aircraft was lost on 8th October 1951 when the No. 4 engine caught fire during a planned in-flight restart attempt. The fire was caused by a leak of burning fuel into the turboprop's nacelle through the joint between the engine and the extension jetpipe. The crew managed a forced landing, escaping from the burning aircraft which was destroyed by the fire.

After completing their trials programmes on the Tu-4LL the TV-2F and the 2TV-2F went on to power two aircraft designed by the Tupolev OKB – the '91' (Tu-91) and the '95-1' heavy bomber (the first prototype of the Tu-95 *Bear-A* strategic bomber) respectively. Both aircraft had an unhappy fate, the '91' falling victim to a whimsical head of state (Nikita S. Khrushchov's predilection towards missiles was the bane of the Soviet aircraft industry's existence) while the '95-1' was lost in a crash caused by an engine fire. This brought about a change of the Tu-95's powerplant to the Kuznetsov NK-12.

One peculiarity of the turboprop testbeds was that the power of the development engine exceeded that of the ASh-73TK by a factor of 2 to 5. This required major structural changes; in particular, the engine bearer(s) had to be designed anew to convey the much higher forces. Also, the development engine(s) ran on kerosene instead of aviation petrol, which meant a separate fuel system had to be provided for the turboprop(s).

In 1954 one more *Bull* was urgently converted into a Tu-4LL in the wake of the crash of the '95-1' bomber. This aircraft served as a testbed for the 12,500-ehp Kuznetsov TV-12 driving Stoopino Machinery Design Bureau AV-60 four-bladed contraprops of 5.6 m (18 ft 4¾ in) diameter. The massive engine was again installed in the No. 3 position, the nacelle protruding far ahead of the wing leading edge so that the propeller rotation plane was almost in line with the flight deck. After passing its test programme the TV-12 was installed in the second prototype of the *Bear* (the '95-2'), subsequently entering production as the NK-12. To this day it remains the world's most powerful turboprop (later versions, starting with the NK-12M, deliver 15,000 ehp for take-off!) and has done sterling service on the Tu-95, Tu-142 *Bear-F* long-range anti-submarine warfare aircraft, Tu-114 *Cleat* long-haul airliner, Tu-126

Moss AWACS, An-22 *Antey* (Antheus; NATO reporting name *Cock*) heavy-lift transport aircraft and Alekseyev A-90 *Orlyonok* (Eaglet) transport *ekranoplan* (wing-in-ground effect craft) for more than 40 years now.

The next Tu-4LL converted by LII for testing turboprop engines in the mid-1950s (identity unknown) was a most unusual testbed. Like the ill-fated Tu-4LL c/n 225402, it had two development engines – initially 4,000-ehp Kuznetsov NK-4s and then identically rated Ivchenko AI-20s driving AV-68 four-bladed propellers – in the Nos. 1 and 4 positions. Both engines had been developed for the An-10 *Ookraïna* (the Ukraine; NATO reporting name *Cat*) and the Ilyushin IL-18 *Moskva* (Moscow, the second aircraft to bear the designation; NATO reporting name *Coot*) and the AI-20 was eventually selected to power both types. The unusual part was that the port turboprop was installed above the wing as on the IL-18, while the starboard turboprop was mounted below the wing as on the An-10! This 'lop-sided' installation served to check the behaviour of the engine in different operating conditions.

The Tu-4LL testbeds with experimental turboprop engines were operated by the Tupolev OKB and LII in 1951-62. Interestingly, LII sources state that three, not four, examples

(including the crashed one) were equipped with turboprops.

Three other Tu-4LLs were used to test new jet engines. Among other things, the Myasishchev OKB together with LII converted an unserialled Moscow-built Tu-4 (c/n 230113) into a testbed for the 5,000-kgp (11,020-lbst) Lyul'ka AL-5 turbojet which was considered as a possible powerplant for the M-4 *Bison-A* strategic bomber at an early development stage. In this guise the aircraft was known at the Myasishchev OKB as the DR-1; D stood for ***dvig**atel'* (engine, a reference to the engine testbed role) while *izdeliye* R was the Tu-4's product code. Later, when the much more powerful Mikulin AM-3 turbojet rated at

Tu-4LL '9 Black' (c/n 220204) was fitted with two Dobrynin VD-3TK radial engines driving AV-28 contra-rotating propellers.

Another view of Tu-4LL '9 Black', showing to advantage the shape of the new nacelles housing the development engines.

The No. 4 NK-4 engine of the Tu-4LL shown opposite. Note the aft position of the oil cooler.

221

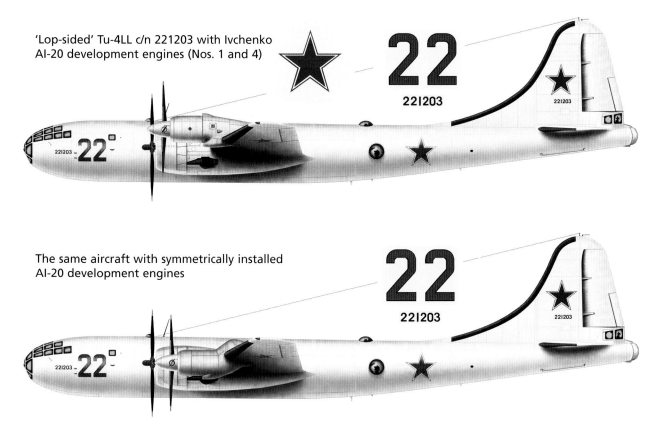

'Lop-sided' Tu-4LL c/n 221203 with Ivchenko
AI-20 development engines (Nos. 1 and 4)

The same aircraft with symmetrically installed
AI-20 development engines

8,700 kgp (19,180 lbst) for take-off was selected for the M-4, this replaced the AL-5 development engine on Tu-4LL c/n 230113 in January 1952 and the in-house designation at OKB-23 was changed to DR-2.

The turbojet was installed in the forward bomb bay in a special nacelle which could move up and down on parallel trailing arms. It was semi-recessed in the fuselage for take-off and landing to give adequate ground clearance, extending clear of the fuselage into the slipstream before start-up; it was also possible to extend the engine on the ground when the aircraft was parked over a special trench with concrete-lined walls and a jet blast deflector for ground runs. When the nacelle was retracted the air intake was blanked off by a movable shutter to prevent windmilling and foreign object damage. In an emergency (for instance, if the hydraulic retraction mechanism failed) the development engine could be jettisoned to permit a safe landing, special pyrotechnical guillotines cutting the fuel and electric lines. The test engineer and his assistant sat in the

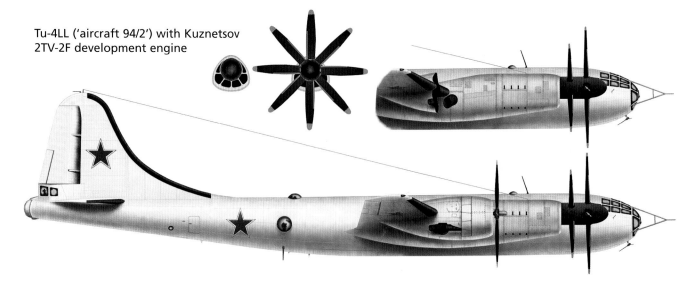

Tu-4LL ('aircraft 94/2') with Kuznetsov
2TV-2F development engine

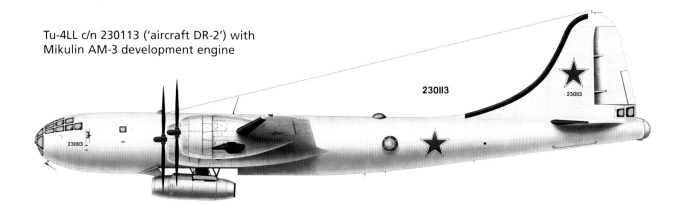

Tu-4LL c/n 230113 ('aircraft DR-2') with Mikulin AM-3 development engine

centre pressurised cabin; the cabin floor incorporated a glazed window for inspecting the nozzle of the development engine.

The testing of the AM-3 involved a lot of defining and refining, but it was worth the effort as it enabled the M-4 and the Tu-16 to quickly complete their test programmes and enter production.

Later two more *Bulls* (one of them was possibly c/n 230314) were converted into Tu-4LLs for testing jet engines. Apart from the AL-5 and AM-3, the three aircraft were used to test a wide range of engines by 1962 – the 2,700-kgp (5,952-lbst) AM-5F, the 3,300-kgp (7,275-lbst) Mikulin AM-9B (RD-9B) afterburning turbojet, the 5,110-kgp (11,265-lbst) Tumanskiy R11-300 afterburning turbojet, the 900-kgp (1,980-lbst) Tumanskiy RU19-300 turbojet, the 6,830-kgp (15,057-lbst) Lyul'ka AL-7 turbojet and its versions – the 7,260-kgp (16,005-lbst) AL-7P and the 9,200-kgp (20,282-lbst) afterburning AL-7F, the 8,440-kgp (18,606-lbst) Klimov VK-3 afterburning turbofan, the 6,270-kgp (13,822-lbst) VK-7 turbojet, the Dobrynin VD-5 turbojet, the 11,000-kgp (24,250-lbst) VD-7 afterburning turbojet and the 5,400-kgp (11,904-lbst) Solov'yov D-20P commer-

cial turbofan. The method of installation and the testing techniques were basically the same as on the DR-1/DR-2, except in the case of the VD-5 and VD-7; these bulky turbojets were positioned too low even when stowed in the bomb bay, necessitating a special non-retractable landing gear with lengthened struts.

Apart from the M-4 and Tu-16, the jet engines tested on the Tu-4LL powered the '98' (Tu-98) experimental bomber, the Tu-110 *Cooker* and Tu-124 *Cookpot* airliners, the Mikoyan/Gurevich SM-1, SM-2, SM-9, MiG-19 *Farmer*, Ye-5, Ye-6, Ye-7, I-1, I-7 and MiG-21F *Fishbed-A* fighters, the Sukhoi Su-7B *Fitter-A* fighter-bomber/Su-7U *Moujik* trainer, Sukhoi T-3, S-1, Su-9 *Fishpot-A* and Su-11 *Fishpot-B* interceptors, the Yakovlev Yak-25 *Flashlight-A*, Yak-27 *Flashlight-C/Mangrove* and Yak-28 *Brewer/Firebar* tactical aircraft family, the Yak-30 *Mantis* advanced trainer, the Ilyushin IL-40 attack aircraft and IL-46 tactical bomber, the Myasishchev 3MS/3MN *Bison-B* and 3MD *Bison-C* bombers which were flight-tested in the 1950s/early 1960s. Many of these aircraft entered production, seeing service with the Soviet Air Force and (in the case of the Tu-124) the Soviet airline, Aeroflot, in the 1960s.

Tu-4LL c/n 220204 with Dobrynin VD-4K development engines (Nos. 1 and 4)

220204

This retouched photo shows how the Tu-12LL testbed for the Bondaryuk RD-550 pulse-jet engine looked.

Tupolev Tu-12LL engine testbed

A single Tu-12 experimental bomber (a turbo-jet-powered, tricycle-gear derivative of the Tu-2) was converted into a testbed for the RD-550 ramjet engine. Developed by Mikhail M. Bondaryuk, the RD-550 was intended to power the LM-15 supersonic target drone designed by Matus R. Bisnovat. The development engine was mounted in an unconventional manner on a forward-swept pylon atop the fuselage. Designated Tu-12LL, the aircraft served with LII in 1949-51.

Tupolev Tu-14LL engine testbed

Similarly, a production Tu-14 *Bosun* twin-turbojet bomber was converted into the Tu-14LL testbed for the Bondaryuk RD-900 ramjet developed for the Lavochkin La-17 subsonic target drone. The tests took place in 1953-56.

Tupolev Tu-16LL engine testbeds (including 17LL-1)

LII's test aircraft fleet included nine engine testbeds called Tu-16LL. These modified *Badger-A*

bombers had the chin-mounted RBP-4 Rubin (*Short Horn*) navigation/bomb-aiming radar and all armament removed and carried the development engine in a special nacelle housed in the former bomb bay. The nacelle was semi-recessed for take-off/landing to provide adequate ground clearance and extended clear of the fuselage on parallel trailing arms by a hydraulic actuator before the engine was started. The nacelle featured an emergency jettison mechanism (in case it failed to retract before landing or the development engine caught fire). A hinged circular cover at the front of the bay closed the air intake when the nacelle was stowed to prevent windmilling and foreign object damage. As a rule, test equipment heat exchangers were mounted on the upper centre fuselage or on the development engine nacelle itself.

The first *Badger* adapted for engine testing was a bomber coded '03' (c/n 1880403) built in Kuibyshev in 1954; it was used to test the Lyul'ka AL-7F-1 afterburning turbojet developed for the Sukhoi Su-7B fighter-bomber. (According to some sources, though, '03' was a Voronezh-built aircraft, c/n 6401403.) The

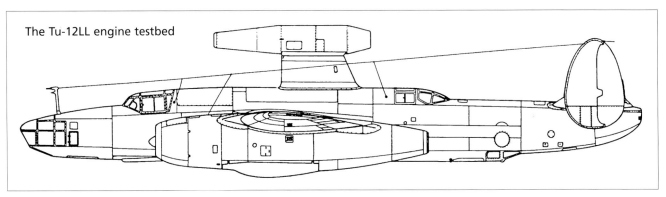

The Tu-12LL engine testbed

One of the Tu-16LL engine testbeds operated by LII – presumably '01 Red' (c/n 6401501) – with a large turbofan engine lowered into running position. Note the test equipment heat exchangers mounted directly on the development engine nacelle.

Another Tu-16LL, '01 Blue' (c/n 6401401; note the differently applied tactical code), with a very similar development engine nacelle – right down to the test equipment heat exchanger placement. Here, the test engine appears to be an afterburning turbofan, which required the rear end of the nacelle to be lengthened accordingly. Note that the Tu-16LLs had the chin-mounted radar removed – presumably to ensure a smoother airflow for the development engine.

The Tu-16LL's bomb bay, looking aft, showing the retractable pylon for the development engine nacelle and its actuation mechanism. Note also the fuel line and the many electrical connectors dangling from the bomb bay.

Another view inside the Tu-16LL's bomb bay, showing the development engine's air intake shutter in retracted position.

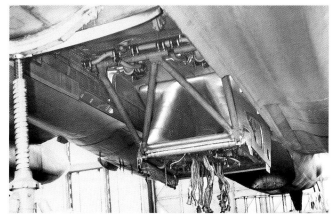

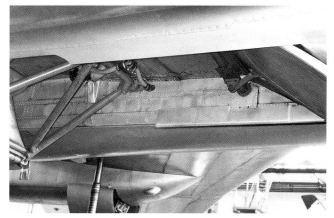

A Tu-16LL testbed is jacked up for maintenance inside LII's hangar, pending installation of the experimental engine.

Above left: The detached development engine nacelle of a Tu-16LL (minus rear fairing) on a dolly in one of LII's hangars, showing the various access panels and cooling air intakes. LII's An-12BK 'Tanker' de-icing systems testbed (CCCP-48974) is visible in the background.

Above: The detachable tail fairing of the development engine nacelle on a dolly alongside Tu-16LL '02 Blue' (c/n 4201002).

Left: The compressor face of a development engine installed in a fully extended nacelle. For ground runs the air intake was closed by a special foreign object damage (FOD) prevention grille.

A fighter engine (the convergent-divergent afterburner nozzle is clearly visible) is prepared for a ground run beneath a Tu-16LL. Note the tarpaulins covering the main gear bogies, the pan collecting any oil dripping from the test engine and the grille erected in front of it as a safety measure. Most of the access hatches have been left open to facilitate adjustments in case of need.

A Lyul'ka AL-7F runs in full afterburner during a ground test on the Tu-16LL.

other Tu-16LLs were Voronezh-built '01 Blue' (c/n 6401401), '08' (c/n 6401408), '41 Blue' (c/n 6401410) and '01 Red' (c/n 6401501), Kazan'-built '02 Blue' (c/n 4201002) and '05 Blue' (c/n 8204105), Kuibyshev-built '10 Red' (c/n 1881110) and an aircraft reported as 'No. 117' (which could be c/n 6203117, 8204117 or 1882117). About 30 Soviet second- and third-generation jet engine types were tested on these aircraft in the course of some 30 years. These included the Dobrynin VD-7M and VD-19 afterburning turbojets, Zoobets RD-16-15 and RD-16-17 turbojets, Kolesov RD36-41 afterburning turbojet and RD36-51 turbofan, Kuznetsov NK-8-2 and NK-8-4 turbofans, Solov'yov D-20P, D-30, D-30K and D-30KP turbofans, D-30F-6 afterburning turbofan, Tumanskiy R11AF-300,

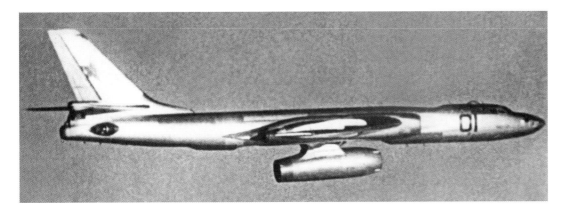

A Tu-16LL in action, probably with a Lyul'ka AL-7F afterburning turbojet. The development engine nacelle is identical to the one in the preceding views.

R11AF2-300, R13-300 and R15B-300 afterburning turbojets, Khachatoorov R27-300 afterburning turbofan, R27V-300 thrust-vectoring turbofan, R29-300 afterburning turbofan and R79V-300 thrust-vectoring afterburning turbofan, Lotarev D-36 turbofan, Lyul'ka AL-7F-2, AL-7F-4 and AL-21F turbojets, AL-31F and AL-41F afterburning turbofans, Ivchenko AI-25TL turbofan, Gavrilov R-95Sh turbofan and Aerosila TA-12 APU.

In the late 1950s and early 1960s two Tu-16LLs served as testbeds for the 13,500-kgp (29,760-lbst) Tumanskiy KR15-300 short-life afterburning turbojet developed for the Tupolev '121' supersonic GLCM and its production derivative – the Tu-123 reconnaissance drone, aka DBR-1 *Yastreb* (Hawk). This engine subsequently evolved into a long-life version – the R15-300 powering the MiG-25.

One of the Tu-16LLs was used in 1978 for testing the 7,000-kgp (15,430-lbst) RD36-51V turbofan intended for the Myasishchev M-17 *Mystic-A* high-altitude 'balloon killer' aircraft. At the Myasishchev OKB this aircraft bore a parallel designation, 17LL-1 – that is, testbed No. 1 under the M-17 programme.

In the 1970s and 1980s Tu-16LL '41 Blue' was modified for performing an extensive test and development programme on the 6,500-kgp (14,330-lbst) D-36 turbofan intended for the Antonov An-72 *Coaler* STOL light transport and the Yakovlev Yak-42 *Clobber* short-haul airliner; at LII this aircraft was known as the Tu-16LL-410. At this time, too, the Tu-16LL found use for testing full-size airframe assemblies together with their engines. For instance, '02 Blue' and '41 Blue' carried complete fuselages of the Yak-36M *Forger* shipboard vertical/short take-off and landing (V/STOL) attack aircraft incorporating a 6,900-kgp (15,210-lbst) R27V-300 lift/cruise engine and a pair of RD36-35V lift-jets; another aircraft reportedly carried a Yak-41 *Freestyle* V/STOL fighter fuselage with an R79V-300 lift/cruise engine rated at 10,500 kgp (23,150 lbst) dry and 15,500 kgp (34,170 lbst) reheat, plus two 4,100-kgp (9,040-lbst) Kolesov RD-41 lift-jets. '10 Blue' (called Tu-16LL-110 at LII) and '02 Blue' carried the fuselage of the Czechoslovak Aero L-39 Albatros advanced trainer with a 1,720-kgp (3,790-lbst) AI-25TL engine, and the like. These tests allowed the effect of the air intake design on the engine's operation to be studied and engine operation at various angles of attack to be verified. As in the case of the engines alone, the aircraft assemblies were

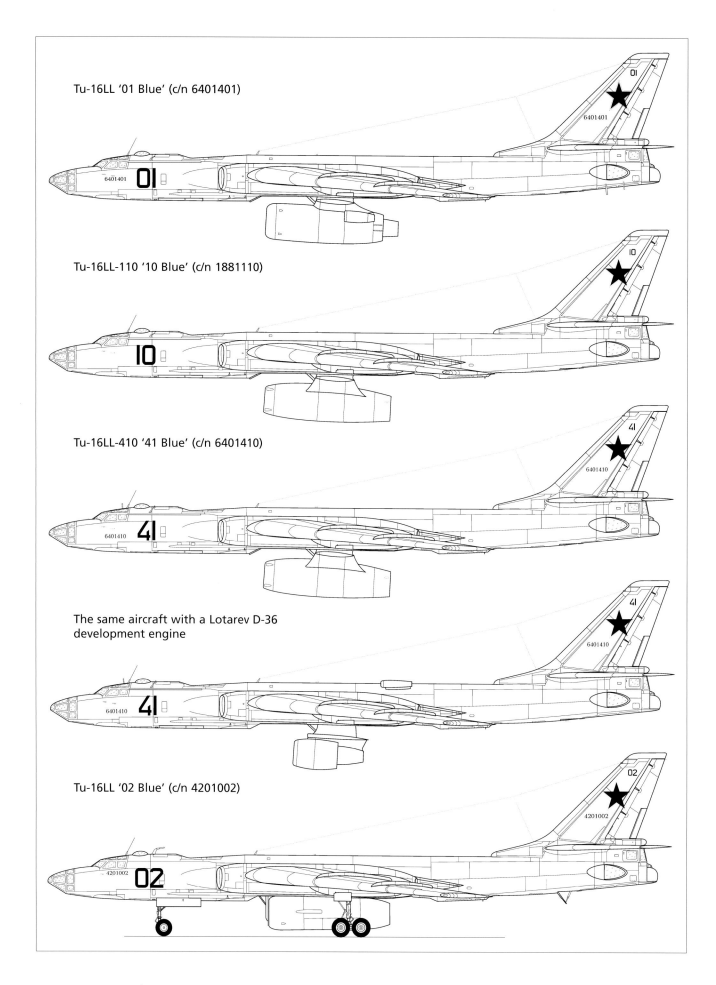

Tu-16LL '01 Blue' (c/n 6401401)

Tu-16LL-110 '10 Blue' (c/n 1881110)

Tu-16LL-410 '41 Blue' (c/n 6401410)

The same aircraft with a Lotarev D-36 development engine

Tu-16LL '02 Blue' (c/n 4201002)

Tu-16LL '01 Blue' (c/n 6401401) makes a low pass with an unidentified development engine in operating position during the Aviation Day flypast at Zhukovskiy on 16th August 1990. LII's airfield was still off limits to the general public then, so demo flights were staged over specially built public grandstands on the bank of the Moskva River.

The same aircraft comes in to land; the development engine is stowed and the intake shutter closed.

housed in the former bomb bay and lowered clear before engine starting.

In 1976-78 a Tu-16LL, sometimes referred to as LL-88, was used to test the Izotov RD-33 afterburning turbofan for the MiG-29 *Fulcrum* tactical fighter rated at 5,040 kgp (11,110 lbst) dry and 8,300 kgp (18,300 lbst) reheat.

Unfortunately the tests did not always end well. One Tu-16LL was lost when the test engineer mistook the glow of the setting sun on the development engine's nacelle for a fire and the entire crew baled out. Another example crashed on 1st February 1971, killing the crew of five captained by the famous test pilot Sultan Amet-Khan, Hero of the Soviet Union; the cause was never found.

By the turn of the century the remaining Tu-16LLs had been retired as time-expired. '41 Blue' was the last of the kind, remaining operational as late as 2001.

For ground runs of the development engine the Tu-16LLs were parked on elevated concrete ramps to increase the ground clearance, as illustrated by '41 Blue' (c/n 6401410). Note the open crew entry hatch.

At one stage of its career '41 Blue' carried a Lotarev D-36 turbofan, complete with a standard Yak-42 engine nacelle and pylon. Note the dorsally mounted test equipment heat exchangers.

Another excellent air-to-air of Tu-16LL '41 Blue' with the D-36 engine running. Note the lack of the dorsal and ventral cannon barbettes.

Tu-16LL '02 Blue' (c/n 4201002) with a stowed development engine sits in front of one of the purpose-built ramps at Zhukovskiy in the 1990s.

Two views of the same aircraft in an earlier config-uration with a different test engine taxies at Zhukovskiy, the nacelle's intake firmly closed by the shutter. Note the difference in the presentation and location of the tactical code and c/n which changed in the course of an overhaul.

Tu-16LL '01 Blue' (c/n 6401401) was part of a static display staged at Zhukovskiy in May 1991 on occasion of LII's 50th anniversary. Note that the design of the nacelle's intake shutter varied on individual aircraft, depending on the type of engine fitted (or, to be precise, its intake diameter).

Close-up of the last engine tested on Tu-16LL '01 Blue'.

Opposite: Tu-16LL '10 Blue' (c/n 1881110) sits on a rain-soaked hardstand. With no development engine installed, the door-less bomb bay creating a concave lower fuselage contour is readily apparent.

Another view of Tu-16LL '01 Blue'; the DK-7 tail barbette is replaced by a fairing.

Engines and other parts missing, time-expired Tu-16LL '05 Blue' (c/n 8204105) sits forlorn on a hardstand opposite LII's hangars. Note the development engine intake cover.

Above and opposite page, top: The end of the road for Tu-16LL '41 Blue'. The retired aircraft sits on the grass next to Tu-154LL CCCP-85024, awaiting disposal.

Another view of the retired Tu-16LL '05 Blue' after it had been towed to the grass area.

The development engine bay of Tu-16LL '41 Blue'; the pylon is in the extended position and the retracted intake cover is just visible.

A fine view of Tu-16LL '01 Blue' parked at Zhukovskiy. Though seemingly intact, this aircraft is also out of service.

The extended development engine pylon of Tu-16LL '01 Blue'.

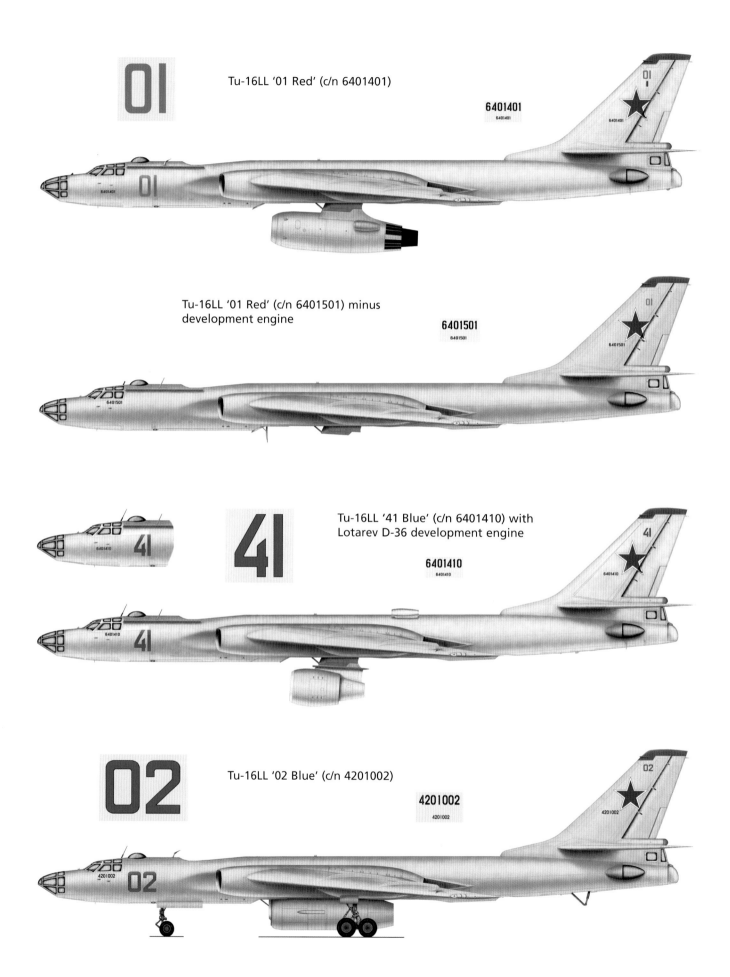

Tu-16LL '01 Red' (c/n 6401401)

Tu-16LL '01 Red' (c/n 6401501) minus development engine

Tu-16LL '41 Blue' (c/n 6401410) with Lotarev D-36 development engine

Tu-16LL '02 Blue' (c/n 4201002)

Tu-16LLs '41 Blue' (above) and '02 Blue' both carried a full-size Yak-36M fuselage for a while, being used for testing the *Forger*'s Tumanskiy R27V-300 lift/cruise engine and Kolesov RD36-35V lift-jet combination.

Another view of Tu-16LL '02 Blue' carrying the 'wish I had wings' Yak-36M powerplant test rig. Note that the RBP-4 radar is still in place.

Tu-16LL '02 Blue' (c/n 4201002) with a Yak-36M powerplant test rig

4201002

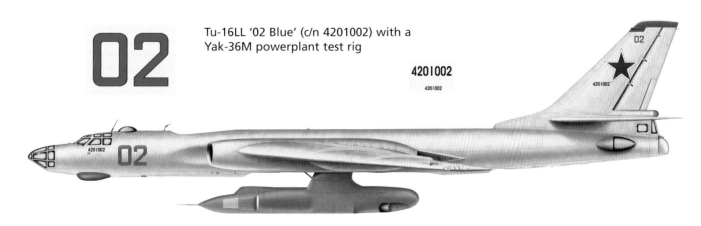

Tu-16LL '10 Blue' (c/n 1881110) was used for testing the Ivchenko AI-25TL turbofan optimised for the Aero L-39 trainer. In order to check the interaction between the engine and the lateral air intakes the engine was installed in a full-size L-39 fuse-lage, seen here in semi-recessed position. Note that the engine covers are marked '110', hence this air-craft's designa-tion Tu-16LL-110.

The same aircraft taxies out for a test mission.

Tu-16LL '10 Blue' with the L-39 fuselage lowered clear of the bomb bay and the test engine running.

238

Tupolev Tu-22M1 engine testbed

As the Tu-22M2 *Backfire-B* 'swing-wing' bomber/missile carrier entered service, improving the thrust/weight ratio became a priority task. However, the Kuznetsov NK-22 afterburning turbofan, an uprated version of the Tu-144's NK-144, had virtually no reserves for further uprating. The ultimate version, designated NK-23, produced 23,000 kgp (50,705 lbst) in full afterburner. A small experimental batch was built and the engine was test flown in a re-engined Tu-22M1 *Backfire-B*, which can thus be regarded as an engine testbed.

Tupolev Tu-22M2Ye engine testbed

OKB-276 Chief Designer Nikolay D. Kuznetsov chose not to go ahead with the NK-23 because he had a new design on the drawing boards. This was the NK-25 rated at 14,300 kgp (31,525 lbst) dry and 25,000 kgp (55,110 lbst) reheat. Unlike the two-spool NK-22, the NK-25 was a three-spool turbofan. It featured a new automatic engine control system based on state-of-the-art semiconductors and had a slightly lower specific fuel consumption at subsonic speeds.

In 1974 a standard Tu-22M2 was refitted with NK-25s for powerplant tests. This engine testbed was known as the Tu-22M2Ye because the engine's product code at the Kuznetsov OKB was *izdeliye* Ye (= E, the sixth letter of the Cyrillic alphabet.)

Concurrently, in 1975-76, the NK-25 was tested on the second Tu-142LL ('043 Black', see below). The engine and its automatic control system worked well but it soon became apparent that simply fitting the NK-25 to the *Backfire-B* was not enough. The new engine was not only more powerful but also heavier. The Tu-22M needed more refined aerodynamics and a weight-trimming diet if an improvement in performance was to be achieved. Tupolev engineers went back to the drawing board, and the result was the Tu-22M3.

Tupolev Tu-22M3 engine testbed

In the early 1980s a production Tu-22M3 *Backfire-C* delivered to LII (c/n 2145345, f/n 4504) was converted into a testbed for the Kuznetsov NK-32 afterburning turbofan rated at 9,400 kgp (30,840 lbst) dry and 25,000 kgp (55,115 lbst) reheat. Though originally developed for the Tu-160 *Blackjack* strategic bomber, this engine was also to power future versions of the *Backfire* – a plan which has since been put on hold.

Appropriately coded '4504 Black', the aircraft differed from standard Tu-22M3s in having four rows of auxiliary blow-in doors instead of three each side (that is, twelve doors instead of nine). The NK-32 could also be identified by its characteristic orange-coloured efflux in full 'burner indicating a high content of nitrous monoxide. Well, it was a military engine and had manners to match.

'4504 Black' took part in the flying display during MosAeroShow '92 (11-16th August 1992) and the MAKS-93 airshow (31st August – 5th September 1993) in Zhukovskiy; at MAKS-95 (22nd-27th August 1995) it was in the static display featuring MERs with 'iron bombs'. It was probably this aircraft that was misidentified by some Western sources as the 'Tu-22M4'; so far Russian sources have not confirmed the existence of such a version.

'4504 Black', the Tu-22M3 refitted with NK-32 engines, at Zhukovskiy with four MBD3-U9 external bomb racks.

Tupolev Tu-95LL engine testbed

When the need arose in the mid-1950s for a flying testbed for verifying new powerful jet engines with a thrust up to 25,000 kgp (55,115 lbst), the Tu-95 was a natural choice. It was large enough to accommodate the test instrumentation, test engineers and relevant mechanical systems; also, the tall undercarriage allowed even the bulkiest engines to be installed under the bomber's belly without any trouble.

Accordingly, on 29th July 1957 MAP issued an order requiring the Tupolev OKB to convert the second prototype Tu-95 into a testbed for the Kuznetsov NK-6 afterburning turbofan rated at 20,000 kgp (44,090 lbst) in full afterburner. This engine was initially regarded as the powerplant for the Tu-105 supersonic bomber (which eventually became the Tu-22 *Blinder*) and several other bomber projects which did not materialise, including the Tu-106 and Tu-135.

Conversion of the '95/2' was performed at the Kuibyshev aircraft factory No. 18 in 1957-58. Designated Tu-95LL, it was ready to fly during 1958.

The development engine was housed in a special nacelle mounted in the bomb bay on a hydraulically-powered trapeze mechanism. A retractable cover closed the engine's air intake to prevent windmilling when shut down. The nacelle was lowered clear of the fuselage before engine starting to avoid any influence of the boundary layer on engine operation (and to prevent the jet exhaust from damaging the fuselage skin). In an emergency the engine could be jettisoned; all piping and wiring would be automatically guillotined to facilitate separation of the nacelle from the bomber.

An automated fuel delivery system was provided; fuel for test engine came from the Tu-95LL's normal fuel tanks. The experimental engine had dual controls which could be operated either by the pilot from the centre control pedestal or by the test engineer from a special control panel. The test instrumentation suite measured 172 parameters at 371 control points.

The NK-6 was beset by serious problems and never reached production status, the Tu-105A (Tu-22) being powered by Dobrynin VD-7 afterburning turbojets instead. However, the ill-fated NK-6 was just one of six new engines to be tested on the Tu-95LL. Later, this aircraft was used for testing and refining the NK-144A afterburning turbofan for the Tu-144 SST, the NK-144-22 for the Tu-22M0 *Backfire-A* and the NK-22 for the Tu-22M1/Tu-22M2, all rated at 15,000 kgp (33,070 lbst) dry and 20,000 kgp (44,090 lbst) reheat.

The Tu-95LL remained in service with LII for some 15 years until it was damaged beyond repair in a major accident. After that, its mission was filled by similarly converted Tu-142 and Tu-142M ASW aircraft (see Tu-142LL).

Tupolev Tu-95LAL nuclear research aircraft

In the mid-1950s serious consideration was given in the USSR to the possibility of developing nuclear propulsion systems for aircraft – prompted, no doubt, by the encouraging experience with nuclear powerplants on surface ships, such as icebreakers. On 12th August 1955, the USSR Council of Ministers approved development of special facilities optimised for prototyping and developing nuclear powerplants for aircraft. The Nuclear Physics Institute headed by Igor' V. Kurchatov initiated studies of such powerplants; the actual design effort was supervised by Aleksandr P. Aleksandrov. Research proceeded on nuclear ramjets, turbojets and turbofans. Reactors with air and intermediate metal cooling were designed; thermal and fast nuclear reactors were studied in great detail. Importantly, particular attention was paid to the problems associated with biological exposure to radiation.

On its part, the Tupolev OKB proposed a twenty-year R&D programme that would lead to the development of a nuclear-powered military aircraft. Both subsonic and supersonic designs were proposed. On 28th March 1956 the Council of Ministers ordered the Tupolev OKB to initiate preliminary design of a nuclear research aircraft based on the Tu-95 bomber. This aircraft would be used to study the effects of radiation on aircraft instrumentation and equipment, as well as on the aircrew and ground support personnel. (A similar testbed,

Below: A ground test rig built to check the placement of the nuclear reactor and its ancillary systems on the Tu-95LAL testbed.

Opposite page: Three views of the Tu-95LAL ('51 Red') as an instructional airframe in Irkutsk, showing the dorsal fairing over the nuclear reactor, its ventral water radiator and the associated coolant tanks under the outer wings.

Top: Close-up of the Tu-95LAL's centre fuselage where the reactor was housed.

The lowered pallet mounting the reactor; the water radiator was mounted on this pallet.

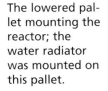

the Convair NB-36H Crusader, had been built in the USA, making 47 flights between 17th September 1955 and 28th March 1957.)

Work on adapting the Tu-95 to the nuclear research role was conducted by the OKB's Tomilino branch headed by Iosif F. Nezval', a very experienced Tupolev principal. The bomber's centre fuselage was redesigned to accommodate the support structure for the reactor. Completely new materials were utilised for the latter, requiring the mastery of totally new manufacturing techniques. Working in close co-operation with chemical industry experts, Tupolev's non-metal technology department under Abram S. Faïnshtein created a structure that was both strong and highly resistant to the effects of nuclear radiation.

During 1958 this support structure was completed and delivered to Semipalatinsk for mating with the reactor. Shortly afterwards, the first airworthy Soviet nuclear reactor (the VVRL-100) was completed and declared ready for shipping, also to Semipalatinsk. The reactor was tested in an underground nuclear facility during late 1959 and then declared ready for the proposed flight test programme.

A production Tu-95M coded '51 Red' (c/n 7800408) was set aside for conversion to meet order No. 247 and, following modifica-

tion during 1961, became the Tu-95LAL research aircraft (*letayushchaya ahtomnaya laboratoriya* – flying atomic laboratory). External recognition features included a large flat dorsal fairing over the reactor immediately aft of the wings, a large ventral fairing with an air scoop further aft at the rear end of the former bomb bay and two large coolant tanks for the reactor under the outer wings. The tanks were teardrop-shaped and adhered directly to the wing undersurface. There was also a small blister fairing on the starboard side of the fin near the top, probably housing a ciné camera.

Following conversion and initial flight tests to determine the effect of the new fairings on the aircraft's aerodynamics, the Tu-95LAL had its reactor installed. From May to August 1961 the aircraft flew 34 test missions related to nuclear propulsion in flight. Flights with the reactor both 'hot' (operational) and 'cold' (shut down) were made. The crew and researchers sat in the forward fuselage in a pressurised compartment. A sandwich-type bulkhead made of lead, polyethylene and ceresin (a special compound) served as a radiation shield for the crew. A radiation level sensor was installed in the bomb bay in close proximity to the reactor. Another sensor was installed in the rear cabin aft of the bomb bay; two more sensors were mounted in the outer wing panels.

The reactor itself was mounted in the bomb bay. It was suspended in a water jacket for cooling purposes and had a coating of lead and resin to reduce hazardous emissions. The water radiator and associated plumbing for the cooling system were located in the above-mentioned ventral fairing. The reactor's control system was hooked up to a test engineer's panel positioned in the forward compartment just aft of the flight deck.

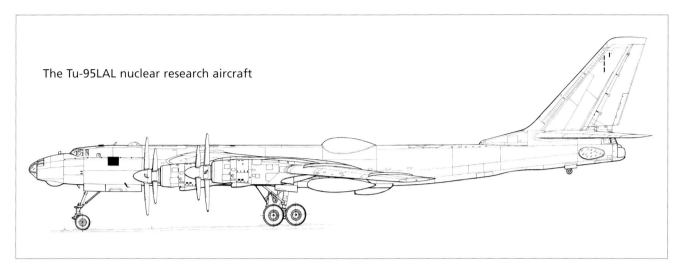

The Tu-95LAL nuclear research aircraft

The Tu-95LAL had a TOW of 145 tons (319,660 lb) and a landing weight of 110 tons (242,500 lb). The nuclear powerplant weighed 33.9 tons (74,735 lb) complete with radiation barriers and support systems. Range was 4,700 km (2,919 miles) and cruising speed 750-800 km/h (465-497 mph). Normal flight duration was 6.4 hours.

The test programme quickly proved the efficiency of the radiation barrier system. Exposure rates for the crew members were extremely low. This gave rise to confidence in the ability of humans to work in close proximity to nuclear reactors.

However, the viability of the nuclear-powered aircraft concept was questionable from an economic standpoint. When it was concluded that the entire national budget for two years would be consumed by the project (!), the Powers That Be decided to terminate the effort and pursue other avenues of bomber development.

Other issues contributing to the programme's demise included the unwieldy nature of the radiation protection systems and associated shielding materials, and the fact the USA had terminated its Aircraft Nuclear Program in 1961 (thus eliminating it as a technology threat). The Tu-95LAL was retired and served as a ground instructional airframe at the Irkutsk Higher Military Aviation Technical School (IVATU – *Irkootskoye vyssheye aviatsionno-tekhnicheskoye oochilishche*) until finally scrapped.

Tupolev Tu-128LL engine/avionics testbed

In 1965 the Tu-128LL development aircraft, a modified *Fiddler* heavy interceptor, entered flight test at Zhukovskiy; this was to be a com-bined engine/avionics testbed. The aircraft was powered by two 13,000-kgp Dobrynin VD-19 afterburning turbojets (this model based on the earlier VD-7 was the precursor of the RD-19R2), with appropriate changes to the air intakes, inlet ducts and rear fuselage structure. It also featured a recontoured radome for the new Smerch-A radar and two strakes under the rear fuselage; the radar itself was still unavailable and had to be substituted by ballast. In effect, the Tu-128LL was to be the prototype of the advanced Tu-28A version forming the core of a new aerial intercept weapons system.

The tests, however, gave disappointing results. The maximum speed in 'clean' configuration (with no missiles) was only 110-120 km/h (68-74.5 mph) higher as compared to the standard aircraft. That said, reaching the 2,000-km/h (1,240-mph) speed target with four R-4 (AA-5 *Ash*) missiles was obviously impossible. The reason for this shortfall in performance was the Tu-128LL's greater fuselage cross-section area due to the new engines' larger inlet ducts, which caused a dramatic increase in drag.

Tupolev Tu-134A-1510 multi-role testbed

In 1987 LII's Tu-134A CCCP-65740 (c/n 2351510), a former Tu-134A 'Salon' VIP aircraft obtained from the Air Force, was converted into a multi-role testbed/research aircraft known as the Tu-134A-1510. Its primary mission was to test small expendable or short-life turbofan engines developed for cruise missiles and unmanned aerial vehicles, such as the 350-kgp (770-lbst) Tumanskiy RDK-300 and the 450-kgp (990-lbst) Saturn Model 36MT turbofan developed for the Kh-59M air-to-sur-

Two views of the first Tu-142LL engine testbed (c/n 4200) during take-off, showing the 12-wheel main gear bogies. The development engine is semi-recessed in the bomb bay.

The first Tu-142LL engine testbed with the development engine running.

The first Tu-142LL during final approach to Zhukovskiy.

face missile. Concurrently, however – or rather between engine tests – the Tu-134A-1510 was used for aerothermophysical and environmental research, such as measuring the heat signature and air pollution levels in the wake of other aircraft (more on this in Chapter 8).

In accordance with its engine testbed role the aircraft gained a large flat-bottomed bulge under the wing centre section which mounted three pylons of different shape. The forward-swept port pylon carried the development engine in a small nacelle, with a sensor shaped like an inverted T immediately aft, while the centre pylon carried a cigar-shaped pod housing unknown equipment.

A hemispherical fairing enclosing heat sensors was mounted dorsally just ahead of the entry door. The tenth cabin window on each side (the one between the emergency exits) had the glazing replaced by emergency

decompression valves with air outlet grilles. These allowed the cabin to be depressurised within two or three seconds prior to bailing out, should the need arise; otherwise the pressure differential would make it impossible to open the doors or rear baggage compartment door, all of which open inwards.

The aircraft's equipment fit and hence appearance changed over the years. By 1992 the original development engine had been replaced by a different model in a larger nacelle (type unknown in both cases), while the cigar-shaped pod gave place to two shorter pods mounted side by side, each of which had a hemispherical front end. A small sensor turret was added at the front of the ventral platform.

CCCP-65740 was usually captained by Merited Test Pilot Yuriy Pavlovich Sheffer, one of LII's best pilots. The engine tests took place between 1987 and 1996.

Tupolev Tu-142LL engine testbed

When the Tu-95LL engine testbed was written off in an accident, the need for a replacement arose. Hence in the early 1970s the first prototype Tu-142 *Bear-F* anti-submarine warfare aircraft (c/n 4200) with 12-wheel main landing gear bogies was converted into an engine testbed designated Tu-142LL for testing high-power turbofan engines. Stripped of all ASW gear and armament, the aircraft was fitted with appropriate test equipment and the development engine mounting trapeze which had been removed from the Tu-95LL.

The first Tu-142LL was used to test three afterburning turbofans: the Kuznetsov NK-25 for the Tu-22M3, the Kolesov RD36-51A for the Tu-144D and the NK-32 for the Tu-160. The chin fairing of the *Gagara* (Guillemot) thermal imager characteristic of the Tu-142 *sans suffixe* was originally retained but removed at a later stage, resulting in a Tu-142M-style nose contour. The aircraft was used until the mid-1980s when it was written off as time-expired.

Later, a second *Bear-F* took over the engine testbed role; this was the first Tu-142MK *Bear-F Mod* ('043 Black', c/n 4243). Interestingly, this aircraft reversed the sequence with the first Tu-142LL by having a chin-mounted weather radar added (looking rather like the RBP-4 radar of the Tu-95 *Bear-A*). In 1990 the Tu-142LL set several time-to-height and sustained altitude records in its class. The aircraft was still based at Zhukovskiy in the mid-1990s but was struck off charge soon afterwards.

The second Tu-142LL (c/n 4243); the four-wheel main gear bogies and the raised flight deck roof are visible.

Tu-142LL c/n 4243 parked at Zhukovskiy over a trench with a jet blast deflector pipe at the rear. This allows the development engine to be run on the ground. Note the traces of the MAD 'stinger' on the fin.

Tupolev Tu-155 cryogenic fuel technology demonstrator

In the mid-1970s, soaring oil prices and the global energy crisis forced scientists to step up research into alternative fuels for industrial applications and transportation. Liquefied natural gas (LNG) and liquid hydrogen (LH2) became prime candidates for the aviation fuel of the future. LNG had a number of valuable advantages over traditional kerosene fuels. The global reserves of natural gas were more abundant than those of oil, which meant that the price of LNG fuel would go down as the oil reserves shrank and the oil prices rose; here it should be noted that the Soviet Union pos-

The second Tu-142LL was retrofitted with a weather radar in a chin radome. The IFR probe, on the other hand, was removed, leaving only a stub.

Close-up of the development engine pod on the second Tu-142LL. Note the shutter closing the turbofan's air intake; it retracts into a special fairing before the development engine is lowered and started up.

The LII crew that established the records in the Tu-142LL in 1990 poses for a photo at Zhukovskiy after the flight.

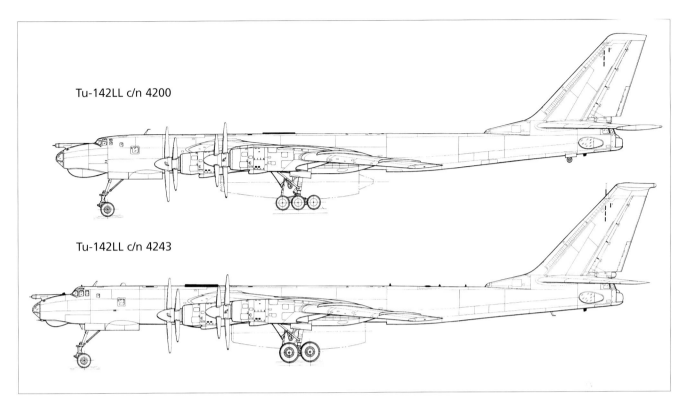

Tu-142LL c/n 4200

Tu-142LL c/n 4243

sessed large supplies of natural gas. Also, the calorific value of natural gas is 15% greater than that of kerosene; last but not least, the toxic emissions are much lower (for example, unlike oil, natural gas has no sulphur content and produces no corrosive sulphurous oxides).

LH2 held even greater promise; it had a specific calorific value three times greater than fossil fuel and was ecologically 'clean', to say nothing of the fact that, unlike fossil fuel, it was replenishable. Therefore the Soviet Academy of Sciences joined forces with a number of industry research and development establishments and design bureaux to work out a programme envisaging large-scale introduction of hydrogen fuel in the national economy. Implementation of this programme was to bring about a marked improvement of the nation's ecology and lay the foundations for the development of future hypersonic aircraft and aerospaceplanes.

In the aircraft industry the programme was code-named **Kho**lod (Cold) as a reference to cryogenic fuels (LNG and LH2). First, in 1974 the Myasishchev design bureau (OKB-23) started work on a hypersonic aerospaceplane with a cryogenic powerplant known as 'project 19'. As a first step towards the aerospaceplane, OKB-23 was to create a propulsion technology testbed designated IL-76Kh (the suffix letter referring to the Kholod-2 programme) – a heavily modified IL-76 transport with a single experimental cryogenic-fuel tur-

bofan and heat-insulated LH2 tanks housed in the freight hold. Later, the Powers That Be intervened; when development was well advanced the Myasishchev OKB was ordered to transfer its know-how to the Tupolev OKB.

Designated Tu-155, the cryogenic fuel technology demonstrator was converted from Tu-154 *sans suffixe* CCCP-85035, c/n 72A035). The extensive conversion involved a whole raft of major structural and external changes. The rear fuselage section between the overwing emergency exits and the rear pressure dome was remanufactured as a windowless bay housing a heavily insulated cryogenic fuel tank with a capacity of 20 m³ (706.29 cu ft); this could hold LH2 chilled to –253°C (–423°F) or LNG chilled to –162°C (–259°F). This installation was necessary because LH2 has a four times lower specific gravity than kerosene and the fuel tanks need to be pressurised to stop the LH2 or LNG from boiling out as the aircraft climbs to high altitude, which makes it impossible to house them in the wings.

The starboard NK-8-2 engine was replaced by an NK-88 turbofan rated at 10,500 kgp (23,150 lbst) – an experimental derivative of the NK-8-2U adapted to run on cryogenic fuels; Chief Designer V. N. Orlov was responsible for this engine. The basic design changes included installation of a cryogenic turbine pump (a two-stage centrifugal pump driven by engine bleed air); it was housed in a special box on top of the engine casing ventilated by

The logo of SNTK Trood (the Kuznetsov OKB) whose engines were tested on the Tu-142LL. The legend reads 'NK engines – high reliability'.

The unique Tu-155 cryogenic fuel technology demonstrator.

Part of the Tu-155's special cryogenic refuelling complex is seen in the foreground.

The faired cryogenic fuel jettison valve at the top of the Tu-155's tail.

air under pressure to prevent accumulation of explosive fuel vapours. An annular heat exchanger was installed downstream of the low-pressure turbine (at the location formerly occupied by the thrust reverser) to convert the fuel to gaseous state before it was fed into the combustion chamber; no attempt was made to feed the hydrogen to the engine in liquid form because vaporisation in the fuel lines made it extremely difficult to maintain a stable flow and regulate combustion. Bench tests of the NK-88 began in February 1980.

The aircraft's fuel system was considerably redesigned to include a cryogenic fuel delivery system, a cryogenic tank pressure monitoring system activating the safety valves if necessary, a recirculation system, a cryogenic tank pressurisation system and an emergency fuel jettison system. The delivery system featured centrifugal and jet (transfer) pumps in the cryogenic tank, heat-insulated fuel lines, a fuel flow regulator and valves.

Three additional systems were installed to operate and monitor the cryogenic fuel complex. These were: a helium system controlling the experimental powerplant units; a nitrogen system which replaced the air in the cryogenic tank bay and fuel system compartments with nitrogen and alerted the crew about any cryogenic fuel leak well before it reached an explosive concentration; and a system monitoring the vacuum in the cryogenic tank's heat insulation. The cylindrical helium bottles were installed ten-abreast on the floor in the centre portion of the former forward cabin, while the spherical nitrogen bottles were mounted

CCP-85035

above the cabin windows in the same area (in lieu of the overhead baggage racks) and in the forward baggage compartment. Test and recording equipment racks were located fore and aft of this 'bottle bay'.

The flight deck was extended aft to accommodate three test engineers' workstations facing the starboard side, and a further workstation was located near the rear entry door, with more test equipment between it and the bulkhead separating the manned cabin from the cryogenic fuel tank bay. An extra large complement of fire extinguishers was installed in the unpressurised rear fuselage near the experimental engine.

The Tu-155's non-standard external features included a small fin-like excrescence atop the vertical tail near the leading edge which incorporated twin (normal and emergency) fuel vapour vents; the drains to these vents were routed along the starboard side and top of the centre engine air intake trunk and the fin leading edge. This required the usual spike housing the radio antenna to be deleted and the upper anti-collision light to be relocated to the fuselage amidships.

Three large conduits enclosing pipelines (one to port and two to starboard) ran along the rear fuselage sides from the pair of windows immediately ahead of the emergency exits. The flattened upper conduits associated with the cryogenic tank bay pressurisation system were located symmetrically above the windows, terminating in line with the fin/centre engine air intake trunk junction. The starboard lower conduit, which housed the heat-

Another view of the Tu-155 stored at the Tupolev JSC's flight test facility.

Parts of the experimental cryogenic refuelling complex at Zhukovskiy.

The experimental NK-88 engine in the Tu-155's starboard nacelle. Note the fairing enclosing the fuel pump and the lack of the thrust reverser.

The pilots' work-stations in the Tu-155. Note the orange-painted throttle knob of the NK-88 engine.

Overall view of the Tu-155's flight deck, with test engineer workstations to starboard. Note the dished seat pans to accommodate parachutes – just in case.

insulated fuel line and was fatter than the others, ran below the windows in such a way as to obstruct the emergency exits (which no one would use anyway) and then curved upwards to mate with the engine pylon. Additionally, two thinner conduits ran from the port wing/fuselage fairing and along the rear fuse-lage underside (offset to port) from the wing centre section to a point in line with the fin leading edge. A fuel jettison pipe curved gently from a ventral fairing offset to starboard to the starboard side of the centre engine nozzle. Finally, the starboard engine nacelle was considerably modified, with no thrust reverser and

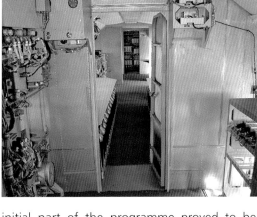

Far left: The forward test equipment bay aft of the flight deck.

The Tu-155's front cabin, looking forward, with helium bottles on the floor and nitrogen bottles under the ceiling.

More test equipment racks aft of the 'bottle bay'.

a boxy fairing with streamlined front and rear ends (offset to starboard) over the NK-88's cryogenic turbine pump housing.

A special cryogenic refuelling installation was built at the OKB's flight test facility in Zhukovskiy for servicing the Tu-155 with LH2 or LNG during the tests. The installation allowed tests involving large quantities of cryogenic fuel to be performed and included the following systems:

• a cryogenic fuelling system utilising special refuelling vehicles based on the KrAZ-257B 6x4 conventional lorry;
• a compressed nitrogen supply;
• an electrical power supply,
• a television monitoring system;
• a gas analysis system;
• a fire-precaution sprinkler system;
• a cryogenic fuel quality monitoring system;
• a vacuum pump vehicle;
• a command and control centre located in an underground bunker to ensure safety in the event of a fire or explosion on the hardstand;
• mobile communications posts.

Because the OKB was breaking new ground and a lot of research was involved, the

initial part of the programme proved to be quite protracted. The Tu-155 made its first flight – the historic first-ever flight of a hydrogen-fuelled aircraft – on 15th April 1988; the aircraft was captained by Merited Test Pilot Vladimir A. Sevan'kayev. This was the beginning of an extensive test programme. Initially liquid hydrogen was used as the cryogenic fuel; in January 1989 the Tu-155 began a new stage of the tests with a modified NK-88 running on liquefied natural gas. The Tu-155's performance was basically similar to that of

A rare shot of the Tu-155 in flight.

251

the Tu-154B; the supply of cryogenic fuel in the 20-m³ tank was sufficient to give an endurance of two hours.

In the course of the tests the Tu-155 established 14 world records. It also made two visits abroad – to Nice (with a stopover in Bratislava) and to Hannover. At the latter location the aircraft participated in the ILA'90 airshow, becoming one of the stars of the show. As a result, Deutsche Airbus (the German division of the Airbus Industrie international consortium) struck a deal with the Tupolev OKB envisaging co-operation in the development of hydrogen-fuelled aircraft, including a version of the A300B4 wide-body airliner. On 11-16th August 1992 the aircraft was displayed statically at the MosAeroShow '92. After that, the aircraft was placed in storage and is still parked on its own hardstand in Zhukovskiy.

The advent of the Tu-155 aroused immense interest in the world's aviation community and aviation press. Inevitably, there were detractors who questioned the safety of cryogenic fuels and dismissed the idea.

The development and testing of the Tu-155 yielded invaluable practical material for further development of cryogenic-fuel aircraft, paving the way for such projects as the Tu-156, Tu-206, Tu-130SPG, Tu-330SPG (**szhizh**ennyy pri**rod**nyy ghaz – LNG) and Tupolev C-Prop. Soviet aircraft designers gained practical experience of operating and testing such aircraft and evolved new testing techniques, as well as fire safety procedures for cryogenic-fuel aircraft.

Tupolev Tu-204-100V engine testbed

As mentioned earlier, the performance and reliability of the PS-90A turbofan left a lot to be desired. Hence, enlisting the help of Pratt & Whitney, the Perm'-based Aviadvigatel' JSC design bureau developed a refined version designated PS-90A2. The new version is to power the Tu-204SM which is to supersede the Tu-204-100 on the production line at the Aviastar-SP factory in Ul'yanovsk.

In order to test the new engine, in May 2009 Aviadvigatel' struck a deal with the Moscow-based charter carrier Red Wings allowing the engine maker to use one of the airline's aircraft, Tu-204-100V-04 RA-64048 (c/n 1450741964048), as a testbed with minimum modifications. The aircraft, which was a 'logojet' wearing an orange overall advertising livery for the travel agency Coral Tour and was undergoing maintenance at Aviastar, had the port engine replaced with a PS-90A-2, making 18 test flights from Ul'yanovsk-Vostochnyy. In the course of the tests the development engine logged a total of 83 hrs 47 min. The flights were conducted by Tupolev JSC test pilots, with the participation of specialists from Aviastar, GosNII GA, Aviadvigatel' and Permskiye Motory (the production plant manufacturing the PS-90A). The trials were successfully completed on 19th November the PS-90A-2 received its type certificate awarded by the CIS Interstate Aviation Committee, and production began in April 2010.

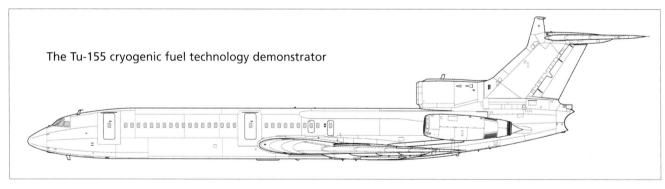

The Tu-155 cryogenic fuel technology demonstrator

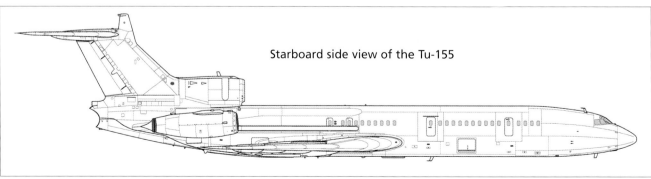

Starboard side view of the Tu-155

This Yak-7B was used as a testbed for DM-4S ramjet boosters (it is often called Yak-7PVRD). Note the non-standard rear canopy glazing.

As for the aircraft, Red Wings chose not to take it back. RA-64048 was reconverted to standard configuration and sold to the North Korean airline Air Koryo, becoming P-633.

Yakovlev Yak-7B propulsion testbeds with DM-4S boosters

One of the unusual tasks tackled by the Yakovlev OKB in the early phase of the Great Patriotic War was the construction of a flying testbed intended for flight development of the experimental Merkoolov DM-4S ramjet booster. The DM-4S had a diameter of 500 mm (1 ft 7^{11}⁄$_{16}$ in), was 2,300 mm (7 ft 6½ in) long and weighed 45 kg (99 lb). Two such engines were suspended under the fighter's outer wings; they ran on gasoline fed from the machine's standard fuel tanks.

The DM-4S was designed in 1941; this was the version that was flight-tested on the I-15*bis* and I-153 fighters before the war. After a period of bench tests two such engines manufactured at the beginning of 1942 were mounted under the wings of a Yakovlev UTI-26 (Yak-7) trainer operated by the 12th Guards IAP. In the spring of 1942 the Soviet Army began a counteroffensive and the frontlines rolled back some distance to the west from Moscow; this made it possible to perform test flights without the risk of being intercepted by Luftwaffe fighters.

The testing proved abortive: installation of the engines led to a significant forward shift of the CG, entailing the risk of the aircraft nosing over. Besides, no special measures were taken to prevent a fire onboard, and vibrations created by the ramjets' operation caused frequent leakage of fuel from the tanks. The flights had to be suspended for almost two years while corrective measures were taken.

In the middle of the war information started coming in about experiments with auxiliary powerplants conducted both by Soviet Union's allies and the Axis powers. For example, the British performed many experiments with ship-borne aircraft – such as Supermarine Spitfire and Hawker Hurricane fighters carried by CAM (Catapult Armed Merchant) ships, while the Germans made use of jettisonable solid-fuel rocket boosters to assist the take-off of overweight bombers from small field airstrips.

In early 1944 certain alterations were introduced into the design of the DM-4S engine with a view to enhancing its reliability. The engines were mounted under the wings of a production Yak-7B; the rear cockpit was outfitted as a workstation for the test engineer. The tests were conducted at LII NKAP, starting on 24th March 1944; on 15th May project test pilot Sergey N. Anokhin switched on the ramjets for the first time.

The boosters guzzled up to 20 kg (44 lb) of gasoline per minute; therefore, the auxiliary powerplant could be put into operation only for shorts spells of time – for example, to shake off an attacking adversary or to chase the target. The flights revealed that the boost-

Close-up of the DM-4 boosters fitted to the Yak-7PVRD.

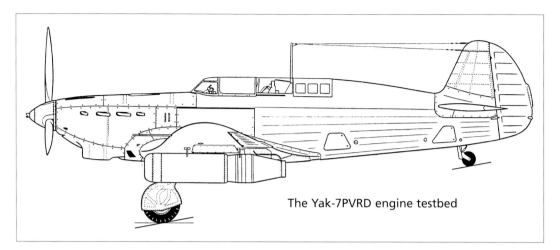

The Yak-7PVRD engine testbed

ers gave an increase of maximum speed amounting to 53 km/h (32.9 mph) versus 22 km/h (13.6 mph) for the I-15*bis*DM and 42 km/h (26.0 mph) for the I-153DM. However, owing to the considerable additional drag of the wings and the massive hollow cylinders attached to them the net speed increase did not exceed 19 km/h (11.8 mph).

These results were found to be disappointing. The work went on until the end of 1944. It encompassed the development of boosted modes of ramjet operation, installation of the ramjets in an aerodynamically refined airframe, making the booster engines easily detachable and so on. Calculations showed that a maximum speed increase could be as high as 100 km/h (62 mph) in the course of ten seconds, but it proved impossible to corroborate these estimates by actual test results.

Yakovlev Yak-3RD experimental fighter

In December 1944 an experimental fighter featuring a mixed powerplant was created on the basis of the Yak-3. In addition to the standard Klimov VK-105PF-2 engine, the machine was equipped with an RD-1 liquid-fuel rocket motor designed by Valentin P. Glushko; it was intended to serve as a booster.

Aleksandr S. Yakovlev entrusted engineer B. Motorin with the design of the unorthodox machine. Motorin installed the RD-1 rated at 300 kgp (661 lbst) in the aft fuselage under the vertical tail on a special mount and enclosed it with an easily detachable cowling faired into the fuselage contours.

In the course of manufacturer's flight test pilot Viktor Rastorgooyev performed 21 flights on this aircraft designated Yak-3RD (*raketnyy dvigatel'* – rocket motor). In eight of these flights he ignited the RD-1. On one occasion,

on 11th May 1945, the aircraft attained a speed of 782km/h (486 mph) at the altitude of 7,800 m (25,584 ft), which was 182 km higher than the speed attained at this altitude by the standard Yak-3 without the rocket booster. However, the automatic devices of the rocket engine were faulty and there were several cases of the engine cutting unexpectedly in flight. Once an incident occurred in which the engine nozzle was severely damaged.

The faults were rectified and repairs were made, whereupon preparations were started for the aircraft to take part in the air display on occasion of the 1945 Air Fleet Day. However, two days before the event, on 16th August, the aircraft suddenly entered a steep dive while performing a scheduled test flight and crashed, killing Rastorgooyev. The cause of the accident was not established, but it is known with certainty that there was no explosion of the rocket booster; neither did the engine disintegrate in flight. Apparently, there was a failure in the control system linkage.

In connection with the emergence of turbojet engines in the post-war years the subject of fitting liquid-fuel rocket boosters to piston-engined aircraft became moot and the work was discontinued.

Yakovlev Yak-3 propeller testbed

A single Yak-3 fighter was fitted experimentally with a three-blade propeller having scimitar-shaped blades. Unfortunately no details are known.

Yakovlev Yak-18T as a ground engine test rig

A time-expired Yak-18T four-seat light aircraft registered CCCP-44310 (c/n 72001313; it was actually painted on this way instead of

The Yak-3R mixed-power fighter. The rocket booster is housed in a conical fairing below the cropped rudder. The aircraft is carefully polished to cut drag and wears no insignia.

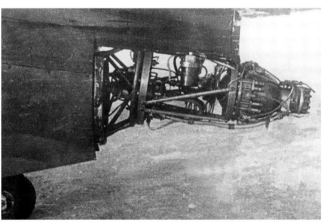

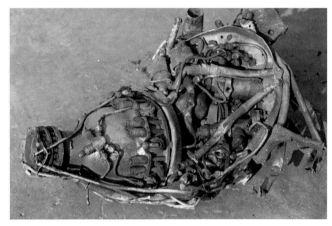

7201313) was converted into a ground test rig for testing the powerplant of the Molniya-1 light executive aircraft. This unusual machine designed by NPO **Mol**niya ('Lightning' Scientific & Production Association) was powered by a Vedeneyev M-14PM nine-cylinder radial driving a three-blade pusher propeller manufactured by the West German company MT-Propeller.

The test rig was intended to simulate the operating conditions of the pusher installation, in particular with regard to the cooling. To this end the aircraft, minus outer wing panels, was fitted with a rather bulky air duct and an electrically-powered centrifugal blower; the latter sent a stream of cooling air through the engine in the reverse direction during engine runs on the ground. A test equipment suite accommodated in a ZiL-131N 6x6 lorry with a van body was hooked up to the aircraft. The installation could be seen at Moscow-Tushino in May 1992.

Above left: The RD-1KhZ liquid-fuel rocket motor with the cowling removed.

Above: The RD-1KhZ recovered from the wreckage of the Yak-3RD after the crash on 16th August 1945.

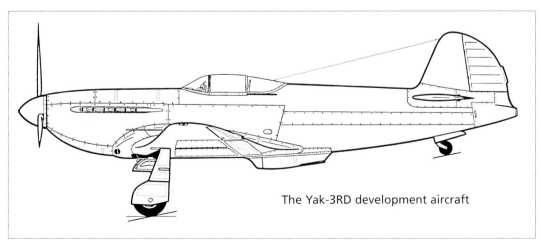

The Yak-3RD development aircraft

This Yak-3 was fitted with an experimental propeller having scimitar-shaped blades.

Yakovlev Yak-18T propeller testbed

The static park of the MAKS-97 airshow (22nd-27th August 1997) included a Yak-18T four-seat cabin monoplane placed on the Aviation Enthusiasts' Federation of Russia register as 01001ФЛА РФ (that is, 01001FLA RF, ex-ФЛА СССР-01001). The aircraft had a non-standard propeller whose blades featured unusually shaped tips, resembling an ancient Nepalese sword called *ram dao*. This example (presumably a testbed) had previously flown with a standard V530TA-D12 propeller. No details are available. The new propeller appears to have remained experimental. The

aircraft was later used by LII in a further research programme, receiving the registration RF-051 (see Chapter 4).

Yakovlev Yak-25M engine testbeds

Several Yak-25M interceptors were converted into engine testbeds. Little is known about these aircraft, except that the development engines were mostly those envisaged for future versions of the *Flashlight*.

Yakovlev Yak-42E (Yak-42LL, Yak-42E-LL) engine testbed

An early-production Saratov-built Yak-42 airliner (CCCP-42525, c/n 11030703) was converted by LII into a testbed for flight-testing the Lotarev D-236T geared propfan. This engine, which had previously been tested on the IL-76LL4 testbed (CCCP-76529, see above), was intended both for the Antonov An-70 transport (which has materialised) and for projected rear-engined propfan airliners, such as the An-180 and Yak-46 (which have not). The latter was to have been a twin-engined derivative of the Yak-42.

The test engine rated at 10,850 shp was mounted in place of the No. 3 (right-hand) Lotarev D-36 turbofan, driving SV-36 contra-rotating tractor propellers of 4.2 m (13 ft 9⅜ in) diameter. To ensure the necessary blade tip clearance, the D-236T was mounted on a new and longer pylon.

The aircraft first flew on 15th March 1991; three months later it was exhibited at the 39th Paris Air Show as the Yak-42E-LL. By August 1997 CCCP-42525 had been withdrawn from use at Zhukovskiy.

The Yak-42LL in an early test flight; the nacelle of the development engine is still unpainted.

4 Avionics testbeds

One of the largest groups of testbeds includes aircraft modified for testing new avionics items and suites, such as navigation systems, targeting systems, electronic intelligence (ELINT) systems and so on. Such aircraft are usually easy to identify by the tell-tale 'bumps and bulges' – non-standard radomes and fairings.

Aero L-39C navigation system testbed

A Czechoslovak-built Aero L-39C Albatros advanced trainer in Russian Air Force colours coded '56 Yellow' (c/n 43 30 26) was modified by LII for testing navigation systems. The aircraft featured an experimental navigation and flight instrumentation suite developed by the private company Rooskaya Avionika (Russian Avionics), including the SOI-30-1 data presentation system (*sistema otobrazheniya informahtsii*). Several experimental inertial and satellite navigation (SATNAV) systems were also fitted – the KompaNav-2, A-737, GG-12 and JGG-30; the latter two designations were rendered in Roman letters in LII materials, indicating the systems' foreign origin.

Wearing a non-standard two-tone green camouflage scheme and the LII logo on the fin, the aircraft was in the static park of the MAKS-2005 airshow (19-24th August 2005). Interestingly, the control surfaces featured deflection angle gauges.

Antonov An-12B ELINT equipment testbed

The majority of An-12 testbeds were used to test new avionics and equipment. One of them was a Voronezh-built An-12B (c/n 402207), which was used for testing ELINT and optoelectronic reconnaissance systems in 1964-1998. Wearing standard overall grey

The logo of the Roosskaya Avionika JSC whose navigation systems were tested on L-39C '56 Yellow'.

L-39C '56 Yellow' in the static park of the MAKS-2005 airshow.

An-12B '08256' (c/n 402207) on the hardstand at Zhukovskiy. These views show the SLAR antenna arrays, the ventral sensor and the tail radome. The combination of a civil registration and Air Force insignia is unusual. Note the An-12BK prototype (CCCP-83962) in the background.

The same aircraft on short finals to runway 30 in Zhukovskiy.

The rear end of An-12B c/n 402207 (the callsign 08256, which became the registration, has yet to be applied). This view shows the rear radome and the unidentified boxy equipment housings extended through the open cargo door.

The logo of the Leninets Holding Co. avionics house.

camouflage and Soviet Air Force star insignia but sporting the non-standard registration 08256 (originally the aircraft's ATC callsign) in lieu of a tactical code, this *Cub* sported long slab-sided fairings with streamlined front and rear ends housing a side-looking airborne radar (SLAR) high on the rear fuselage sides. The fairing incorporated dielectric panels, which were flat (except for an elliptical bulge at the front) and canted slightly outward.

The DB-65U tail turret was supplanted by a long ogival dielectric fairing looking like a supersonic fighter's radome; the tail gunner's station probably accommodated one of the test engineers. Additional boxy fairings were located low on the starboard side of the forward fuselage and beneath the starboard SLAR fairing; the rear

'box' was much larger and incorporated what appears to be two rows of eight four-round flare launchers. Two oblong square-section boxes with unidentified equipment could be extended aft through the open cargo doors. Finally, a bulged observation blister was built into the forward starboard emergency exit. The aircraft was based at the LII airfield in Zhukovskiy.

Antonov An-12A/An-12B ASW/ELINT equipment testbed

A quasi-civil grey-painted An-12A or B registered CCCP-11417 (c/n unknown) was used for testing anti-submarine warfare (ASW) or ELINT equipment – most probably by the All-Union Electronics Research Institute (VNIIRA –

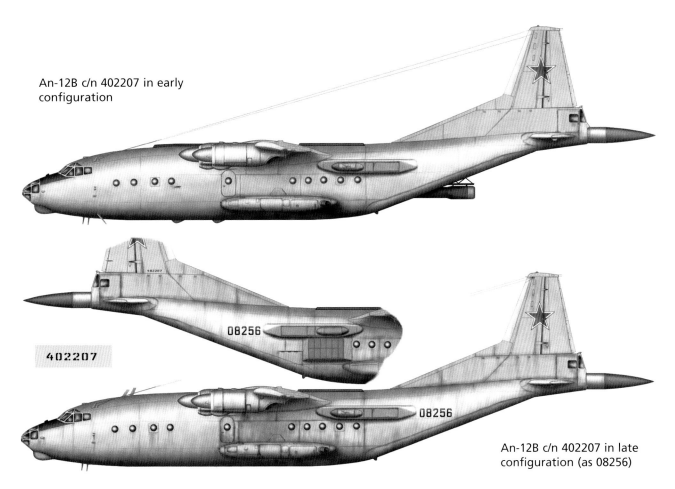

An-12B c/n 402207 in early configuration

402207

An-12B c/n 402207 in late configuration (as 08256)

Vsesoyooznyy naoochno-issledovatel'skiy instituot rahdioelektroniki), aka LNPO Leninets (Leninist), in Leningrad. This establishment, a division of the Ministry of Electronic Industry (MRP – *Ministerstvo rahdioelektronnoy promyshlennosti*), was one of the Soviet Union's leading avionics houses. (LNPO = *Leningrahdskoye naoochno-proizvodstvennoye obyedineniye* – Leningrad Scientific & Production Association. LNPO Leninets is now known as the Leninets Holding Company.)

Originally large cylindrical pods with cropped conical ends were mounted on the forward fuselage sides beneath the cabin windows; these housed a **Sa**blya (Sabre, *izdeliye* 122) monobloc SLAR of the type fitted to the Mikoyan MiG-25RBS *Foxbat-D* reconnaissance/strike aircraft. A smaller rectangular box located ventrally on the centreline in line with their aft ends accommodated a Bulat (Damask steel, pronounced *boolaht*) radar. A fairly large rounded radome mounted on a hinged frame was installed at the rear of the freight hold, protruding through a cutout in the starboard cargo door segment; it was semi-recessed in the fuselage for take-off and landing and fully extended in flight. A small angular fairing – probably a camera housing – projected

downwards from the rear cargo door segment.

The first test flights performed by GNIKI VVS crews gave disappointing results. The lateral pods generated powerful vortices and the turbulent airflow from the port pod was caught by the propeller of the No. 2 engine, striking the entry door. As a result, the door would vibrate, producing a deafening roar that was absolutely unbearable. Besides, the mounting frame of the rear radome was not rigid enough and the radome started swaying dangerously. To remedy the situation the SLAR antennas were attached directly to the fuselage sides and enclosed by huge teardrop fairings, and the said frame was stiffened.

Later the aircraft had an *Oospekh* (Success) search radar in a large quasi-spherical radome

An-12B CCCP-11417, one of the avionics testbeds operated by LNPO Leninets, in its revised form. The aircraft served for testing ASW systems.

259

SOVIET AND RUSSIAN TESTBED AIRCRAFT

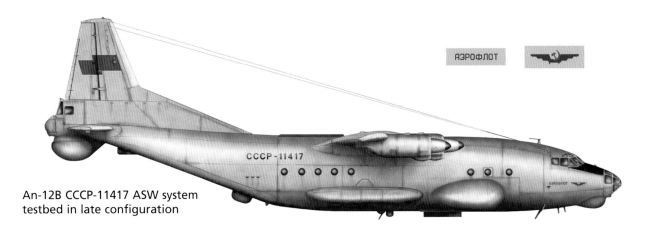

An-12B CCCP-11417 ASW system
testbed in late configuration

An-12A
CCCP-11790, an
ECM or ELINT
testbed operated
by the Yermolino
Flight Test
Centre. The
antenna arrays
replacing the tail
turret and
obstructing the
cargo door are
clearly visible.

installed beneath the tail gunner's station; this
radar had been developed for the Kamov
Ka-25Ts *Hormone-B* shipboard over-the-
horizon (OTH) targeting helicopter. The test
missions flown by CCCP-11417 took it over
the Barents Sea where it was intercepted by
Royal Norwegian Air Force Lockheed F-104
Starfighters in the summer of 1984.

Antonov An-12A ECM/ELINT equipment testbed

Another quasi-civil grey *Cub*, an early-production Voronezh-built An-12A (CCCP-11790,

c/n 1400302), served as an electronic countermeasures (ECM) or ELINT testbed with the
Yermolino flight test centre near Moscow in
the early 1980s. It featured a large boxy structure incorporating five dielectric panels of different size attached to the aft fuselage underside, supplanting the rear cargo door segment,
and a fairly long square-section tapered
'stinger' fairing protruding from the tail gunner's station in lieu of a turret. The aircraft was
later stripped of all non-standard appendages,
serving as an ordinary transport with the
Yermolino Flight Test & Research Enterprise
(YeLIIP – *Yermolinskoye **lyot**no-ispy**tah**tel'-
noye is**sled**ovatel'skoye predpri**yah**tiye*) as
RA-11790; the DB-65U turret was reinstated.

Antonov An-12AP ECM/ELINT equipment testbed

Yet another avionics testbed operated by the
Yermolino flight test centre was An-12AP
CCCP-11916 (c/n 2400901). In its days as a
testbed it also wore grey Air Force colours, featuring a similar 'stinger' fairing replacing the
tail turret and a large cylindrical 'proboscis'
with a ventral dielectric insert and a wiring
conduit to starboard, which required most of

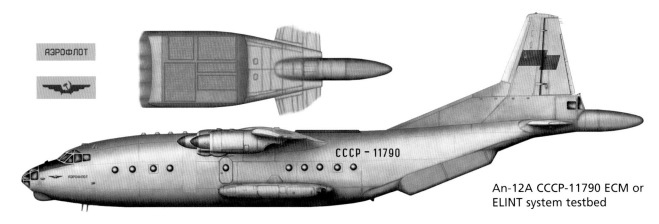

An-12A CCCP-11790 ECM or
ELINT system testbed

An-12AP CCCP-11916 was another testbed operated by the Yermolino test centre and featured a similar tail radome, plus a 'thimble' with a dielectric insert supplanting the navigator's station glazing.

the navigator's station glazing to be removed. Again, the aircraft was reconverted to standard configuration by 1993 and was still operated by YeLIIP as RA-11916 in 1998.

Antonov An-12BP ECM/ELINT equipment testbed

A Tashkent-built An-12BP, CCCP-11819 (c/n 6344009), wearing a version of Aeroflot's 1973-standard livery with a blue/white vertical tail served as an ECM or ELINT equipment testbed. The greater part of the navigator's station glazing was replaced by a long conical fairing with a rounded tip and a ventral excrescence near the tip; a smaller upturned conical fairing replaced the tail turret, and a small ventral canoe fairing was located immediately ahead of the wings. This aircraft, too, was reconverted to standard configuration and transferred

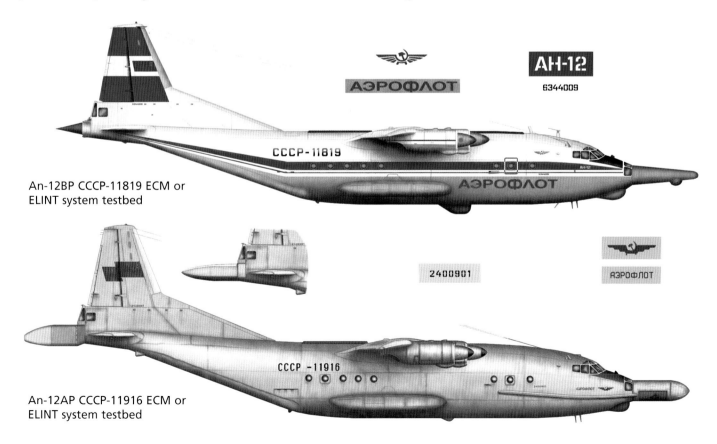

An-12BP CCCP-11819 ECM or ELINT system testbed

An-12AP CCCP-11916 ECM or ELINT system testbed

An-12BP CCCP-11819 was an avionics testbed with a 'proboscis', a conical tail fairing and a ventral canoe fairing. Note the non-standard livery with a blue/white tail.

plethora of aerials led many observers to believe the aircraft was an airborne command post version designated An-12VKP Zebra (*vozdooshnyy komahndnyy poonkt* – ABCP).

In 1975 the aircraft served as a testbed for the mighty BLT-5 winch driven by a ram air turbine; the winch was used to deploy a 2,500-m (8,200-ft) trailing wire aerial (TWA) with a stabilising drogue at the end. The unit had been developed for the Tu-142MR *Bear-J* communications relay aircraft maintaining very low frequency (VLF) communications between submerged submarines and shore-based or airborne command posts in the event of a nuclear attack. The An-12 also featured a whole network of wire aerials stretched from the tail unit to the wings, looking almost like a cobweb.

Antonov An-12A/B SATCOM equipment testbed

An unidentified An-12A or B in Air Force colours was fitted out for testing what appears to be satellite communications (SATCOM) equipment. The aircraft sported a large teardrop fairing housing antennas on top of the forward fuselage.

to the transport department of the Khar'kov aircraft factory No.135.

Antonov An-12A antenna/feeder system testbed

In 1960-88 an uncoded Irkutsk-built An-12A (formerly '19 Red', c/n 9900902) was converted for testing aerials (antenna/feeder systems) designed for communicating with submerged submarines. The aircraft was characterised by three cigar-shaped fairings at the wingtips and atop the fin. Additionally, long 'towel rail' aerials ran along the upper and lower fuselage sides from the wing trailing edge to a point in line with the front end of the fin fillet. This

Antonov An-12A SLAR testbed

A grey-painted Tashkent-built An-12A with the non-standard registration CCCP-13321 (c/n 2340301) had been used by the Moscow-based avionics house NPO Vzlyot (Take-off), another Ministry of Electronics Industry division, as a SLAR testbed at one time. When first noted at Zhukovskiy in August 1992 the aircraft appeared perfectly standard at first glance, but careful inspection of the rear fuselage sides showed traces of antenna fairings similar to those of An-12B 08256.

An-12A c/n 9900902 (often misidentified as the An-12VKP Zebra airborne command post) was an antenna system testbed with wingtip- and fin-mounted antenna pods and various aerials on the centre fuselage.

Antonov An-12BP radar testbed

Another *Cub* with a non-standard registration, An-12BP CCCP-13402 (c/n unknown) in 1973-standard Aeroflot livery was also an avionics testbed – probably operated by LNPO Leninets. This aircraft had a huge quasi-spherical radome housing an Oospekh radar instead of the normal chin-mounted RBP-2; the thing looked like the vocal sac of a tree frog! As in the case of CCCP-11417, the rear cargo door segment carried a camera fairing.

Antonov An-12BK radar testbed

A Russian Air Force An-12BK (identity unknown) was seen at one of the Russian airbases with a curious truss-type structure on top of the fuselage; it was made of thin metal tubes and looked almost like a construction scaffolding. This aircraft was reportedly used to test the mock-up installation of some radar antenna, although it obviously could not have flown in this guise.

Antonov An-12A spectrometric equipment testbed

A Soviet Air Force An-12A coded '15 Red' (c/n 1340105, the fifth Tashkent-built *Cub*) was used by LII to test spectrometric and radiometric equipment. The test equipment sensors were installed in lieu of the DB-65U tail turret and gun ranging radar.

Antonov An-12 Koobrik avionics testbed

In 1969 a Soviet Air Force An-12A or B coded '77 Red' (c/n unknown) was extensively modified as the An-12 **Koob**rik (crew quarters on a ship) – a testbed for thermal imaging systems and IR sensors designed for detecting targets on water, land and in the air. The aircraft looked positively hair-raising. The navigator's station glazing was almost entirely supplanted by a cylindrical metal adapter mounting a conical radome at the front and a small teardrop radome underneath; the adapter also incorporated two small air scoops. A large boxy dorsal canoe fairing ran nearly the full length of the forward fuselage from just aft of the flight deck to the wing leading edge, carrying a large cylindrical 'smokestack' near its aft end. A small bullet-shaped pod was carried on a short forward-swept pylon installed under the port forward emergency exit.

The rear end was non-standard as well. The tail gunner's station was moved aft appreciably by inserting a 'plug' between it and the rest of the airframe. It carried a PRS-4 Krypton (*izdeliye* 4DK) gun ranging radar with a distinctive boxy radome (hence the NATO codename *Box Tail*; PRS = *pri**tsel rah**diolokatsion-nyy strel**kov**yy* – gunner's radar sight); the Afanas'yev/Makarov AM-23 cannons were replaced by some kind of sensor. Finally, a camera fairing protruded from the rear cargo door segment.

Antonov An-12BK navigation system testbed

The An-12BK prototype (CCCP-83962 No. 2, c/n 402210) was used in 1966-72 to test the

The rear end of An-12A '15 Red' (c/n 1340105) with radiometric and spectrometric equipment supplanting the DB-65U tail turret.

first Soviet automated navigation system, SAU-1T-1 Polyot-1 (Flight-1; SAU = *sistema avtomaticheskovo oopravleniya* – automatic control system), and the **Koo**pol (cupola – or rather, in this context, parachute canopy) precision paradropping system. The latter was developed for the An-22 Antey (*Cock*) heavy military transport.

Antonov An-12B/BP navigation system testbed

One more *Cub* served for long-range radio navigation (LORAN) systems development in 1977-82; the research culminated in the Alpha LORAN system fitted to some Soviet combat

aircraft. The aircraft was referred to as 'An-12 No. 3108' – that is, either Voronezh-built An-12B c/n 403108 or Tashkent-built An-12BP '15 Red' (c/n 5343108). If the latter machine was involved, it was later transferred to the airlift regiment at Kubinka AB.

Antonov An-12 thermal imager testbed

Another An-12 served as a testbed for the *Prostor* (wide expanse, or ample space) thermal imager in 1968-70. The unit was designed for infrared mapping and locating fires.

Antonov An-24T Troyanda avionics testbed

In the mid-1960s the US Navy was bolstered by new nuclear-powered submarines having a much lower acoustic signature compared to traditional diesel-electric subs. The Soviet Union countered by launching development of new ASW systems using new principles of detecting submerged submarines. These systems were to be installed on new heavy ASW aircraft; for instance, the Antonov OKB proposed an ASW version of the An-22 heavy transport, which remained on the drawing board.

The An-12B Koobrik avionics testbed has few competitors for the title of the Most Bizarre *Cub* Ever! Note the tail gunner's station moved aft by inserting a plug; the PRS-4 Krypton gun ranging radar was developed for the IL-76.

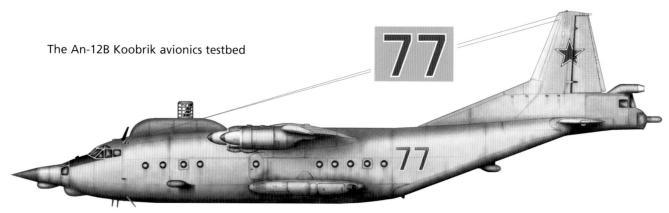

The An-12B Koobrik avionics testbed

The bizarre An-24T Troyanda ('04 Red') was purpose-built for testing new anti-submarine warfare systems. The huge 'cheek pouches' house infra-red sensors for detecting submerged nuclear submarines.

Meanwhile, the mission equipment had to be tested, and the assorted testbeds involved included a highly modified An-24T *Coke* tactical transport allocated the codename *Troyanda* ('rose' in Ukrainian). The aircraft's mission avionics were developed by VNIIRA (LNPO Leninets).

Unlike most aircraft of a similar nature, the bizarre An-24T Troyanda was not converted from a standard An-24T; it was custom-built to a specification drawn up by VNIIRA and delivered 'green' for outfitting at the institute's flight test facility in Pushkin, Leningrad Region. The fuselage featured major structural changes. Huge teardrop fairings were grafted onto the forward fuselage sides, making the aircraft look like a chipmunk which has stuffed its cheek pouches full of food; the undersides of these fairings incorporated large circular antenna panels associated with infra-red sensors. Since the starboard fairing would have obstructed the entry door, this was eliminated completely, access to the aircraft being solely via the ventral cargo hatch.

The wing/fuselage fairing was extended aft, resulting in a large flat-topped structure whose top was level with the top of the wing airfoil; it mounted a 'towel rail' aerial and incorporated hinged access panels. Camera windows were provided in the fuselage underside. The cabin housed six mission equipment operators working the IR sensor system, cameras and the sonobuoy receiver (SPARU – *samolyotnoye priyomnoye avtomaticheskoye rahdio'ustroystvo*, 'airborne automatic radio receiver device') which detected and processed the signals generated by sonobuoys.

Coded '04 Red' (c/n unknown) and wearing the overall medium grey scheme which was standard for the An-24T, the aircraft underwent a lengthy test programme.

However, only part of the equipment tested on the An-24T Troyanda eventually found its way to operational ASW aircraft.

Antonov An-24B avionics testbed

A late-production Kiev-built An-24B with the non-standard An-26 style registration CCCP-26196 (c/n 17307303) was another avionics testbed operated by LNPO Leninets, featuring some kind of antenna or sensor array on the port side immediately aft of the wings. A vertical camera of unknown type was housed in the rear fuselage, the lens being closed by a prominent teardrop fairing with clamshell doors located between the ventral fins. It is not known what purpose this aircraft served.

This full frontal of the An-24T Troyanda emphasises the size of the lateral fairings.

Antonov An-26LL-PLO avionics testbed

The 1970s saw a new generation of US Navy *Ohio* class nuclear-powered submarines and British Royal Navy *Resolution* class submarines armed with Trident sea-launched ballistic missiles; each missile could carry ten warheads and cover a distance of 10,000 km (6,200 miles) from the launch site to the target. These missile subs posed so serious a threat to the Soviet Union that a crash programme was initiated with a view to upgrading the Soviet ASW arsenal. Between 1972 and 1993 a testbed based on the An-26 *Curl* tactical transport – specifically, the sole An-26A (c/n 0901) – took part in that work.

Designated An-26LL-PLO (*le**ta**yushchaya labora**tor**iya protivo**lod**ochnoy obo**ron**y* – ASW flying testbed), the aircraft was used primarily for developing new ultra-sensitive magnetic anomaly detectors (MAD). It was believed that these devices, which until then had not been noted for their high precision and discriminating ability, would be able to replace the extremely costly sonobuoys (one Soviet RGB-3 sonobuoy, for example, cost more than a GAZ-24 Volga D-segment saloon car in terms of 1973 prices!).

The operation of onboard MADs, which detected the magnetic field of a submarine moving at a great depth and took its bearings, was hampered by the aircraft's own electromagnetic field. For that reason the MAD sensors detector were to be placed as far from eventual sources of interference as possible. Hence the An-26LL-PLO was equipped with two double-kinked booms, each 5 m (16 ft 5 in) long, mounted beneath the cargo ramp; one of them housed the sensor of a reference MAD, while the other boom carried an experimental system. The sensors of two other MAD systems were towed on cables 170 m (560 ft) long. During the test flights three systems functioned in combat mode and one, as noted above, was used as a standard for comparison.

(Note: The above description, together with photo evidence to match, suggests that An-26 c/n 0901 wore the non-standard registration CCCP-29113. However, this registration is consistently quoted in the Western press with the c/n 1301!)

The aircraft was converted by the Taganrog Machinery Plant named after Gheorgiy Dimitrov (TMZD) with the participation of LII and the organisation responsible for the equipment, NPO Roodgheofizika. The testing continued from the mid-1980s until the early 1990s, but the onset of *perestroika* (restructuring), when the funding for a number of military programmes was reduced to a trickle, made it impossible to complete the work. The entire fleet of the principal Soviet ASW aircraft types – the Ilyushin IL-38 *May* and Tupolev Tu-142 *Bear-F* – was placed in a difficult situation. Moreover, continuing the programme was pointless because the plans to launch production of the Beriyev A-40 Albatross (NATO codename *Mermaid*) ASW amphibian were abandoned. Subsequently the An-26LL-PLO was used for ecological monitoring. It was later stored at Zhukovskiy as RA-29113 with all non-standard equipment removed.

Antonov An-26P ('Prozhektor') avionics testbed

In the late 1960s the USSR initiated development of guided weapons intended to replace the not overly successful Zvezda Kh-66 and Zvezda Kh-23M air-to-surface missiles with radio command guidance (both types had the NATO codename AS-7 *Kerry*) in the arsenal of the Soviet Air Force's tactical arm. New air-to surface missiles were to be fitted with a semi-active laser homing warhead, and their carriers were to be equipped with laser target designators.

In 1973 the An-26P avionics testbed was converted from a production aircraft as a joint effort of LII and NPO Gheofizika. In this case the P suffix stood for *Pro**zhek**tor* (searchlight), as the new weapons programme was codenamed; this testbed should not be confused with the An-26P waterbomber, where the P stood for *po**zhar**nyy* (firefighting, used attributively). The aircraft carried a podded Prozhektor laser designation system and a Type 24N1 laser seeker head from the Zvezda Kh-25L (AS-10 *Karen*) ASM. Test flights conducted in 1973-74 served for studying the peculiarities of the laser beam's reflection from typical backgrounds and maritime or land objects, including road and railway bridges, and for studying the system's ability to work during sandstorms. In-flight experiments with weapons suspended under the aircraft enabled the researchers to study the ability of the laser homing warheads to withstand interference and investigate the possibility of guiding the Kh-25L and Vympel Kh-29L (AS-14 *Kedge*) missiles to targets illuminated by a ground laser designator.

The work was conducted for the benefit of the Sukhoi OKB with a view to developing a weapons system for the Su-17MKG fighter-bomber. Although the target designation system proved to be not quite up to the mark, it was, nevertheless, introduced into service and adopted for the production Su-17M2.

Antonov An-26K ('Kaïra') avionics/weapons testbed

Soviet combat aircraft equipped with a laser rangefinder/marked target seeker (LRMTS) of the Klyon (Maple) family could use the thing only in a dive because, if released in level flight, the guided bomb quickly began to lag behind the aircraft, entering a 'blind spot' out of reach of the ray of the aircraft's quantum generator. To solve this problem, development of the Kaïra-23 (Model 16S-1) and Katoon'-BI laser/TV sighting systems was initiated. (**Kaï**ra is Russian for 'great auk' while *Ka**toon'*** is the name of a Russian river.) This was a radically different kind of equipment. Its development involved, among other things, the An-26K testbed (the K initially stood for Kaïra, the name of the programme).

This time the main customer was the Mikoyan OKB which was working on the MiG-27K *Flogger-J2* fighter-bomber. The Kaïra-23 LRMTS was accommodated in a fairing under the nose of the An-26K; two Kh-25 or Kh-29 missiles were carried externally on both sides of the fuselage.

The conversion job ran seriously behind schedule – in fact, so much that, as related in the memoirs of Merited Test Pilot Valeriy Ye. Menitskiy, by the time when the testbed started flying at long last, all the tests and live missile firing trials, as well as drops of laser- and TV-guided bombs, had already been conducted on the actual fighter-bomber. Flights within the framework of the Kaïra's test programme were performed by a crew captained by NIIRP (*Na**ooch**no-is**sled**ovatel'skiy insti**toot** rah**d**iopri**bor**ostro**yen**iya* – Radio Appliance Design Research Institute) project test pilot L. Tetsman; the Mikoyan OKB was represented by test navigators Leonid S. Popov and Valeriy S. Zaïtsev. The Kaïra system was adopted for service. In addition to the MiG-27K, it was used on the Sukhoi Su-24M *Fencer-D* tactical bomber, which had a version of this system designated LTPS-24).

Antonov An-26K ('Kaplya') avionics testbed

After the completion of flights associated with the Kaïra programme the An-26K was used for other work. Within the framework of the

*Kap*lya (Drop [of liquid]) research programme Soviet designers created an optical correlation guidance system for air-to-surface missiles, development of which also involved the use of the suitably re-equipped An-26K; the K now conveniently stood for Kaplya.

On 13th March 1989 the An-26K crashed, killing the crew captained by LII test pilot Igor' V. Borisov. During a night test flight at low altitude over the Azov Sea the aircraft collided with a wild goose. The big bird took out the entire windshield, the pilot in command was wounded by the flying splinters and involuntarily pushed the control wheel forward…

Antonov An-26 'Polyot' avionics testbed

The 1960s saw the beginning of the introduction of the Unified Air Traffic Control System in the USSR. For this system Soviet designers developed the aforementioned Polyot-1 navigation suite which was based on the KSB navigation/attack system of the ill-starred Myasishchev M-50 *Bounder* supersonic strategic bomber. Several dozen aircraft were involved in the development of this system; one of them was a testbed based on the An-26 and designated An-26 'Polyot'.

Test flights of this machine within the framework of the Polyot programme started at LII in 1968. The work on the KSB and Polyot systems resulted in the development of the TKS-P precision compass system (***toch**naya koorso**va**ya sis**tem**a*), a vertical gyro of the TsGV family, the TGV-P precise vertical gyro (***toch**naya **ghee**roverti**kahl'***), the SVS-72 air data system (*sis**tem**a voz**doosh**nykh sig**nah**lov*), the '*Strela*' (Arrow) Doppler speed/drift sensor, the Koors-MP (Heading-MP) compass system and RSBN-2 short-range radio navigation system (***rah**diotekh**nich**eskaya sis**tem**a **blizh**ney navi**gah**tsiï – SHORAN), the SOM-64 ATC/SIF transponder, the NV-PB navigation computer, the SAU-1T automatic control system and other items of equipment. The Polyot-1 system found wide use on Soviet military and civil aircraft.

Subsequently the same testbed was used for testing the BNK-1P basic navigation instrument set (***bah**zovyy navigatsi**on**nyy kom**plekt***) for civil aviation. This work was conducted as a joint effort of the design bureau developing the equipment, the Antonov OKB and LII; Aleksandr Yu. Garnayev was the programme's project test pilot.

Antonov An-26 automatic landing system testbed

When the tests of the first prototype An-26 (c/n 0201) had been completed, the aircraft became an avionics testbed used in 1982-1993 for developing elements of automatic landing approach systems (ALS). This included the E-326 range/azimuth beacon, the *Plats**darm*** (Bridgehead) microwave landing system and the first automated landing approach instrument set involving the use of GPS. The flights were performed by test pilot Ye. A. Lebedinskiy, navigators Yu. M. Goobarev and N. I. Anisimov. The results obtained were used for perfecting the avionic suites of the Yak-40 and Yak-42 aircraft.

Antonov An-26 avionics testbed

The assorted fleet of Elf Air, the now-defunct commercial flying division of NPO Vzlyot, included an all-white An-26 with the non-standard registration RA-13398 (c/n 2607). A close look at this aircraft revealed an elongated fairing immediately aft of the rudder, the absence of the first two windows on each side (these were faired over) and an elongated metal plate just aft of the entry door. This plate was clearly intended to reinforce the skin where some item of equipment had previously been installed.

Antonov An-26 avionics testbed (?)

Sometime in the 1980s LNPO Leninets obtained an early-production An-26. Registered CCCP-26648 (c/n 06-09), this was probably a navigation systems testbed – or, equally possibly, a navaids calibration aircraft (flight checker). Outwardly it was distinguishable by a dorsal strake aerial on the forward fuselage and two small unpainted metal fairings located in tandem beneath the centre fuselage, offset to port. In 2001 the aircraft was still in use by the Leninets Holding Co. as RA-26648.

Antonov Antonov An-30B with SLAR

In the late 1980s two grey-painted Soviet Air Force An-30B *Clank* photo reconnaissance aircraft coded '07 Red' (c/n 0405) and '86 Red' (c/n 0806) were equipped with an unidentified SLAR of West German origin. The SLAR antenna arrays were housed in compact cylindrical pods on the lower centre fuselage sides; the pylons were of a different type than hitherto,

with small elongated bulges on the fuselage sides immediately above them. '07 Red' and '86 Red' were based at the LII airfield in Zhukovskiy.

Antonov An-30 navigation systems testbed

Several An-30s were used by different organisations as flying testbeds and research aircraft. One of them was the abovementioned An-30B '07 Red' which belonged to LII. The aircraft was used for testing new navigation systems (more exactly, the so-called correlation-type navigation systems for extreme conditions) between 1982 and 1994. The results of these tests were used for the introduction of theses navigation systems on military aircraft and for evolving the methods of testing such systems.

Antonov An-72 navigation system testbed

A production An-72 *Coaler-A* STOL tactical transport coded '03 Red' (c/n 365.720.60.610, f/n 0603) was converted by LII for testing advanced automatic landing systems and SAT-NAV systems. The aircraft had worn a different tactical code (possibly '24') before transfer to the institute.

Outwardly the only non-standard features were two whip aerials in tandem and a small antenna blister low on the port side of the forward fuselage. The new avionics and test/data recording equipment were mounted in a small cubicle at the front of the freight hold; a navigation display was installed in the flight deck. Unusually, a 'Danger, air intake' warning triangle was applied ahead of the APU air intake. The testbed was displayed statically at the MAKS-97 airshow in Zhukovskiy on 19-24th August 1997.

Antonov An-72 data link system testbed

An unspecified An-72 (possibly the same aircraft) was used by LII in 1993-97 for testing data link systems.

Antonov An-74 avionics testbed

Towards the end of its flying career the An-74 *Coaler-B* commercial transport prototype (c/n 003), by then registered CCCP-72003, was converted into an avionics testbed with a magnetic anomaly detector (MAD) boom projecting aft from the tailcone.

ВИСХАГИ

The logo of the State Project & Research Institute of Land Cadastre Photography (VISKhAGI) worn by **Accord-201** 02695 ФЛА РФ.

Avia Accord-201 navigation system testbed/survey aircraft

By mid-2005 the prototype of the Accord-201 four-seat twin-engine light utility aircraft developed by Nizhniy Novgorod-based NPO Avia Ltd. and registered 02695 ФЛА РФ (that is, 02695 FLA RF, ex-RA-03010; c/n A33 00 001?) was modified by LII for testing navigation systems. (FLA RF = *Federahtsiya lyubiteley aviahtsii Rosseeyskoy Federahtsii* – Aviation Enthusiasts' Federation of Russia, the amateur pilot association.) The tests included precision navigation along a pre-programmed route with an error margin of less than 0.1 m (3¹⁵⁄₁₆ in) and aircraft-to-aircraft navigation tests. The cockpit was equipped with reprogrammable head-up and head-down displays.

Additionally, 02695 ФЛА РФ was outfitted as a multi-mode survey aircraft for performing photo mapping, thermal imaging and gravimetric survey with precise determination of the coordinates via SATNAV. To this end the back seat was removed to enable installation of a vertical camera and sensors; these were protected by a sliding ventral door during take-off and landing. The modification was done by the Nature/Technical Systems Safety Safety Centre JSC (TsBPTS – *Tsentr bezopahsnosti prirodno-tekhnicheskikh sistem*) based in Zhukovskiy. In this configuration the aircraft was in the static park at the MAKS-2005 airshow, sporting the logos of the TsBPTS and the State Project & Research Institute of Land Cadastre Photography (VISKhAGI) in whose interests it was operated.

Boeing B-17 avionics testbeds

In the closing stage of the Second World War a number of US Air Force Boeing B-17 Flying Fortress bombers force-landed in Eastern Europe and were abandoned by their crews after being damaged by German air defences during trans-Atlantic shuttle raids. Some of them were salvaged and restored to airworthy condition by the advancing Soviet forces, seeing limited service with the Soviet Air Force. Yet, such aircraft were not operated exclusively by first-line units; some found use in Soviet research and development programmes. For example, five B-17s were used by the NII-17 avionics house as flying testbeds for testing different items of radio equipment. In 1952 one of them, a B-17G (Boeing Model 299-Z), was fitted with the brand-new RP-6 *Sokol* (Falcon) aerial intercept radar developed for a number of Soviet all-weather interceptors (RP = *rahdiopritsel* – radio sight, the Soviet term

for AI/fire control radars). The radome supplanted the glazed bomb-aimer's station.

Cessna 172S navigation system testbed

A new-production Cessna 172S Skyhawk operated by the Omega Light Aviation Centre and serialled '054 Black' (c/n 172S9054) was converted into a navigation system testbed by LII. This included verifying SATNAV applications for general aviation – air traffic control at low/ultra-low altitudes with the help of GPS navigation and GSM mobile phone channels, as well as evaluation of the ADS-B (Automatic Dependent Surveillance-Broadcast) co-operative surveillance technique utilising the VDL-4 digital data transfer protocol. According to the data placard at the MAKS-2005 airshow where the aircraft was displayed, the tests included real-time processing and recording of the aircraft's movement parameters and 'exploring the possibility of real-time tracking and control of aircraft beyond visual range, using data transfer via mobile phone channels'.

Ilyushin IL-4 radar testbed

In 1944 an IL-4 (DB-3F) long-range bomber was fitted experimentally with the Gneys-2M radar intended for detecting surface vessels.

The prototype Avia Accord-201 (02695 ФЛА РФ) converted into a navigation system testbed and photo survey aircraft was displayed at the MAKS-2005 airshow.

Cessna 172S '054 Black', a navigation system testbed, at the MAKS-2005.

The IL-14SLL avionics testbed (CCCP-48106) sits on a rain-soaked hardstand, showing the missile seeker head mounted under the nose radome.

CCCP-58641, the IL-14IRE avionics testbed with the avionics pod suspended under the fuselage.

Here is IL-18A CCCP-75643 at Moscow/Vnukovo-1 in June 1965. It was converted into the IL-18SL testbed for the IL-38's Berkoot search radar; the radome aft of the nose gear is visible.

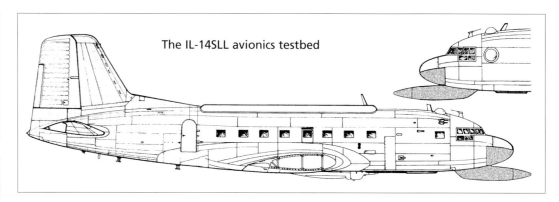

The IL-14SLL avionics testbed

Ilyushin IL-4 IFF system testbed

Another IL-4 was used for experiments with the SCh-3 identification friend-or-foe (IFF) transponder in the immediate post-war years. SCh stands for *svoy/choo**zhoy*** – in this context, friend/foe.

Ilyushin IL-14 automatic landing system testbed

An IL-14 *Crate* airliner (identity unknown) served as a testbed for developing elements of an automatic landing approach system. The tests took place at LII in 1957.

Ilyushin IL-14SLL avionics testbed

A Tashkent-built IL-14TG with the non-standard registration CCCP-48106 (c/n 8344001) was converted by NPO Vzlyot into the IL-14SLL (*spetsi**ahl'**naya le**ta**yushchaya labora**tori**ya* – 'special flying laboratory'). It featured an extended parabolic nose whose shape resembled the radome of the IL-18 airliner; the lower half appeared to be dielectric. A cigar-shaped pod – apparently the seeker head of an air-launched missile – was attached to the underside of the nose, protruding beyond the tip of the radome. Furthermore, a wide, shallow bulge was wrapped around the upper fuselage, running all the way from the wing leading edge to the dorsal fin. The aircraft was subsequently stripped of mission equipment and withdrawn from use but subsequently sold to a private owner as RA-02299 and restored to airworthy condition, retaining the the dorsal bulge and featuring an IL-18 style parabolic radome.

Ilyushin IL-14IRE avionics testbed

A further IL-14 with the non-standard registration CCCP-58641 (c/n unknown) was used by the Moscow Institute of Electronics and Automatic Systems (MIREA – *Mos**kov**skiy insti**toot rah**dioelek**tron**iki i avto**mah**tiki*) for testing electronic systems of an unknown nature (possibly ECM or ELINT). The equipment was carried in a large cylindrical pod with ogival front and rear ends under the fuselage. The testbed was designated IL-14IRE.

Ilyushin IL-14 ECM (?) equipment testbeds

A Soviet Air Force IL-14T coded '21 Red' (c/n unknown) had angular horizontal aerials built into the sides of the nose below the forward equipment bay access cover. This aircraft was in all probability an ECM equipment testbed.

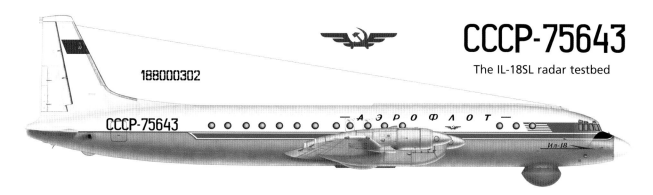

CCCP-75643

The IL-18SL radar testbed

188000302

CCCP-75643 — АЭРОФЛОТ — Ил-18

Ilyushin IL-18SL (SL-18A?) radar testbed

By June 1965 IL-18A CCCP-75643 (c/n 188000302) was converted into a testbed for the **Berk**ut (Golden Eagle) 360° search radar developed for the IL-38 anti-submarine warfare aircraft. The radar, a product of the Leningrad-based NII-131 (LNPO Leninets), was installed in identical fashion to the IL-38 in a large quasi-spherical radome immediately aft of the nose gear unit on a faired mounting ring. An unidentified tubular pod was fitted ventrally on a short pylon just aft of the wings. By then the aircraft had been retrofitted with a TG-16 APU.

CCCP-75643 has been referred to as the IL-18SL (*samolyot-laboratoriya* – laboratory aircraft). However, in keeping with LNPO Leninets's system of designating its testbeds the aircraft was probably designated SL-18A.

Ilyushin IL-18SL navigation system testbed

In the early 1990s CCCP-75643 became a testbed for the GLONASS satellite navigation system (*globahl'naya navigatsionnaya spootnikovaya sistema*, the Soviet/Russian counterpart of GPS). The search radar was removed (with no sign of its ever being there) and large cruciform markings were applied to the nose radome; these were possibly some kind of angle reflectors. By May 1992 the aircraft was retired and preserved at the 32nd Kilometre railway station a short way from Pushkin (Leningrad Region), home of LNPO Leninets's flight test centre – only to be broken up later.

Ilyushin IL-18P (IL-18 'Polosa') navigation system testbed

In 1965-66 (some sources say 1968) the Experimental Machinery Plant headed by Vladimir M. Myasischchev (formerly OKB-23) operated an aircraft designated IL-18 'Polosa'

(Stripe) or IL-18P. This was probably a modified IL-18B in pre-1973 Aeroflot colours with the non-standard registration CCCP-06180 No. 2 (ex-CCCP-75647, c/n 188000401). It had small cheek fairings immediately aft of the radome and small square fairings on the sides of the fin about halfway up; these incorporated slot aerials. Four L-shaped aerials were carried on the nose above and below the said fairings, plus one more on the flight deck roof. Incidentally, the registration had previously been used on the second prototype Mil' Mi-2 *Hoplite* utility helicopter (c/n 0102); it was later 're-reused' for a production Mi-26 *Halo* heavy-lift helicopter (c/n 226205) owned by the Mil' OKB.

The IL-18P served for testing an experimental navigation system permitting automatic flight along a predesignated route and automatic approach/landing. The test programme was performed by pilots Ghennadiy N. Volokhov and Aleksandr M. Tyuryumin. The aircraft was later converted into the IL-18RTL (aka IL-18SIP) – the prototype of the IL-20RT telemetry relay aircraft used during missile

The logo of MIREA whose avionics were tested on the IL-14IRE and IL-18IRE.

IL-18B CCCP-06180 was probably the IL-18 'Polosa' (IL-18P) used by the Myasishchev OKB for navigation system tests. Note the fairings on the sides of the nose and fin housing slot antennas and the L-shaped aerials on the nose.

The IL-18P

CCCP-06180

CCCP-06180

tests – and reregistered CCCP-27220; however, this version lies outside the scope of this book.

Ilyushin SL-18V radar testbed

IL-18V CCCP-75786 (c/n 181003905) became the SL-18V testbed with LNPO Leninets. The aircraft was used to test the *Obzor-K* (Perspective-K) navigation/attack radar developed for the Tupolev Tu-160 *Blackjack* missile carrier. This supplanted the RPSN-1 Emblema weather radar and was enclosed by a long conical radome whose large diameter required the use of a special adapter with a ventral cutout for the nose gear doors. Small blister fairings were located low on the aft fuselage sides in line with the fin fillet/fin junction.

A special pylon was installed under the fuselage in line with the forward entry door, per-

CCCP-75786, the SL-18V testbed, in pre-1973 colours. Note the Kh-45 fixed acquisition round under the fuselage.

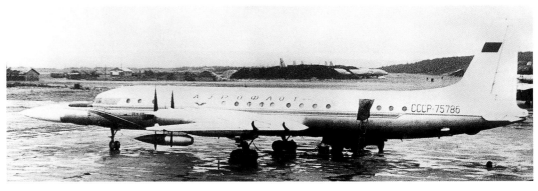

The SL-18V at Leningrad-Pulkovo airport after repaint in 1973-standard livery.

By 1996 the SL-18V had been retrofitted with an APM-60 MAD boom, making it the longest *Coot* ever. It is seen here stored at Pushkin in August 2001 as 75786.

The SL-18V radar testbed in original colours
with fixed Kh-45 acquisition round

mitting carriage of fixed missile acquisition
rounds for systems integration purposes. For
instance, at an early stage of the programme
the SL-18V carried a large acquisition round
simulating the Kh-45 **Mol**niya (Lightning) air-
to-surface missile. Originally developed by the
Sukhoi OKB as the main weapon for the Sukhoi
T-4 bomber (development was later continued
by NPO Raduga in Doobna north of Moscow),
this weapon was a candidate to arm the Tu-160
for a while but was rejected in 1977.

Ilyushin SL-18V (modified)
At a late stage of its test career the SL-18V (by
then reregistered RA-75786) was retrofitted
with an APM-60 **Or**sha MAD (named after a
Russian town) and associated IL-38-style boom
removed from the IL-18D-GAL geophysical
survey aircraft (CCCP-74267, see Chapter 7).
The purpose of the modification is unknown.
Thus RA-75786 became the longest IL-18
ever! By 1995 the aircraft was retired in
Pushkin; it was still sitting there engineless in
August 2001 as 75786 (with no nationality
prefix, as befits an experimental aircraft).

Ilyushin SL-18D avionics testbed
An IL-18D with the out-of-sequence registration CCCP-75713 No. 2 (c/n 186009403) was
yet another LNPO Leninets avionics testbed
designated, quite logically, SL-18D. (Previously

The nose of the
SL-18V radar test-
bed used for
testing the
Tu-160's radar.

the registration had belonged to IL-18V
c/n 180001804, which was retired in 1974.)

The SL-18D was used in several test programmes and the end result was rather
bizarre. The RPSN-1 weather radar gave way
to a bulbous nose fairing accommodating a
ground mapping radar developed for the
Antonov An-124 Ruslan (*Condor*) heavy military transport. This was compounded by a

This in-flight shot
of the SL-18D
shows well its
many excres-
cences.

The SL-18D
avionics testbed
(CCCP-75713
No. 2) parked at
Pushkin.

273

Another view of the SL-18D at Pushkin.

The latest incarnation of the SL-18D, with a standard nose and no dorsal/ventral radomes but with an APM-60 MAD. The aircraft wears NPP Mir colours; note that the RA- prefix is carried on the wings but omitted on the fuselage.

The logo of NPP Mir operating the Leninets Holding Co's aircraft.

Berkoot search radar aft of the nose gear unit; hence a mudguard was fitted to the nose-wheels and the nose gear doors bulged accordingly. (Interestingly, the SL-18A had no such mudguard.)

A small canoe fairing with forward-looking infra-red (FLIR) or laser equipment was mounted on the flight deck roof, with a larger dorsal canoe fairing over the wing centre section and a small ventral canoe fairing ahead of the wing leading edge. The underside aft of the wings bristled with small blade and 'hockey stick' aerials and dielectric blisters. Finally, as on the SL-18V, two small fairings were located low on the aft fuselage sides.

In 1993 the SL-18D was stripped of experimental equipment and leased to Djibouti-based Daallo Airlines as RA-75713, retaining the bulbous nose. Today it is operated by NPP Mir, the commercial flying division of the Leninets Holding Co., as 75713. All non-standard fairings on the forward/centre fuselage have been

The SL-18D avionics testbed in original configuration

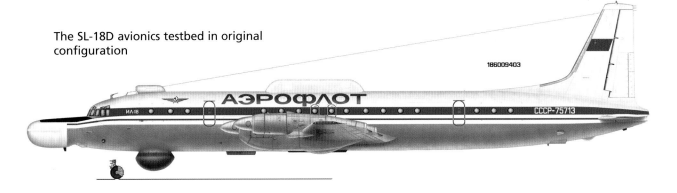

НПП МИР

The SL-18D in current configuration

removed and the normal nose radome has been reinstated; on the other hand, an IL-38 style MAD boom has been added.

Ilyushin SL-18I (SP-T18I) radar testbed

IL-18V CCCP-75804 (c/n 182004305) became yet another avionics testbed with LNPO Leninets. Since the V suffix letter was by then allocated to CCCP-75786, this testbed was designated SL-18I or SP-T18I. Again, the aircraft participated in several test programmes, each one leading to more tell-tale 'bumps and bulges'. Firstly, the SL-18I was fitted with a non-standard nose radar in an elongated and downward-curving radome; the type and application is unknown.

Secondly, once again a Berkoot search radar was installed (with appropriate modifications to the nose gear unit), but this time a large perfectly cylindrical pod with a hemispherical rear end was grafted on to its radome from behind (!); this housed a Mech (Sword) SLAR. Next, a fairly large dorsal teardrop fairing with a hemispherical cupola on top was mounted in line with the overwing emergency exits; this was presumably associated with satellite navigation. Finally, the aircraft had the small blister fairings on the aft fuselage which were common to almost all of the SL-18 series.

In 1993 the SL-18I was likewise stripped of experimental equipment and leased to Daallo Airlines (initially as 75804, with no nationality prefix, and then as RA-75804). Curiously, all

The SL-18D tucks up its landing gear as it departs on a mission.

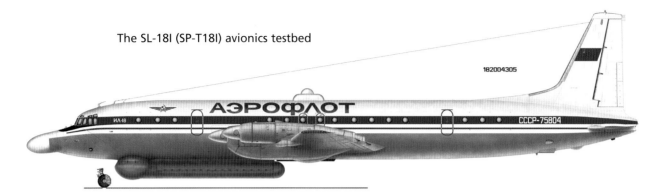

The SL-18I (SP-T18I) avionics testbed

of the non-standard fairings except the dorsal dome were retained, causing many a raised eyebrow wherever the aircraft went; the eye-catching SLAR 'sausage' even caused one Polish writer to mistake 75804 for an IL-20 *Coot-A* ELINT aircraft! Later the ventral radomes were removed during an overhaul but the non-standard nose radome and the mounting for the Berkoot radar remained.

Ilyushin SL-18P radar testbed

The last of the SL-18 series, in Russian alphabetical order, was the SL-18P converted from the last IL-18E, CCCP-75411 (c/n 186009205). Once again, the aircraft had at least four configurations; unlike the SL-18D and SL-18I, however, it carried equipment associated with only one programme at any one time.

The first configuration dated back to 1966 when the brand-new aircraft was converted to

CCCP-75804 was one of the most exotic-looking LNPO Leninets. testbeds, the SL-18I or SP-T18I. These views show the combined search radar/ SLAR fairing, the dorsal dielectric cupola and a 'Roman nose' housing a development radar.

The flight line at Pushkin in the early 1990s with some of LNPO Leninets's test-beds, including (left to right) the SL-18D, the SL-18I and An-26 CCCP-26648

276

The SL-18P (CCCP-75411) as originally flown with the Taïfoon-M radar for the Su-15TM. Note the shiny factory finish and the weather radar 'egg' above the flight deck.

The SL-18P in its third configuration with a late version of the Su-24 bomber's PNS-24 Tigr navigation/attack suite. Note the ventral sensor fairings.

test the Taïfoon-M (Typhoon-M) fire control radar developed for the Sukhoi Su-15TM *Flagon-F* interceptor; this may explain the P suffix (for *pere***khvat***chik* – interceptor). Inevitably, the radar (enclosed by a long ogival radome) was installed on an adapter supplanting the standard RPSN-1 radar. A small weather radar 'egg' was mounted on a short pylon on the flight deck roof.

The second and third configurations of the SL-18P (in pre-1973 and post-1973 colours respectively) were very similar, being used to test two versions of the PNS-24 Tigr (Tiger) navigation/attack system (*pritsel'no-navigatsi-***on***naya sis***tem***a*) designed for the Su-24 *Fencer* tactical bomber. In both cases the short, almost rectangular-section radome terminated in a pitot – a horizontal H-shaped structure (as fitted to the delta-winged Sukhoi T6-1 prototype) on Version 2 and the characteristic F-shaped pitot (popularly known as the

'goose' among Su-24 crews) on Version 3; the radome itself was dark green and white respectively.

An optical sensor offset to starboard and covered by a rotating hemispherical guard was fitted aft of the nose gear on the second version; on the third this was replaced by a different optical system in a small teardrop fairing. Additionally, a short ventral canoe fairing was located just ahead of the wing leading edge *à la* SL-18D. The 'egg' over the flight deck was removed

The final version of CCCP-75411 was a testbed for the **Korshoon** (Kite, the bird) 360° search radar developed for the Tu-142M *Bear-F Mod* ASW aircraft. The large teardrop radome was fitted, IL-38 style, aft of the nose gear unit; the RPSN-1 weather radar was reinstated and a mudguard added to the nose gear unit. In this guise the aircraft flew until the early 1990s when the radar was removed;

The SL-18P testbed in original configuration with a Taïfoon-M radar

CCCP-75411

The final test configuration of the SL-18P. The RPSN-2 weather radar has been reinstated and a Korshoon search radar for the Tu-142M fitted aft of the nose gear.

the mounting for the radome remained for a while but was deleted during the next overhaul.

Currently RA-75411 is operated by NPP Mir. The Korshoon radar underwent further tests on a converted IL-38.

Ilyushin IL-18B radar testbed

IL-18B CCCP-75703 (c/n 189001505) was converted into an avionics testbed of some sort, probably by NPO Vzlyot. The aircraft had a shallow boxy fairing immediately aft of the port wing root (probably a synthetic-aperture radar), with two round antennas mounted in line with it closer to the fuselage centreline. The shape of this fairing was changed later in the aircraft's career. A flat-bottomed bulge aft of the nose gear unit revealed that CCCP-75703 had been fitted with a Berkoot search radar which was subsequently removed and the bottom of the mounting ring closed with sheet metal. A small blade aerial was mounted dorsally ahead of the forward entry door and another one ventrally aft of it.

CCCP-75703 was based in Zhukovskiy. Despite its age, the aircraft stayed operational long enough to see the break-up of the Soviet Union, being last noted at Moscow-Bykovo as RA-75703 in June 2000.

The SL-18P, version 4 with Korshoon search radar
Bottom: The SL-18P, version 3 with PNS-24 nav/attack suite

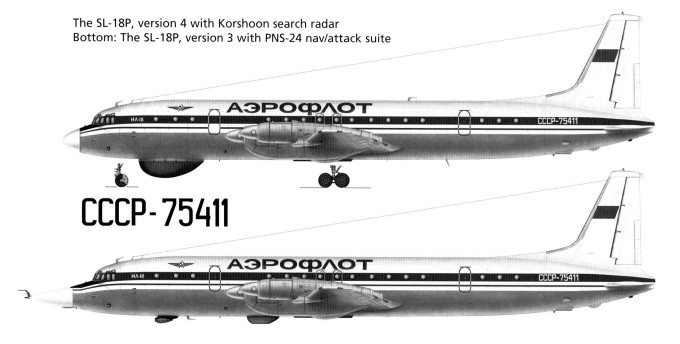

CCCP-75411

IL-18B CCCP-75703, an avionics testbed of unknown purpose, at aircraft repair plant No. 402 (Moscow-Bykovo) in the late 1980s. Note the antenna array aft of the port wing root (probably a synthetic-aperture radar).

Ilyushin IL-18V missile guidance systems testbed

In 1984 NPO Vzlyot added IL-18V CCCP-75851 (c/n 182005501) to its aircraft fleet; this was an aircraft re-imported from Cuba, having served with Cubana de Aviación since 1964 as CU-T832. The machine became a testbed for missile guidance systems. The weather radar was replaced by an adapter terminating in a

IL-18V CCCP-75851 was used by NPO Vzlyot in the mid-1980s to test missile guidance systems. Note the missile tracker head replacing the radome, the ventral and rear sensor fairings and the ventral pod.

IL-18V CCCP-75851 missile guidance system testbed

Close-ups of the
nose adapter for
experimental
avionics fitted to
CCCP-75851.

The ventral
acquisition round
pod on the same
aircraft, seen
from the rear.

cone with a dielectric tip and a pronounced 'chin'. The adapter was made to fit instead of the standard radome without requiring any additional modifications to the airframe.

A large detachable pod with the missile's seeker head and a parabolic rear end was pylon-mounted ahead of the wings, and small angular fairings housing additional equipment were located aft of the nose gear unit and on the fuselage tailcone. By August 1992 CCCP-75851 was reconverted to standard and transferred to Elf Air as an IL-18Gr package freighter, becoming RA-75851.

Ilyushin IL-18REO avionics testbed

Sometime before 1973 IL-18V CCCP-75811 (c/n 182004504) was converted into the IL-18REO avionics testbed (*rah*dioelek**tron**noye obo**roo**dovaniye – electronic equipment) by MIREA. In its original configuration the

nose radome gave place to a long cylindrical metal 'plug' terminating in a dielectric fairing which looked like an outsize match (to use the *mildest* of comparisons). A large cylindrical pod with a hemispherical front end and a parabolic rear end (both of them dielectric) was installed under the forward fuselage. The nature of the equipment tested is unknown.

Later, CCCP-75811 was fitted with a Berkoot search radar, a small dielectric blister over the wing centre section and a small ventral canoe fairing aft of the wings. Interestingly, the search radar was located noticeably farther aft than on the SL-18D and SL-18I. In 1996, long after this equipment had been removed, the aircraft was operated by the Ilyushin OKB's own airline, Ilavia, as RA-75811.

Ilyushin IL-18B avionics testbed

An IL-18B with the non-standard registration CCCP-48093 No. 1 (ex-CCCP-06187 No. 1; originally CCCP-75660, c/n 188000604) was yet another obscure avionics testbed used by LII, with a large flat-bottomed fairing under the extreme aft fuselage. Upon retirement in 1982 the aircraft was preserved in a pioneer camp near St. Petersburg, still in its original pre-1973 colours; the registration was reused in 1989 for an Antonov An-32 *Cline* transport (c/n 0703).

CCCP-75811 The IL-18REO avionics testbed in late configuration

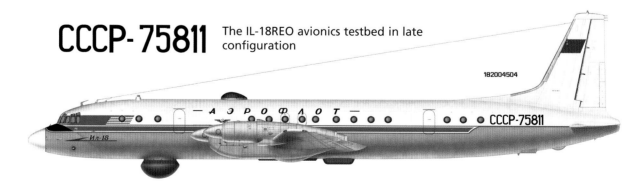

CCCP-75811, the IL-18REO avionics testbed, in early configuration. No comments on the shape of the nose please...

CCCP-75811 in later guise. The Berkoot search radar is located further aft than on the IL-18SL and SL-18D. The dorsal dielectric blister and ventral 'bathtub' are barely visible.

Ilyushin IL-18V CCCP-75894 avionics testbed

A former Soviet Air Force IL-18V with the out-of-sequence registration CCCP-75894 (c/n 182004801) was converted into an avionics testbed by NPO Vzlyot before 1992. Logically it should have been CCCP-75820 but this registration already existed as IL-18 *sans suffixe* c/n 187000103 (ex-CCCP-Л5820). The aircraft featured a large canoe fairing (with a downward-projecting 'thimble' at the rear) under the forward fuselage and a smaller ventral 'bathtub' aft of the wings, both incorporating large dielectric panels. It was still active in this condition in 2002 as RA-75894.

Ilyushin IL-18V navigation system testbed

IL-18V CCCP-78732 (ex-CCCP-75794, c/n 181004103) presumably operated by NPO Vzlyot was yet another testbed for the Polyot-1 automated navigation system. It was also used for testing an autothrottle, the BSU-3P automatic landing approach system (*bortovaya sistema oopravleniya* – on-board control system) and a LORAN system. The non-standard registration is noteworthy, since the 787xx series is allocated to Ilyushin IL-76MD *Candid-B* transports and IL-78/IL-78M *Midas-A/B* tankers. The aircraft had a small cylindrical pod with a hemispherical front end and a

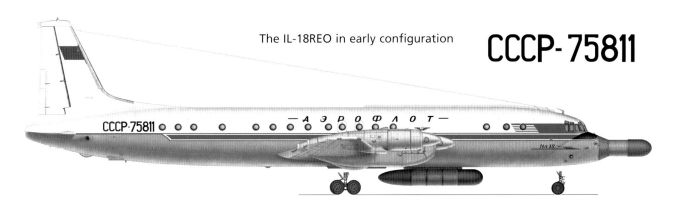

The IL-18REO in early configuration

CCCP-75811

CCCP-48093

The IL-18B CCCP-48093
avionics testbed

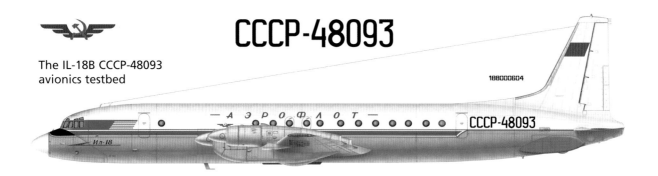

188000604

— А Э Р О Ф Л О Т — CCCP-48093

pointed rear end under the forward fuselage, plus a small flat-topped dorsal fairing near the aft entry door. The tests took place in 1960-68 (*sic* – as per LII sources, even though the aircraft was built in 1961!).

The tests made it possible to develop and refine automatic landing approach systems for later versions of the IL-18, as well as for the Tupolev Tu-124 *Cookpot* and Tu-134 *Crusty* twinjet short-haul airliners. The IL-18 testbed was last noted at Zhukovskiy as RA-78732 in March 1994 minus the ventral pod and was reportedly operated by Volare, LII's own airline, at the time.

Ilyushin IL-28LL radar testbed

One IL-28 *Beagle* bomber (identity unknown) was converted in 1952 for testing the RP-6 Sokol radar and designated IL-28LL. This radar

with a 30-km (16.2-nm) detection range had been developed by NII-17 under the direction of Ghedaliy M. Koonyavskiy for two interceptors – the Yakovlev Yak-120, which entered production and service as the Yak-25 *Flashlight-A*, and the Lavochkin La-200B, which remained a prototype. As noted earlier, initial tests were performed on a converted Boeing B-17G. When it transpired that development of the Yakovlev fighter was taking longer than predicted and that the La-200 would be the first to receive the new radar, Semyon A. Lavochkin suggested that a heavy aircraft but a faster one than the B-17 be used to bring the radar up to scratch. The IL-28 jet bomber was the obvious choice.

To accommodate the radar the glazing of the navigator/bomb-aimer's station was cut away at fuselage frame 2 and substituted by a cylindrical metal structure (part of the Yak-120's nose incorporating the avionics bay). The huge antenna dish of the RP-6 was enclosed by a large glassfibre radome which had an almost hemispherical front end rather rather than the usual pointed or parabolic shape. The conversion work was done by Lavochkin OKB specialists under the supervision of the Ilyushin OKB (which was not directly interested in the project but held responsibility for the IL-28 anyway).

The famous test pilot Mark L. Gallai flew the IL-28LL, with R. A. Razumov as test engineer; the latter was the worst off, sitting in a dark and extremely cramped bay aft of the radar set – all that remained of the navigator's station. A total of 33 flights were made without any problems; the test programme, which ended in December 1952, included simulated interception of real aircraft. By the end of 1953 the radar had been perfected and was fitted to the late-production Yak-25M from 1954 onwards, replacing the RP-1D Izumrood-1 (Emerald) radar fitted to the Yak-25 *sans suffixe* as a stopgap measure.

IL-18V RA-75894, an obscure avionics testbed belonging to NPO Vzlyot, takes off from runway 12 at Zhukovskiy. The large ventral canoe fairings are clearly visible.

Ilyushin IL-28 missile targeting system testbed

In 1960 the Ministry of Defence's Central Research Institute No. 30 (TsNII-30 – *Tsen-trahl'nyy na**ooch**no-is**sled**ovatel'skiy insti-**toot***) joined forces with NII-2 and the Research Institute of the State Committee for Electronics (NII GKRE – *Na**ooch**no-is**sle-d**ovatel'skiy insti**toot** Gosu**dar**stvennovo komi**tet**a po **rah**dioelek**tron**ike*) to develop active radar homing systems for anti-shipping missiles. (Note: GKRE was the name of the former MRP during the Khrushchov era.) To this end it was necessary to analyse the characteristics of the radar pulse reflected from surface ships. Thus an IL-28 and a Lisunov Li-2 *Cab* transport were converted into avionics testbeds equipped with two experimental radars and special recording equipment.

The measurement and recording system (MRS) developed by NII-2 was housed in the *Beagle*'s bomb bay. It included a high-speed ciné camera capturing the radar pulses reflected from the ship and appearing as lines on the radar display. The two testbeds made more than 50 flights from Kirovskoye airbase on the Crimea Peninsula, using Black Sea Fleet cruisers, destroyers and minesweepers as 'targets'. The ships were either anchored on the roadstead at Feodosiya or moved on predetermined headings. Measurements were made in 38 flights at 2,000-5,000 m (6,650-16,400 ft) and 110-167 m/sec (360-550 ft/sec) at 10 to 50 km (5.4 to 27 nm) range.

43 measurements were made with the cruisers, 64 with destroyers and 40 with minesweepers at various sighting angles in various sea states. The results were analysed by a computer, making it possible to develop algorithms for determining the class of a ship in a group; this helped to develop guidance systems for stand-off anti-shipping missiles.

Ilyushin IL-28U automatic landing system testbed

An IL-28U *Mascot* bomber trainer coded '18 Blue' was converted into a testbed for an automatic landing system, sporting several non-standard aerials under the forward and rear fuselage. The tests took place in 1957.

Ilyushin IL-38 radar testbed

As mentioned above, the Korshoon search radar developed for the Tu-142M underwent tests on a converted IL-38 ASW aircraft (identity unknown).

Ilyushin IL-62 navigation systems testbed

In 1978 an early-production IL-62 *sans suffixe* (*Classic*) long-haul airliner powered by Kuznetsov NK-8-4 engines (CCCP-86674, c/n 80304) was converted by LII into a testbed for LORAN systems and inertial navigation systems (INS). Outwardly it was virtually a standard IL-62, featuring only an additional dorsal

IL-18V CCCP-78732 was a testbed for the Polyot-1A1 navigation system. Note the small cylindrical pod under the forward fuselage.

The IL-28LL radar testbed

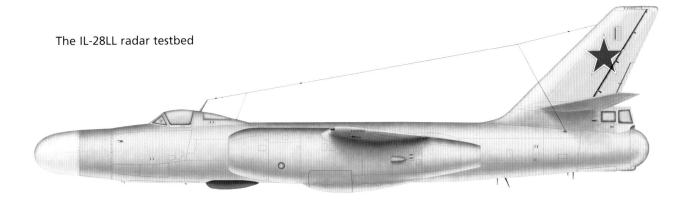

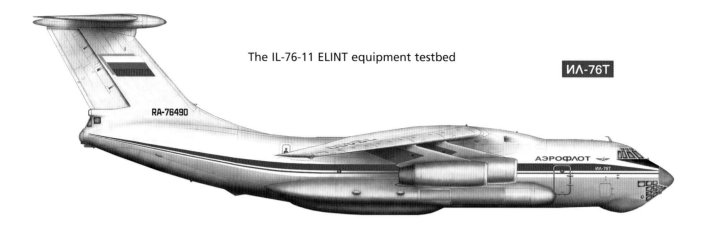

The IL-76-11 ELINT equipment testbed

ИЛ-76Т

RA-76490

The IL-76-11 avionics testbed (RA-76490) sits engineless at Zhukovskiy in August 1993. The ELINT systems fairing is just visible ahead of the starboard main gear fairing.

Close-up of the ELINT systems fairing on the IL-76-11, showing the dielectric panels of unequal size.

strake aerial offset to starboard immediately ahead of the wings. The results of the work performed on this aircraft were used for perfecting the methods of testing inertial navigation systems and LORAN systems. According to some sources, the machine was also used for testing an automatic landing approach system. The testbed was reportedly used until 1985, but the aircraft stayed active until 1993, gaining the Russian nationality prefix and becoming RA-86674.

Ilyushin IL-62M navigation systems testbed

Later, between 1984 and 1992, LII used an IL-62M powered by Solov'yov D-30KU engines for researching and testing LORAN and

autonomous navigation systems. The aircraft was converted from an IL-62M 'Salon TM-3SUR' VIP jet, CCCP-86515 (c/n 2138657), retaining this version's distinctive fat fuselage spine fairing associated with the Surgut-T secure communications suite. It took over the duties of its predecessor – the abovementioned IL-62 CCCP-86674.

In addition to the IL-62M's standard avionics suite permitting domestic and international flights, CCCP-86515 was equipped with LORAN, SATNAV and INS; a large flat-topped fairing with an astrodome at the front was added on top of the forward fuselage for testing astro-inertial navigation systems. Means of recording the standard-setting flight parameters (an aerial camera and flight trajectory measurement sensors) were installed, as were an on-board time synchronisation system and a data recording system. The aircraft could stay airborne for up to ten hours.

A special software package called *Etalon* (Standard) was used for detailed post-flight processing of the flight research and test materials. Quick approximate analysis of the test materials could also be performed in real time on board the aircraft, using an automated processing system. Detailed analysis of the flight test materials could be completed in two or three days after the flight.

The test results obtained with CCCP-86515 facilitated the introduction of LORAN systems on civil and military aircraft (the Tu-154 *Careless* medium-haul airliner, the IL-86 *Camber* wide-body medium-haul airliner, the MiG-31 interceptor and so on). It is probable that this aircraft was also used in experiments with blind landing systems (a report about such experiments conducted on an IL-62 in 1988 did not specify the aircraft).

In 1993 the testbed, reregistered RA-86515, was stripped of its mission equip-

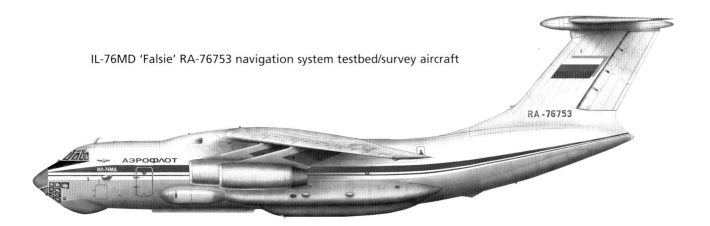

IL-76MD 'Falsie' RA-76753 navigation system testbed/survey aircraft

ment and sold to Moscow Airways for use as a passenger aircraft. Later it saw service with the Moscow-based cargo carrier Airstars as a freighter (still retaining the fat fuselage spine); it is now in storage at Moscow-Domodedovo.

Ilyushin IL-76-11 ELINT equipment testbed

NPO Vzlyot converted IL-76T RA-76490 (ex-CCCP-76490, ex-Iraqi Air Force IL-76M YI-AKO, c/n 093416506, f/n 1307) into an ELINT equipment testbed called IL-76-11 in some sources. Interestingly, despite wearing 'IL-76T' nose titles, the aircraft had a tail gunner's station as befits an IL-76M; therefore one might call it an IL-76T 'Falsie'.

The ELINT antennas were housed in a large semi-cylindrical fairing on the starboard side only which blended smoothly into the main gear fairing. The antenna fairing was unpainted and thus plainly visible; it had a cut-off front end to clear the entry door and incorporated three dielectric panels of different size.

The aircraft was first seen at Zhukovskiy in this configuration during the MAKS-93 airshow. By August 1995 it had been withdrawn from use, sitting engineless at Zhukovskiy. By August 1999 RA-76490 had been reconverted to standard configuration and returned to service with Elf Air, the flying division of NPO Vzlyot.

Ilyushin IL-76MD and IL-76T 'Falsie' SATCOM equipment testbeds

IL-76MD CCCP-76790 (c/n 0093496903, f/n 7306) and an unidentified IL-76T 'Falsie' were converted into satellite communications and data link equipment testbeds for the Ilyushin/Beriyev A-50 *Mainstay-A* airborne warning & control system (AWACS) aircraft. Both aircraft had a large dielectric fairing over

SATCOM antennas ahead of the wing centre section; CCCP-76790 was demilitarised. In 1992 both aircraft were sold to the Yekaterinburg-based airline SPAir; the experimental avionics had been removed by then but the dielectric fairing remained. This fairing vanished from RA-76790 after an overhaul and conversion to IL-76TD standard.

Ilyushin IL-76MD 'Falsie' navigation system testbed/ geophysical survey aircraft

In December 1987 the Moscow-based Fine Instruments Research Institute (NIITP – *Na**ooch**no-is**sled**ovatel'skiy insti**toot toch**-nykh pri**bor**ov*) took delivery of brand-new IL-76MD 'Falsie' CCCP-76753 (with 'IL-76MD' nose titles but an IL-76TD-style tailcone; c/n 0073481461, f/n 6206), which was immediately converted into a navigation systems testbed. The aircraft is equipped with an IK-VR centimetre-waveband synthetic-aperture SLAR; the designation probably stands for *izme**rit**el'nyy **kom**pleks vysok**ovo** razre**sh-en**iya* – high-resolution measurement suite. The SLAR antennas are located on both sides of the fuselage ahead of the main gear fairings in flattened semi-cylindrical fairings which hinge open fully for maintenance.

Apart from navigation systems testing the aircraft can also be used for geophysical and ecological survey. To this end it is equipped with *Mala**khit*** (Malachite), ***Po**isk* (Search, or Quest) and NP-50 infra-red scanners; their sensors are located in square flat projections atop the wing centre section fairing and under the forward fuselage just aft of the emergency exits. An unswept blade aerial is added just aft of the flight deck. The conversion job took the better part of a year and the first post-conversion flight took place on 4th October 1988.

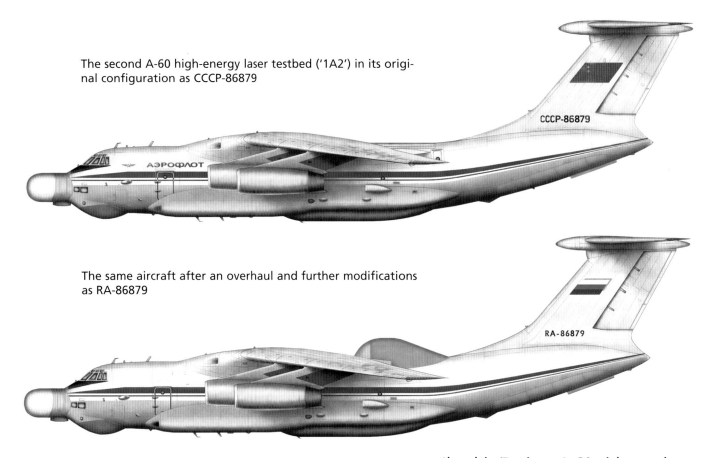

The second A-60 high-energy laser testbed ('1A2') in its original configuration as CCCP-86879

The same aircraft after an overhaul and further modifications as RA-86879

The test equipment is located in a compact cubicle in the forward part of the freight hold, leaving the rest of the hold free, so that the aircraft can carry cargo and generate revenue as well! Reregistered RA-76753, the aircraft was in the static park at the MAKS-95 airshow in Zhukovskiy which is its home base.

Ilyushin/Beriyev A-60 airborne laser laboratory (*izdeliye* 1A)

When US President Ronald Reagan announced the Strategic Defence Initiative (SDI) supposed to protect the US from Soviet intercontinental ballistic missiles, US weapons makers set to work inventing new weapons systems, including laser weapons. Perhaps it was the laser bit that caused the SDI to be nicknamed Star Wars.

One unusual aircraft used in the Star Wars programme was the Boeing NKC-135ALL (Airborne Laser Laboratory) converted from KC-135A-02-BN 55-3123 (c/n 17239, f/n T0006). Photos of this aircraft with its characteristic dorsal canoe fairing incorporating a high-energy laser (HEL) turret were sometimes

The second A-60 laser weapons testbed ('1A2'), CCCP-86879, at the Beriyev facility in Taganrog. The bulbous nose radome and the cheek fairings housing turbine power packs are clearly visible.

used by the Soviet press for propaganda purposes (to illustrate articles exposing the 'evil Yankee militarists preparing to unleash war in space'). What these articles did not tell, however, was that the Soviet Union had a similar aircraft up its sleeve.

A photo of the prototype Kamov Ka-32 *Helix-D* helicopter (CCCP-31000) published by the TASS news agency in June 1991 showed two of the four Ilyushin IL-80 *Maxdome* national emergency airborne command posts in the background. On seeing this photo, the writers at *Flight International* perceived the apparent similarity of the *Maxdome's* SATCOM fairing with the NKC-135ALL's superstructure and raised a ballyhoo, claiming a 'Soviet laser experiment revealed'! In reality, however, the Soviet laser testbed was a much-modified IL-76. The aircraft was modified extensively enough to warrant a separate designation, A-60; another designation was *izdeliye* 1A.

The reader might argue that such an aircraft belongs in the weapons testbed chapter. Yet, since laser weapons are nowhere near entering service, the authors, on some deliberation, have chosen to place the A-60 here.

The Beriyev OKB and the co-located Taganrog aircraft factory (TMZD) did the conversion job; the *A*/**maz** (Diamond) Central Design Bureau was the main contractor responsible for the HEL. The appearance of the A-60 could easily be described as hair-raising. The weather radar was replaced by a bulbous fairing reminiscent of the Boeing C-135N ARIA (Apollo Range Instrumentation Aircraft), if rather smaller. This probably housed a target acquisition/tracking antenna dish. The extensive navigator's glazing was almost entirely faired over, leaving only two small windows each side; the result was a vicious scowl instead of the *Candid's* familiar big grin.

The laser used a tremendous amount of electric power which the standard engine-driven generators could not provide. Hence two mighty gas turbine power units were installed on the forward fuselage sides in huge fairings which blended into the main gear fairings, making the aircraft look like a chipmunk. Each unit was a 2,820-ehp (2,103-kW) Ivchenko AI-24VT turboprop engine (the type powering the An-24, An-26 and An-30) driving four 90-kilowatt GT-90PCh6A stable-frequency three-phase AC generators. The fairings had large circular air intakes at the front and downward-angled jetpipes at the rear.

The power pack fairings overlapped the entry doors, which had to be modified (a small

part of the fairing swung open together with the door and the handle was relocated). The port forward emergency exit was deleted. The TA-6A auxiliary power unit was moved to the rear portion of the port main gear fairing, receiving a dorsal 'elephant's ear' intake and a downward-angled exhaust *à la* A-50.

The HEL gun turret was buried in the rear fuselage, popping up through dorsal doors immediately aft of the wings. The cargo doors were removed and faired over to add structural stiffness to the rear fuselage; the cargo ramp remained but was non-functional.

The first A-60 ('1A', identity unknown) entered flight test on 19th August 1981, captained by Beriyev OKB test pilot Yevgeniy A. Lakhmostov. Unfortunately no more details are known, except that the aircraft is said to have been written off in 1989 after a ground fire at Chkalovskaya AB. Ten years later, on 29th August 1991, the second A-60 ('1A2') converted from IL-76M 'Falsie' CCCP-86879 (c/n 0013430893, f/n 2304) made its first flight, captained by Beriyev OKB test pilot Vladimir P. Dem'yanovskiy. Originally it had the same fully retractable HEL turret; later the aircraft was reregistered RA-86879 and modified to feature an HEL turret in a large dorsal fairing.

Work on the A-60 is continuing in response to the USA's repudiation of the Anti-Missile Defence treaty and the development of the AL-1 ballistic missile interceptor – a modified Boeing 747-400F freighter armed with an HEL.

The same aircraft in flight with the Russian registration RA-86879.

The logo of the Beriyev OKB which co-developed the A-60.

A-60 RA-86879 at Taganrog in modified form. Note the new dorsal fairing for the laser gun turret.

The IL-114LL arrives at Zhukovskiy for the MAKS-2007 airshow. Note thew Russian Navy flag aft of the 'Radar' (that is. Radar-MMS) titles.

Ilyushin IL-114LL surveillance/ targeting systems testbed

An IL-114 regional airliner (RA-91003, c/n 2053800109) was custom-built for the St. Petersburg-based avionics house NPP Radar-MMS, which outfitted it as an avionics testbed designated IL-114LL for developing ELINT and surveillance systems for the Russian Navy. A small pod with an ARGS-14E active radar seeker head (*aktivnaya rahdiolokatsionnaya golovka samonavedeniya*) for an anti-shipping missile at the front was located aft of the nose gear unit; this pod also housed a retractable gyrostabilised optoeletronic imaging 'turret' protected by clamshell doors. Two detachable conformal pods housing SLAR antennas could be installed under the wing roots; the SLAR operated in the metre and

centimetre wavebands. Twelve mission operator workstations were outfitted in the cabin, with a crew rest area at the back.

The aircraft was unveiled in model form at the IMDS-2004 (the annual International Maritime Defence Show) in St. Petersburg in July 2004; that same year Radar-MMS placed an order for the actual airframe with the Tashkent Aircraft Production Association manufacturing the IL-114. Just over a year later, the real aircraft was on display at the MAKS-2005 airshow but the SLAR pods were not yet fitted. At the MAKS-2007 the IL-114LL showed up again – this time with the pods in place. However, at the IMDS-2010 and the Hydro Aviation Show-2010 in Ghelendjik on the Black Sea the aircraft was again seen flying minus SLAR pods.

The IL-114LL makes a flypast at the MAKS-2007, showing the SLAR antenna pods and the pod with the missile seeker head.

Kamov Ka-25 landing approach system testbed

In 1967-78 a modified Kamov Ka-25PL *Hormone-A* shipboard ASW helicopter was used by LII for testing automatic landing approach systems facilitating shipborne operations. The results of the tests were used in developing the approach system for the Ka-27 *Helix*-A shipboard ASW helicopter and Ka-29 *Helix*-B shipboard transport/assault helicopter.

Kazan' Helicopters Ansat-LL targeting systems testbed

In 2005 NPP Radar-MMS converted an early-production Kazan' Helicopters Ansat light utility helicopter (the name means 'easy' in the Tatar language) for testing a new navigation system with naval aviation applications. The company had placed an order for the airframe in 2000. Designated Ansat-LL, the testbed was on display at the MAKS-2005; in May 2006 it was formally delivered to the Russian Navy.

Registered 20440 (c/n 050101), the helicopter differed outwardly from the standard Ansat in having a modified nose fairing incorporating a radar seeker head for an anti-shipping missile, with a dorsal bulge aft of it, and a gyrostabilised optoeletronic imaging 'turret' under the nose. Also, the Ansat-LL had no cabin door on the port side, showing it was custom-built for the testbed role.

Lavochkin La-200B as a radar testbed

When the main part of the RP-6 Sokol radar's test programme on the modified IL-28 ended, testing continued on the second prototype of the La-200 two-seat all-weather interceptor (La-200B) which, after being rejected by the Air Force, was transferred to NII-17 in December 1952. This version had been designed to take the RP-6, featuring a huge radome around which three air intakes feeding the two engines were arranged; therefore no additional modifications to the airframe were necessary.

Manufacturer's tests of the RP-6 lasted throughout the first half of 1953, with Mark L. Gallai as project test pilot. State acceptance trials of the radar on the La-200B at GK NII VVS began on 26th June. Eventually the manufacturer succeeded in ironing out the radar's main 'bugs' and simplifying its operational and maintenance procedures. The La-200B made a total of 109 flights in the radar testbed role.

By mid-1953 the radar had reached an adequate reliability standard. Hence in November 1953 the tests of the radar on the Lavochkin interceptor were terminated and the go-ahead was given to install the radar in the Yak-120. The result was the Yak-120M, known as the Yak-25M in production form.

Lisunov Li-2 guidance system testbed for KS cruise missile

Development of the guidance system for the first Soviet anti-shipping cruise missile, the Mikoyan KS-1 (*izdeliye* K; NATO codename AS-1 *Kennel*), was conducted initially on two converted Li-2 aircraft. One of the modified transports emulated the cruise missile, while the other posed as the missile carrier aircraft. This was followed by experiments with the MiG-9L and MiG-17SDK testbeds described below, and only thereafter by tests of the missile proper.

Lisunov Li-2 guidance system testbed for K-10S cruise missile (Li-2IRE, Li-2LL IRE?)

On 3rd February 1955 the Soviet Council of Ministers issued a directive initiating the development of the K-10 anti-shipping missile system (code-named *Loo*ga-S after the River Looga in the Leningrad Region), which was ultimately destined to remain in the inventory of the Soviet Naval Aviation for more than 30 years. The system was built around the K-10S supersonic cruise missile (*izdeliye* 352; NATO

20440, the Ansat-LL testbed, at the MAKS-2005. Again, the helicopter wears Radar-MMS titles and the Russian Navy flag. Note the nose radome and optoeletronic 'turret', as well as the lack of the port cabin door.

The logo of NPP Radar-MMS whose missile guidance systems were tested on the IL-114LL and the Ansat-LL.

A still from the movie *Bar'yer neizvesnosti* showing the Li-2IRE testbed for the K-10 missile's guidance system.

codename AS-2 *Kipper*). The missile was designed for destroying surface ships with a displacement of 8,000-10,000 tons and designed to be launched by the Tu-16K-10 *Badger-C* naval missile carrier.

Development of the missile began at the Mikoyan OKB's Section K, aka OKB-2-155, in November 1957 under the leadership of Mikhail I. Gurevich. (It subsequently became a separate entity called MKB **Raduga** – 'Rainbow' Machinery Design Bureau.) Test work in support of the programme was conducted simultaneously on four aircraft, including a Li-2 converted into an avionics testbed. This aircraft featuring the seeker head of the K-10 missile on a conical adapter attached to the fuselage nose (the result looked like a sharpened pencil) and fitted with some other additional devices flew test missions from the airfields of the Black Sea Fleet Air Arm.

This machine was accidentally caught on camera and a glimpse of it could be seen in

the Soviet feature film *Bar'yer neizvesnosti* (The Barrier of the Unknown) released by Lenfilm in 1961, provoking a lot of speculations on what it could be. In some publications this aircraft was called Li-2LL IRE or Li-2IRE; the IRE acronym denoted *Instituut rahdioelektroniki* – that is, MIREA to which the aircraft allegedly belonged, although no information was published to substantiate this claim.

Lisunov Li-2 avionics testbed with Oval radar
The OKB-30 avionics design bureau led by A. P. Golubkov was tasked with preparing technical documentation for the installation of the Oval radar on a Li-2. One production Li-2 was converted to this configuration.

Lisunov Li-2 avionics testbed with Proton-M radar
Six production Li-2s were equipped with the Proton-M radar; this version was likewise developed by OKB-30.

Lisunov Li-2 with Azimuth SLAR
A Tashkent-built Li-2T coded '08 Red' (c/n 33444205) was fitted with the Azimuth side-looking aircraft radar; the radar set was accommodated in the cargo cabin. The aircraft was also fitted with equipment for remote IR sensing of the Earth, which was located near a special hatch on the port side of the rear fuselage. This equipment was developed in 1958 and flight-tested in 1959.

Li-2REO '08 Yellow' (c/n 23440603) seen near Leningrad in 1958. Note the ventral radome.

Lisunov Li-2 in experiments to study data characteristics of targets

In late 1950s intensive work was conducted in the Soviet Union on various air-launched missile systems. In order to create new radar-based airborne target designation and guidance systems intended for guiding cruise missiles to high-priority maritime surface targets, it was necessary to analyse the characteristics of the radar echo created by the target. To this end, in 1960 the MoD's Central Research Institute No. 30 (TsNII-30) together with NII-2 (Research Institute No. 2), NII GKRE (Research Institute of the State Committee for Radio-electronics, an establishment responsible for the development of radar target designation and guidance systems) and NII-4 VMF (Research Institute No. 4 of the Soviet Navy) organised and conducted test flights intended to examine these parameters within the framework of the framework of the Saksaul (saxaul, a desert plant) R&D programme.

Two types of target illumination radars and experimental equipment for recording the parameters of the radar echoes were installed in a modified IL-28 bomber and a modified Li-2 transport. The role of targets was played by Black Sea Fleet cruisers, destroyers and minesweepers which were moored on the roadstead of Feodosiya (a seaport in the east of the Crimea Peninsula) or moved along a predesignated route. The aircraft took a course which was at right angles to the movement of the ships. Recording of characteristics was performed in 38 flights at altitudes between 2,000 and 5,000 m (6,560-16,400 ft) and at flight speeds of 110-167 m/sec (360-548 ft/sec), the distance between the aircraft and the target ranging from 10 to 50 km (6-31 miles). 43 recordings of reflected signals from a cruiser, 64 signals from a destroyer and 40 signals from a minesweeper were used for the purposes of processing. The recorded signals were processed on a computer.

Other Lisunov Li-2 avionics testbeds

Other Li-2s were used as flying testbeds by various organisations dealing with the development of avionics for military purposes. One of them, registered CCCP-04368 (in the registration block reserved for Polar Aviation aircraft), was used for developing the bomb-aiming radars which were planned for installation on the Tu-4 bomber; the radar was housed in a large teardrop radome under the wing trailing edge.

There was also a Li-2 coded '08 Yellow' (c/n 23440603) which was referred to in some documents as the Li-2REO (*rahdioelektronnoye oboroodovaniye* – electronic equipment). Again, it featured a ventral radome (this time of hemispherical shape) aft of the wings; it was used for developing various types of radio equipment for military use. The aircraft was later registered CCCP-13386.

Mikoyan/Gurevich I-320 (*izdeliye* R) as a radar testbed

In the late 1940s the Mikoyan OKB developed the I-320 (*izdeliye* R) two-seat twin-turbojet all-weather interceptor equipped with the *Toriy-A* (Thorium) AI radar. It was a direct rival to the broadly similar La-200 and entered flight test in April 1949. However, by trying to get ahead of the competitor by all means the Mikoyan OKB did itself a disservice. The first prototype (designated R-1) turned out to be suffering from a rash of teething troubles; all attempts to cure them were in vain, and in November 1949 the manufacturer's flight tests were abandoned. The R-1 was transferred to NII-17, which used it as an avionics testbed for debugging the Toriy-A radar.

Mikoyan/Gurevich MiG-9L (*izdeliye* FK) missile guidance system testbed

A major issue during the development of stand-off air-to-surface missiles is the testing of the guidance system. The general problem with missiles is that test launches result in the destruction of the prototypes and thus do not yield as much information for their further refinement as in the case of aircraft. The obvious solution is to use a piloted analogue which can be landed safely, allowing test data to be analysed.

Hence a single-seat MiG-9 *Fargo* fighter (*izdeliye* FS – that is, *izdeliye* F se*reey*noye, production version) was converted into an avionics testbed for the purpose of verifying the guidance system of the KS-1 anti-shipping cruise missile. Looking rather like a scaled-down MiG-15P*bis* (*izdeliye* SP-1) experimental interceptor minus cockpit canopy, this weapon had been developed by OKB-2-155 under the Kometa (Comet) programme; KS stood for *Kometa snaryad* (Comet missile). Hence the heavily modified *Fargo*, which bore the designation MiG-9L (for [*le*ta*yushchaya*] labora*tor*iya), had the product code *izdeliye* FK –

The bizarre-looking MiG-9L (*izdeliye* FK) missile guidance system testbed. The nose and tail radomes, the wing-mounted antennas and the second cockpit for the guidance system operator are clearly visible.

Close-up of the MiG-9L's fin-mounted antenna pod.

that is, modified for tests under the Kometa programme.

The conversion job was completed on 14th May 1949. The aircraft had a crew of two – the pilot and the guidance system operator (the latter sat in a separate unpressurised cockpit with a bubble canopy immediately aft of the wing trailing edge). Like the KS-1 cruise missile, the FK was equipped with two radars. A K-1M target illumination radar (used during the terminal guidance phase) was installed in a

large bullet-shaped radome above the air intake, protruding perceptibly above the nose contour. The radar echo was received by two antennas accommodated in small cylindrical pods mounted on the wing leading edge. A second, aft-looking transceiver radar was installed in a large cigar-shaped fairing on top of the fin; it was intended for testing the systems providing mid-course guidance from the missile's carrier aircraft, as well as the guidance system of the missile itself.

Development of the Kometa system's avionics on the MiG-9L (*izdeliye* FK) testbed began in 1949 and went on for four years. Other test aircraft involved in the programme included the MiG-17K (MiG-17SDK) cruise missile emulator, which had the same guidance system, and the KSK experimental aircraft which was a manned version of the actual KS-1 (these are described below).

In the course of the *Kennel*'s trials, prototype missiles were launched from a Tupolev Tu-4KS missile carrier aircraft. Eventually the missile passed its state acceptance trials successfully in 1952-53 and was included into the Soviet Navy inventory, the Tu-4KS and later the Tu-16KS *Badger-B* acting as the carrier aircraft.

Mikoyan/Gurevich MiG-15*bis* avionics testbed with SRD-3 Grad gun ranging radar

On 6th October 1951 a US Air Force/4th Fighter Interceptor Wing North American F-86A-5-NA Sabre (49-1319/'FU-319', c/n 161-313) was shot down and captured by the Soviet forces during the Korean War, being delivered to the Soviet Union for detailed study. This was an invaluable prize for the Soviet aircraft industry, providing it with a wealth of information on the potential adversary's technologies and aircraft systems. Many equipment items from the Sabre were studied, tested, reverse-engineered and put into production. These included the AN/APG-30 gun ranging radar, whose Soviet copy was designated SRD-3 Grad (Hail; pronounced *grahd*); SRD stood for *samolyotnyy rahdiodal'nomer* – aircraft-mounted radio rangefinder.

In mid-1952 the Mikoyan OKB converted a production MiG-15*bis Fagot-B* into a testbed for the SRD-3. No separate designation has been quoted for this development aircraft. Little is known except that the forward fuselage was redesigned up to frame 4; the radome was built into the intake upper lip, causing the S-13 gun camera to be relocated to the starboard side of the intake. The SRD-3 was later tested on a modified MiG-17 designated *izdeliye* SG (see below).

Mikoyan/Gurevich MiG-17 (*izdeliye* SG) avionics testbed

The equipment items sourced from the captured F-86A also included the A-1C computing optical gunsight copied as the ASP-4N Sneg (Snow; ASP = *avtomaticheskiy strelkovyy pritsel* – automatic gunsight). It could project the aiming reticle and target range data on the bulletproof windscreen, thus acting as a head-up display, which was very convenient.

In mid-1952 the Mikoyan OKB developed a version of the MiG-17 *sans suffixe* (*Fresco-A*) equipped with the ASP-4N gunsight (in lieu of the standard ASP-3N) and the SRD-3 gun ranging radar. The aircraft was converted from the 14th Gor'kiy-built example ('114 Red', c/n N54210114) and completed in October 1952, entering flight test early next year. The manufacturer's designation was *izdeliye* SG – that is, *izdeliye* S [s *sistemoy*] Grahd.

The forward fuselage of the *izdeliye* SG was redesigned up to frame 4, representing a curious blend of Sabre and MiG. The aircraft

had the F-86's characteristic 'beak' of the gun ranging radar radome on the intake upper lip; this caused the S-13 gun camera to be relocated to the starboard side of the intake. However, the dorsal avionics/equipment bay cover ahead of the windshield characteristic of the early MiGs (up to and including the MiG-21 *Fishbed*) and the standard cannon armament were retained. The windshield was lengthened to accommodate the ASP-4N gunsight, featuring two curved sidelights and an upper glazing panel *à la* MiG-17P/PF, and the bulletproof windscreen was more sharply raked, Sabre-style (30° instead of 37°). The DC battery was relocated to the ventral fin to make room for the gun ranging radar and the instrument panel was suitably redesigned. The standard 3-kW GSR-3000 generator was replaced by a 6-kW GSR-6000 unit, and the RSIU-3 radio and the ARK-5 Amur automatic direction finder (*avtomaticheskiy rahdiokompas* – ADF) were powered by a common MA-250 transformer.

In addition to the standard armament the fighter carried two pods with eight 57-mm (2.24-in) ARS-57 folding-fin aircraft rockets (FFARs) each on short unswept pylons. Rocket launch was controlled from a panel in the cockpit and automatically disabled when the landing gear was down.

The conversion was completed on 28th October 1952. On 16th November, after passing its manufacturer's flight tests, the aircraft was transferred to NII-17 for equipment tests. Meanwhile, a further two MiG-17s were custom-built to *izdeliye* SG standard at the Gor'kiy aircraft factory No. 21. In the first six months of 1953 the three aircraft underwent state acceptance trials at GK NII VVS with the purpose of verifying the ASP-4N gunsight and the SRD-3 gun ranging radar.

Both items were recommended for production. However, the SRD-3 could not be brought up to scratch and was eventually rejected in

The forward fuselage of the *izdeliye* SG avionics testbed ('114 Red') with the avionics bay cover removed, showing the SRD-3 gun ranging radar.

favour of the more refined indigenous Kvant (Quantum) gun ranging radar which found widespread use on Soviet fighters. The Kvant was also installed on new-build MiG-17s and retrofitted to earlier *Frescos* in service.

Mikoyan/Gurevich MiG-17 (*izdeliye* SG-5) avionics testbed

Later, '114 Red' was refitted with the ASP-5N gunsight, retaining the SRD-3 gun ranging radar; in this configuration the aircraft was known as *izdeliye* SG-5. Conversion work was completed in December 1953 and the aircraft commenced manufacturer's flight tests on 11th January 1954.

Once again the gunsight and gun ranging radar were tested by NII-2 and TsKB-589. Shortly afterwards the ASP-5N gunsight was removed for modifications and returned on 26th April 1954 for re-installation. On 12th May the SG-5 commenced state acceptance trials, passing them satisfactorily.

Mikoyan/Gurevich MiG-17 radar warning receiver (?) testbed

One MiG-17 coded '32 Red' featured non-standard probe-like excrescences on the intake upper lip (almost like a refuelling probe) and on top of the fin (facing aft). The aircraft was probably an avionics testbed for a radar warning and homing system.

This MiG-17 coded '32 Red' was an avionics testbed, possibly for testing a SHORAN system. Note the bullet fairings on the nose and tail and the flush antennas in the fin.

Mikoyan/Gurevich MiG-17P (*izdeliye* SP-8) avionics testbed

In 1953 NII-17 led by Viktor V. Tikhomirov developed the RP-5 Izumrood-5 AI radar. This was a derivative of the RP-1 Izumrood-1 with increased detection range (12 km/6.48 nm), higher resistance to jamming and a field of view of ±60° in azimuth and +26/–14° in elevation (some sources say +26/–16°). It was only logical that this radar should be fitted to the MiG-17PF *Fresco-D* all-weather interceptor. A Council of Ministers directive to this effect appeared on 9th August 1954, followed by an appropriate MAP order on 23rd August; the aircraft was to begin state acceptance trials in December.

At the end of the year a very early MiG-17P with 0.522-m² (5.61-sq ft) airbrakes and two NR-23 cannons was converted into a testbed for the RP-5. Known in-house as *izdeliye* SP-8, the unserialled aircraft was also used, together with the SG-5, to evaluate the SRD-3 Grad gun ranging radar (which in this case was linked to the RP-5 radar).

Mikoyan/Gurevich MiG-17P (*izdeliye* SP-11) avionics testbed

A production *Fresco-B* was converted into a testbed for the experimental **Vstre**cha-1 (Rendezvous) aiming radar linked to an SRD-3 Grad gun ranging radar and an ASP-4NM gunsight. Designated *izdeliye* SP-11, the aircraft also featured an SIV-52 infrared sight (*samolyotnyy infrakrasnyy vizeer* – aircraft-mounted IR sight [developed in 1952]) designed for night operations in clear weather only. The IR sight was mounted atop the instrument panel shroud and could slide back and forth along a special guide rail. In the event of ejection it was retracted by a pneumatic actuator triggered by the ejection seat mechanism in order to prevent pilot injury. The aircraft was completed on 5th December 1954.

Mikoyan/Gurevich MiG-17PF avionics testbed with Globus-2 system

A production MiG-17PF *Fresco-D* was used as a testbed for the Globus-2 system (no separate '*izdeliye* SP-something-or-other' designation is known). A DDV-1 aerial was mounted on the fin and two identical aerials were installed on the search antenna radome.

Mikoyan/Gurevich MiG-17PF avionics testbed with Yupiter FLIR

In 1956 a production MiG-17PF was used as a testbed for the *Yupiter* (Jupiter) forward-looking infra-red (FLIR) sensor; once again, no separate designation is known. The FLIR weighed 16 kg (35 lb) and was installed in place of the RP-5 radar's tracking antenna radome on the air intake splitter.

Mikoyan/Gurevich MiG-17PF avionics testbed with Aïst radar

The same aircraft was previously used to test the *Aïst* (Stork) radar mounted in the same location.

Mikoyan/Gurevich MiG-17K (MiG-17SDK, *izdeliye* SDK or SDK-5) missile guidance system testbed

Upon completion of the initial test phase involving the abovementioned modified Li-2 transports and MiG-9L, the Mikoyan OKB developed a missile emulator version of the MiG-17 for testing the KS-1 missile's guidance system. It was known as the MiG-17K or MiG-17SDK (*samolyot-dooblyor Komety* – 'doubler aircraft', that is, analogue, of the Comet missile; as mentioned earlier, until the mid-1960s the term *dooblyor* was commonly used in Soviet aviation to mean 'second prototype', but this meaning is obviously not applicable here). The manufacturer's designation was *izdeliye* SDK-5, though some documents contain the abbreviated form '*izdeliye* SDK'.

The aircraft had the K-1M guidance radar installed in a large bullet-shaped radome on the air intake upper lip protruding perceptibly above the nose contour, and the S-13 gun camera was relocated to the starboard side of the intake. Thus the shape of the forward fuselage was similar to that of the SP-1, except that the

radome protruded even further ahead of the intake. A large cigar-shaped fairing installed on top of the fin housed an aft-looking antenna for receiving mid-course guidance signals from the mother aircraft. The armament was deleted to make room for the autopilot and there was a flat-bottomed fairing offset to port under the centre fuselage over test equipment. Several MiG-17s were converted to *izdeliye* SDK-5 standard; known serials are '06 Red' (c/n N54211006?) and '007 Red' (c/n N54211007).

The trials technique was as follows. The MiG-17SDK was suspended under the wing of a bomber by means of a lug on the centre fuselage, just like a real missile; the aircraft took off and proceeded to the target area. The crew completed all preparations for missile launch as if they were working with the real thing, with the exception of engine starting which was performed by the SDK's pilot. Having located the target, the navigator/radar operator switched the Kobal't-N radar to automatic tracking mode; after making sure he had a good lock-on, he gave the OK to start the engine. The MiG-17SDK was then released by the bomber and guided to the target by the bomber's radar and then by its own K-1M radar at the terminal guidance phase; the pilot sat back and did not touch the controls. If all went well and the emulator stayed on its intended course, the pilot took over at 500-600 m (1,640-1,970 ft) from the target and brought the aircraft home where test equipment readouts would be analysed.

Later, the Mikoyan OKB built several KSK aircraft which were piloted versions of the KS-1 missile with a bicycle landing gear and a cockpit instead of the explosive charge. This was the final step towards 'the real McCoy' (see below).

The KS-1 passed its state acceptance trials successfully in July 1952 – January 1953 and was included into the Air Force and Naval Aviation inventory. After that, the MiG-17SDK emulator aircraft were used for further refine-

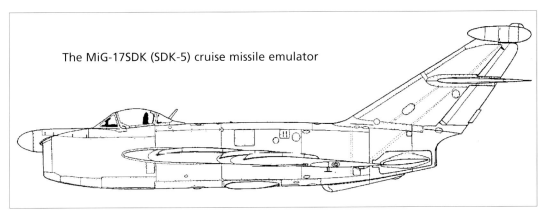

The MiG-17SDK (SDK-5) cruise missile emulator

The first MiG-17SDK (*izdeliye* SDK-5) missile guidance system testbed ('007 Red', c/n N54211007) shows the nose radome of the K-1M radar and the rear antenna for mid-course guidance.

ment of the missile. Pursuant to the Soviet Navy HQ's directive No. 53280 issued on 30th August 1955 the 124th Long-Range Heavy Bomber Regiment was formed in the Black Sea Fleet air arm. The unit operated twelve Tu-4KS missile strike aircraft and several support aircraft, including two MiG-17SDKs. MiG-17SDK operations continued in the Black Sea Fleet and the North Fleet at least until 1958 along with practice launches of KS-1 missiles; for instance, 77 simulation flights were made between 1955 and 1958 in the North Fleet alone.

Mikoyan/Gurevich MiG-17K (MiG-17SDK, *izdeliye* SDK-5TG) missile guidance system testbed

One MiG-17 was similarly converted into a cruise missile simulator with a different guidance system. The K-1M guidance radar was replaced by a Sputnik-2 (Companion, or Satellite) IR seeker head linked to an S-3 flight control system; hence the aircraft was designated *izdeliye* SDK-5TG, the TG standing for *teplovaya golovka* [*samonavedeniya*] – IR homing system. The conversion was performed by plant No. 256 in 1958.

Mikoyan/Gurevich MiG-17K (MiG-17SDK, *izdeliye* SDK-7 and SDK-7A) missile guidance testbed

Three more MiG-17s were converted to KS-1 missile emulator aircraft for the Air Force in 1953-54, plus a fourth aircraft for the Navy in July-August 1954; these aircraft were known as *izdeliye* SDK-7. Four more conversions were undertaken in the first quarter of 1956; designated *izdeliye* SDK-7A, all four aircraft were delivered to the Air Force in February-June 1956. This version probably differed from the original SDK-5 in having a different control system.

Mikoyan/Gurevich MiG-17K (MiG-17SDK, *izdeliye* SDK-7TG) missile guidance system testbed

This was an IR-homing derivative of the SDK-7. Like the SDK-5TG, it featured a Sputnik-2 IR seeker head but a K-1M guidance radar was also fitted. The conversion was performed by plant No. 256 in December 1957.

Mikoyan/Gurevich MiG-17K (MiG-17SDK, *izdeliye* SDK-15) missile guidance system testbed

A single MiG-17 was converted for testing the guidance system of the P-15 *Termit* (Termite, *izdeliye* 4K40) maritime cruise missile developed by MKB Raduga under Aleksandr Ya. Bereznyak. The conversion was performed by plant No. 256 in 1957 for KB-1.

Mikoyan/Gurevich MiG-19 avionics testbed with ARK-544 Ilim ADF

In 1956 a production MiG-19 *sans suffixe* (*Farmer-A*) was fitted experimentally with an ARK-544 Ilim automatic direction finder (pronounced *eelim*; named after a Russian river). Three different ADF aerial configurations were tried – a built-in aerial the fin, a loop aerial buried in the fuselage and two strake aerials on the fuselage sides.

Mikoyan/Gurevich MiG-19 avionics testbed with RUP-4 ILS computer

In 1956 another production *Farmer-A* was used to test the RUP-4 ILS computer (*reshayushcheye oostroystvo posahdki* – landing computer) coupled with the ARK-5 Amur ADF and GIK-1 gyro flux-gate compass. It enabled the aircraft to navigate and make automatic landing approaches, using marker beacons.

Mikoyan/Gurevich MiG-19 navigation system testbed

In 1956 an unidentified MiG-19 served as a testbed for the navigation system of the aforementioned Lavochkin Boorya intercontinental GLCM.

Mikoyan/Gurevich MiG-19 (*izdeliye* delta SM-9) with ASP-5N automatic gunsight

To ensure accurate firing and bombing a production MiG-19 was fitted with an ASP-5N computing gunsight linked to an SRD-1 gun ranging radar. The sight had two modes for firing the cannons or launching unguided rockets respectively (the two types of weapons could not be fired simultaneously). The aircraft was known as *izdeliye* delta SM-9.

Mikoyan/Gurevich MiG-19S development aircraft with VK-65 remote astrocompass

One production MiG-19S *Farmer-C* was experimentally fitted with a VK-65 remote astrocompass for determining true heading and following predetermined headings. The compass worked between 40° and 90° northern latitude if the sun was visible and at up to 70° above the horizontal plane of the aircraft.

Mikoyan/Gurevich MiG-19S (*izdeliye* SM-9/3) development aircraft with AP-28 autopilot

In 1957-58 the Mikoyan OKB tested a MiG-19S fitted experimentally with an AP-28 autopilot installed in place of the centre (fuselage-mounted) NR-30 cannon. The autopilot facilitated flying, especially in instrument meteorological conditions. The pilot could switch on the autopilot which would maintain the current flight mode (that is, climb, descent or level flight); this could be changed by means of the autopilot's formation-keeping control knob. The autopilot also incorporated a 'panic button' feature, automatically bringing the aircraft into straight and level flight from any attitude, should the pilot become disoriented.

The yaw, pitch and roll damping and autostabilisation features were tested and found satisfactory; the aircraft's handling had improved in all modes, including aerobatics. Originally the AP-28 was rigidly connected to the control circuits; later, isodromic feedback (with automatic dampers) was introduced in the pitch and roll channels. The revised AP-28A autopilot successfully passed its state acceptance trials in 1959 and was fitted to production aircraft from that year onwards.

Mikoyan/Gurevich MiG-19 (*izdeliye* SM-20 and SM-20P) cruise missile guidance system testbeds

In the mid-1950s the Soviet Union began a massive design effort aimed at creating air-launched cruise missile systems. New bombers and stand-off missiles were developed in parallel. One of these missiles was the turbojet-powered Kh-20 (AS-3 *Kangaroo*) supersonic missile developed as part of the K-20 weapons system which also included the Tu-95K-20 *Bear-B* bomber. The Kh-20 was developed by the Mikoyan bureau's Section K (OKB-2-155); it was based on a heavily-modified airframe of the Mikoyan I-7U experimental swept-wing interceptor.

To verify the missile's guidance system developed by KB-1, two early-production MiG-19 *Farmer-As* were obtained from the Air

Another MiG-17SDK with a post-1955 tactical code, '06 Red'. The aircraft is fitted with drop tanks.

'105 Red' (c/n N59210105), the SM-20/1 (aka SM-20P) manned testbed for the Kh-20 missile's guidance system. The guidance system fairing and enlarged spine with attachment lug are clearly visible, as are the MiG-19S-style high-set slab stabilisers.

Force and converted into avionics testbeds designated *izdeliye* SM-20. Perhaps it would be more accurate to call them missile simulators, since they were carried and launched by a Tu-95, just like the real thing. The first aircraft, SM-20/1 ('105 Red', c/n N59210105), was a manned version, hence the alternative designation SM-20P (*peelo**tee**rooyemyy* – flown by a pilot). The second aircraft, SM-20/2 ('425 Red', c/n N59210425), was remote-controlled.

The aircraft were lightened by removing the armament, radio, IFF transponder, gun ranging radar, marker beacon receiver, radio altimeter, RHAWS and NI-50IM navigation display. Both SM-20s were fitted with the YaK autopilot developed for the Kh-20 missile,

which received target information from the bomber's YaD (NATO *Crown Drum*) target illumination radar. The SM-20P retained the standard fighter's manual control system but could be controlled by the autopilot. An intercom was installed so that the fighter pilot could communicate with the bomber crew prior to separation. Conversion of the SM-20/1 and SM-20/2 was completed in February and September 1956 respectively.

Outwardly the SM-20 could be identified by a large teardrop fairing aft of the nose gear unit, which housed part of the autopilot's modules, and by a prominent square-section fuselage spine with a hefty lug for the bomber's BD-206 missile rack amidships. The

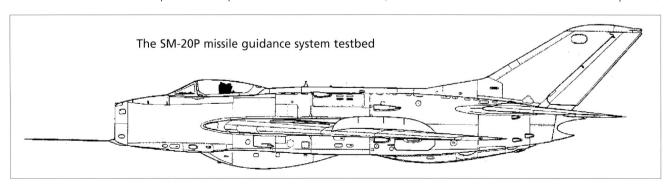

The SM-20P missile guidance system testbed

fighter was hooked up to the bomber via a special adapter which could remain attached to either aircraft after separation. The attachment lug and accompanying local reinforcement of the fuselage necessitated a reduction in the capacity of the No. 1 fuel cell.

The *Farmer-A*'s conventional stabilisers and inset elevators were replaced by slab stabilisers *à la* MiG-19S *Farmer-C* but the early-model vertical tail with a small root fillet was retained. The dielectric fin cap was more rounded than usual; it probably housed the antenna receiving mid-course guidance signals from the bomber. Another non-standard feature was a downward-pointing whip aerial on the starboard side of the nose.

The two Tu-95K prototypes converted from Tu-95 *Bear-A* bombers (c/ns 5800401 and 5800404) served as 'mother ships' for the SM-20. They were equipped with the YaD radar replacing the original glazed nose and chin-mounted RBP-4 Rubin navigation/bomb-aiming radar; the new radome was wider than the bomber's fuselage, resulting in a characteristic 'duck bill' nose. The bomb bay had a special cutout at the rear to accommodate the fighter's fin; on at least one of the bombers, small windows were added to admit some light into the bay when the SM-20/1 was semi-recessed in the fuselage. A retractable fairing was added immediately forward of the bomb bay to stop the fighter's engines from wind-milling during cruise. Ciné cameras were installed in three teardrop fairings under the inner wings and the starboard wingtip to capture the separation sequence.

Conversion of the SM-20/1 was completed in mid-1956, and tests began in the same year. Two stages of the manufacturer's flight tests were held on 4th August – 13th September, using both Tu-95K prototypes but the fighter was not yet disengaged from the bomber at this stage. The pilotless SM-20/2 was completed in late 1956, joining the test programme early next year; it differed from the first aircraft in test equipment fit.

Disengagement and guidance tests began in 1957; the SM-20/1 was flown by LII test pilot Sultan Amet-Khan and V. G. Pavlov of the Special Machinery Research Institute (an MRP division). A. I. V'yooshkov was the programme's project engineer.

The SM-20/1 was used to verify the Kh-20's guidance system at the terminal guidance stage. The aircraft was suspended under the Tu-95K with the pilot sitting in the cockpit. The engines were not started until the mother

ship's pilot gave the OK via intercom when the altitude was right. After disengagement the SM-20/1's pilot directed the aircraft towards the target 'painted' by the Tu-95's powerful radar and monitored the operation of the guidance system. When the system got a lock-on the pilot switched it off and flew the aircraft manually, landing at the home base in the usual way.

Some difficulties were experienced with engine starting at high altitude. Therefore, the SM-20/1 was refitted with modified engines featuring a carburetted ignition system.

By the end of 1957 the two SM-20 testbeds had made 27 test flights involving conventional take-off and 32 drops from the Tu-95K 'mother ship'. In October 1958 they were transferred to GK NII VVS in order to test the guidance system for electronic counter-measures (ECM) resistance. (It has to be said here that ECM resistance was not one of the Kh-20's strong points.) The SM-20/1 made nine flights at GK NII VVS, including seven from the Tu-95K.

After making more than 150 sorties between them the SM-20s were withdrawn

The SM-20/1 semi-recessed in the belly of the Tu-95K 'mother ship'. The latter aircraft featured six small windows to admit some light into the bomb bay.

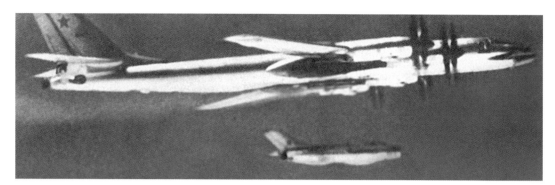

from use as time-expired. The initial test phase was deemed successful; the guidance system was removed from the aircraft and fitted to the prototypes of the Kh-20 missile.

Mikoyan/Gurevich MiG-19SMK (*izdeliye* SM-K/1 and SM-K/2) cruise missile guidance system testbeds

As mentioned earlier, on 3rd February 1955 the Council of Ministers ordered the development of the K-10 weapons system built around the K-10S anti-shipping cruise missile. As part of the design effort, in 1957 the Mikoyan OKB converted two MiG-19s – this time *Farmer-Cs* (c/ns N61210418 and N61210419) – into MiG-19SMK avionics testbeds (cruise missile emulator aircraft). These were used to test the guidance system of the K-10S cruise missile. The two aircraft were designated *izdeliye* SM-K/1 and SM-K/2 respectively.

The MiG-19SMK was fitted with the YeS guidance system working with the Tu-16K-10 missile carrier's YeN target illumination radar. (The second letter shows where the component is installed; S = *snaryad* – missile, N = *nositel'* – carrier aircraft.) The system comprised the YeS-2 mid-range guidance receiver, the YeS-1 terminal guidance (radar homing) unit and the YeS-3R autopilot. Test instrumen-

tation was also fitted. To make room for all this the cannons, ASP-5N gunsight, radio, radar warning receiver, marker beacon receiver, radio altimeter, gun camera and No. 2 fuel cell had to be removed. The hydraulics and electric system were modified.

Part of the guidance system was carried in the fuselage and the wing cannon bays; the remainder was housed in special underwing pods replacing the usual drop tanks. The control system featured hydraulic servo actuators controlled by the YeS-3R autopilot.

The aircraft commenced manufacturer's flight tests in March and October 1957 respectively. Upon completion of ground and flight tests the two MiG-19SMKs were turned over to GK NII VVS (some sources state OKB-1) in August and December 1957 respectively. Tests of the guidance system continued until late 1960; between September 1959 and November 1960 the two missile emulators made 62 simulated launches against Soviet Navy/Black Sea Fleet ships. The tests were successful, and the Tu-16K-10 and the K-10S missile entered service with the Soviet Naval Aviation in 1960.

Mikoyan/Gurevich MiG-19 ARM guidance system testbed

Another production MiG-19 was a testbed for the guidance system of the first Soviet anti-radar missile (ARM) – the Raduga Kh-58U (AS-11 *Kilter*) carried by the MiG-25BM *Foxbat-F* air defence suppression ('Wild Weasel') version. Unfortunately, no details are available of this particular testbed referred to in Russian documents as *imitahtor* (emulator).

Mikoyan/Gurevich MiG-19/APB-1 autopilot testbed

Between 1958 and 1962 a production MiG-19 (exact model unknown) was used by LII as a testbed for an automatic landing system built around the APB-1 autopilot (*avtopeelot bombardirovshchika* – bomber autopilot).

Mikoyan/Gurevich Mikoyan KSK (*izdeliye* K) missile emulator

As mentioned earlier, to speed up the testing of the KS-1 anti-shipping cruise missile the Mikoyan OKB brought out the KSK (*izdeliye* K) simulator aircraft – a manned version of the actual missile. It had a retractable bicycle landing gear, wing flaps and an unpressurised cockpit (equipped with an ejection seat, a set of controls and a very basic instrument fit) instead of the explosive charge. The KSK designation stood for *KS s kabinoy* (KS with a

Front and rear views of the KSK missile emulator aircraft on the ground.

Another view of the KSK. Note attachment lug aft of the cockpit for hooking up to the Tu-4KS. The size of the cockpit gives an idea of how small the aircraft was.

On the ground the KSK had a marked nose-up attitude.

The Tu-4KS prototype (c/n 224203) carrying two KSK missile emulators; one is the fourth example (K-4). Note the cutouts in the missile pylons to accommodate the cockpit canopies.

An interesting aspect of a KSK as it falls away from the 'mother ship', showing the attachment lug and the small span and sharp sweep of the wings and stabilisers.

cockpit). (On reflection, had Western experts been aware of the KSK's existence, they would undoubtedly have decided the crazy Russians were developing suicide bombs!)

Four examples of the emulator aircraft (designated K-1 through K-4) were built. The attachment lug for hooking the missile up to the bomber was immediately aft of the cockpit, therefore the missile pylons of the Tu-4KS prototype (no serial, c/n 224203) had to be modified by providing cutouts to accommodate the KSK's cockpit canopy. As in the case of the real missile, the engine of the KSK was

started by the crew of the mother ship; unlike the actual KS-1, however, the aircraft had a standard RD-500 turbojet with a normal throttle and fuel control unit instead of the expendable RD-500K (*korotkoresoorsnyy* – with a short service life).

In 1951-52 the Tu-4KS prototype passed manufacturer's flight tests at Chkalovskaya airbase and the Bagerovo test range near Kerch on the Crimea Peninsula, in the course of which the system was brought up to scratch. In numerous flights the KSK aircraft was carried aloft and launched by the Tu-4KS. Two

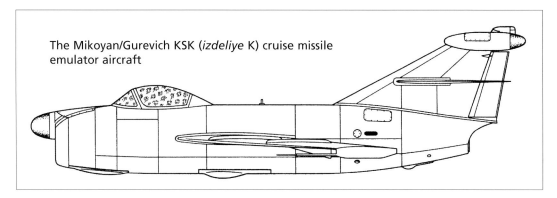

The Mikoyan/Gurevich KSK (*izdeliye* K) cruise missile emulator aircraft

'manned missiles' were always carried but only one was launched at any one time. The bomber's propeller discs were uncomfortably close to the cockpit and the pilots were reluctant to gun the engine immediately after separation, fearing a collision; as a result, the KSK would drop 600-800 m (1,970-2,625 ft) below the bomber's flight path, which made it difficult to catch the mid-course guidance beam.

After 150 manned flights had been made, the appropriately modified K-4 commenced a series of unmanned (remote-controlled) launches in May 1952. Further testing was done with the actual KS-1 missiles.

Mikoyan/Gurevich Ye-7M development aircraft

The improvements made in the course of the MiG-21PF *Fishbed-D* (and subsequently the MiG-21PFS *Fishbed-D/F*) interceptor's production concerned not only the structure but the interceptor's equipment as well. The Ye-7SPS, as the MiG-21PFS was known at OKB-155, was followed by the Ye-7M (*modernizeerovannyy* – updated) featuring an upgraded targeting system. The interceptor was fitted with a compact ASP-PF computing sight, which was integrated with the RP-21 radar and a *Samotsvet* (an old Russian word for 'gem') forward-looking infra-red (FLIR) seeker. The ASP-PF sight received target range inputs from the radar, allowed the pilot to fire his missiles accurately and deliver rocket or bomb strikes against ground targets in daylight visual meteorological conditions. The FLIR detected the target at night and permitted targeting regardless of the flight level.

The Ye-7M was converted from the Ye-7SPS prototype (c/n N76210103), which had completed its state acceptance trials by then and was available for modification. First of all, the test and recording equipment housed in the forward avionics bay was removed and the RP-21 radar was reinstated. The bulletproof windscreen inside the canopy was deleted to make room for the Samotsvet FLIR and the ASP-PF sight; the latter supplanted the standard radarscope, which was moved to a new location. This required the main instrument panel and the side control consoles to be redesigned; changes were also made to the transverse beam in the cockpit supporting the instrument panel.

The Ye-7M was fitted with the K-51 weapons system. Thus, the standard BD3-60-21U wing pylons could carry either APU-13 launch rails for K-13 (R-3S; AA-2 *Atoll*) IR-homing AAMs or PU-12-40 launch rails for RS-2-US (AA-1 *Alkali*) AAMs with radio command guidance. Alternatively, APU-7 launch rails could be fitted.

The avionics included an SRZO-2M Khrom-Nikel' IFF interrogator/transponder and an *Iskra* (Spark) SHORAN system. The rear avionics bay was enlarged by cropping the upper portion of the No. 1 fuselage fuel tank to accommodate the additional equipment, and an extra access hatch was provided to starboard symmetrically to the existing one on the port side. Finally, the canopy featured a rear view mirror.

The modification work was completed on 17th October 1962. Just over two months later, on 27th December, the Ye-7M was submitted for joint state acceptance trials. Concurrently the aircraft participated in the trials of the K-51 weapons system, the greater part of which was performed on the first production MiG-21PF (c/n N76210101).

The trials revealed certain deficiencies in the RP-21 radar and the ASP-PF sight, which had to be modified. Nevertheless, the results were generally encouraging and the Ye-7M was recommended for production in 1964.

MiG-21PF c/n N76210103 (the Ye-7SPS development aircraft) was further converted into the Ye-7M with an upgraded weapons control system.

303

The logo of the Phazotron-NIIR avionics house whose radars were tested on Mikoyan fighters.

MiG-21SM '44 Red' served as an equipment testbed for the projected MiG-21DP UAV, being identifiable by the wiring conduit below the cockpit and the ventral fairing aft of the nose gear unit. It was displayed at Moscow-Khodynka in August 1989 (bottom) and at Zhukovskiy in August 1992.

Mikoyan MiG-21MF development aircraft with Volk gunsight

In the late 1960s Maj.-Gen. Protopopov, who headed one of the chairs at the Air Force Academy named after Nikolay Ye. Zhukovskiy (VVA – *Voyenno-vozdooshnaya akademiya*) in Moscow, proposed a new targeting method to be used against aerial and ground targets. This method involved dual superimposition of the sighting reticle on the target. The concept of the new computing gunsight using this method was developed by the Academy's research laboratory under the guidance of V. G. Katyushkin; the sight was code-named Volk (Wolf) and incorporated an analogue processor. A prototype of the Volk gunsight was tested on the Su-7B fighter-bomber; this first version could be used only against ground targets, but flight tests of an air-to-air capable version began shortly afterwards.

A MiG-21MF *Fishbed-J* tactical fighter refitted with the Volk sight was initially tested by GNIKI VVS in Akhtoobinsk. In the summer of 1973 the tests continued at the Tactical Aviation's 4th TsBP i PLS (*Tsentr boyevoy podgotovki i pereoochivaniya lyotnovo sostava* – Combat Training & Aircrew Conversion Centre) in Lipetsk, with I. I. Fyodorov as project test pilot. A MiG-21*bis Fishbed-L* fighter flown by A. Smetanin, another 4th TsBP i PLS pilot, initially served as a practice target; the two fighters often pulled up to 7 Gs as they manoeuvred during mock dogfights.

Later, the testing was again transferred to GNIKI VVS where the MiG-21MF attacked target drones with cannon fire, using armour-piercing rounds. The results were encouraging; at a range of some 800 m (2,620 ft) the target could be destroyed at the expenditure of just four rounds, notwithstanding the fact that the range had to be entered manually.

Mikoyan MiG-21*bis* navigation system testbed

A single MiG-21*bis* (identity unknown) became a testbed for the Polyot-1 navigation suite comprising the onboard components of the SHORAN system, a Romb-1K (Lozenge) analogue processor, an SKV-2N attitude and heading reference system (*sistema koorsovertikali* – AHRS) and an air data system. The test programme was undertaken by LII at Zhukovskiy.

Mikoyan MiG-21 AAM guidance system testbed

As part of the effort to develop the MiG-31 *Foxhound* heavy interceptor, in 1970 a MiG-21 (version and identity unknown) was converted into a testbed for the guidance system of the Vympel K-33 long-range radar-homing AAM (known as the R-33/NATO AA-9 *Amos* in production form).

Mikoyan MiG-21SM ('44 Red') testbed

By the early 1980s many MiG-21s were approaching the limit of their service lives. Apart from the obvious application as target drones, such aircraft could be converted into unmanned combat aerial vehicles (UCAVs). Therefore the Mikoyan OKB developed the technology for converting high-time MiG-21s (in particular, MiG-21*bis* fighters) to MiG-21DP (*distantsionno peeloteeruyemyy [letatel'nyy apparaht]* – remotely piloted vehicle) or *izdeliye* 21-14 standard – actually an optionally piloted vehicle. Two *Fishbeds* were converted into testbeds for the purpose of verifying the INS and the new autopilot. One of them was a MiG-21SM coded '44 Red'; this aircraft had the radar and cannon removed to make room for the new equipment. Part of the electric wiring had to be relocated to a large conduit running along the starboard side of the fuselage.

The modified aircraft was shown publicly for the first time at the aviation display which

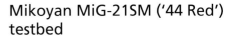

took place at Moscow-Khodynka in late August 1989. Subsequently '44 Red' appeared in the static park at the MosAeroShow '92.

Mikoyan MiG-21-93 as an avionics testbed

Upon completion of the principal test programme the prototype of the upgraded MiG-21-93 became an avionics testbed, being used for verifying the compact Phazotron-NIIR *Osa* (Wasp) fire control radar intended for light combat aircraft and for mid-life upgrades of old aircraft.

Mikoyan MiG-23S automatic landing system testbed

In 1970-75 MiG-23S *Flogger-A* (tactical code unknown, c/n 1501) was used by LII for evolving control algorithms during automatic landing approach and testing new duplicated yaw dampers.

Mikoyan MiG-23M avionics testbed

A MiG-23M *Flogger-B* in overall grey air superiority camouflage coded '07 Red' served as a testbed for an avionics item code-named *Ozon* (Ozone). This may be a new IFF transponder, as a photo of the aircraft shows it to have a non-standard dorsal blade aerial ahead of the windshield.

Upon completion of the trials the MiG-21-93 was converted into a testbed for the Phazotron Osa phased-array radar. Here the radome is removed, exposing the antenna. The piece of linoleum is placed on top merely to stop the falling snow from getting in.

(*letayushchaya laboratoriya* [*po programme izdeliya*] 9.15). Altogether, these trials lasted three years.

Mikoyan LL-915 avionics testbed

The second prototype MiG-27K *Flogger-J2* fighter-bomber (*izdeliye* 32-26, originally known as MiG-23BK) coded '362 Blue' was outfitted with the experimental SN-29 navigation system in September 1978. As its designation suggests, the system was intended for a version of the MiG-29 *Fulcrum* tactical fighter designated MiG-29M (*izdeliye* 9.15). Hence the avionics testbed was designated LL-915

Mikoyan MiG-25P IFF system testbed

A production MiG-25P *Foxbat-A* interceptor coded '76 Blue' served as a testbed for the new SRO-1P *Parol'*-2 (Password-2) IFF system comprising the *izdeliye* 632-1 interrogator and *izdeliye* 620-20P transponder. The presence of the new equipment was revealed by the distinctive triangular blade aerials replacing the equally characteristic triple rod aerials of the then-standard SRO-2 Khrom-Nikel' IFF system

This MiG-23M in overall grey air superiority camouflage coded '07 Red' served as a testbed for an IFF transponder codenamed Ozon.

(Chromium-Nickel; SRO = *samolyotnyy rahdi-olokatsionnyy otvetchik* – aircraft-mounted radar responder) that had given rise to the NATO codename *Odd Rods*.

Mikoyan MiG-25 cruise missile control system testbed

A single MiG-25P was converted into a testbed for verifying the control systems of cruise missiles developed under the guidance of General Designer Vladimir N. Chelomey.

Mikoyan LL-1104 avionics testbed

In 1977 a standard production MiG-25P (tactical code unknown, f/n 1104) was converted into an avionics testbed designated LL-1104. The aircraft served for verifying the SAU-155PD automatic flight control system (*sistema avtomaticheskovo oopravleniya*), the updated ground controlled intercept (GCI) guidance system and other avionics intended for the upgraded MiG-25PD *Foxbat-E* interceptor.

Mikoyan MiG-25PDSL development aircraft

A modified MiG-25PDS *Foxbat-E* interceptor coded '94' was designated MiG-25PDSL, entering flight test in 1985; the L stood for [*letayushchaya*] *laboratoriya* – testbed. It differed from the MiG-25PDSG development aircraft in having an L006LM *Beryoza-LM* (Birch) radar homing & warning system (RHAWS) and a podded Gardeniya-1FU (Gardenia) active jammer; the KDS-155 chaff/flare dispensers were the same as on the MiG-25PDSG.

Mikoyan MiG-25RB navigation system testbed (Project *Trapetsiya*)

The MiG-25RB *Foxbat-B* reconnaissance/strike aircraft proved a very convenient platform for

testing various electronic equipment and systems thanks to the large camera pallet in the forward fuselage. This allowed the ELINT gear and cameras to be easily replaced with experimental avionics and test instrumentation.

LII used MiG-25RBs in various trials programmes. The first of these was Project *Trapetsiya* (Trapeze) under which the aircraft was used to test the navigation system of a new cruise missile.

Mikoyan MiG-25PU avionics testbed

A MiG-25PU *Foxbat-C* interceptor trainer was modified in 1975 for testing the SAU-155MP automatic flight control system and KN-25 navigation suite (*kompleks navigahtsii*) intended for the MiG-31 (originally known as the MiG-25MP).

Mikoyan MiG-29 (*izdeliye* 9.14) development aircraft

In 1984 the Mikoyan OKB developed a version of the 'fatback' MiG-29 (*izdeliye* 9.13) *Fulcrum-C* with enhanced strike capability. Known in-house as *izdeliye* 9.14, it differed from the standard *Fulcrum-C* in having a new weapons control system which included a *Ryabina* (Rowan) low light level TV/laser designator pod. This had been developed for the upgraded MiG-27M *Flogger-J* fighter-bomber, as well as for projected light strike/counter-insurgency (COIN) versions of the MiG-101 twin-turboprop utility aircraft (which remained a 'paper aeroplane') and the Yakovlev Yak-58 five-seat piston-engined light aircraft.

The weapons range was expanded to include Zvezda Kh-25M laser-guided air-to-ground missiles, Vympel Kh-29T/Kh-29L TV-guided or laser-guided AGMs and KAB-500 'smart bombs' (*korrekteeruyemaya aviabomba* – guided bomb). The maximum bomb load was 4,500 kg (9,920 lb) with eight 500-kg bombs

This MiG-25P coded '76 Blue' was a testbed for the Parol'-2 IFF system, as revealed by the small 'bump' below the radome.

The *izdeliye* 9.14 development aircraft ('07 Blue') at Zhukovskiy.

The same aircraft in later guise as '407 Blue'. Note the Mikoyan logo on the air intake trunk.

on multiple ejector racks (MERs) under the wings and one on the fuselage centreline. (Note: The Zvezda weapons design bureau is not to be confused with Guy I. Severin's NPP Zvezda specialising in ejection seats and in-flight refuelling systems.)

The *izdeliye* 9.14 prototype, '07 Blue' (c/n 2960507682, f/n 0407), was a converted *izdeliye* 9.12 *Fulcrum-A*. Defying superstition, it flew for the first time on 13th February 1985 with Toktar O. Aubakirov at the controls. The tactical code was later amended to '407 Blue'.

The Ryabina LLLTV/laser designator system was still under development at the time, so in the summer of 1985 a dummy version was fitted for the initial flight tests. At this stage the aircraft made several test flights with the maximum ordnance load of nine FAB-500 free-fall bombs. However, live weapons tests with *izdeliye* 9.14 never began because in the mid-1980s the Mikoyan OKB was already working on the radically improved MiG-29M (*izdeliye* 9.15) which could deliver pinpoint strikes with precision-guided munitions. The *izdeliye* 9.14 programme was terminated and the one-off prototype became a 'dogship', serving for bomb armament tests, as well as for control system tests, aerodynamics research and performance testing within the *Fulcrum-C*'s state acceptance trials.

Later, when the 'fatback' had completed its trials programme, '407 Blue' was used in the MiG-29S (*izdeliye* 9.13S) upgrade programme. In 1991-92 it participated in several international airshows. In the mid-1990s the fighter was still operational at the 929th GLITs in Akhtoobinsk; in late 1997 it was transferred to LII for use in a short take-off research programme involving the MT-1 mobile ski jump. By then '407 Blue' had more than 800 flights to its credit.

Mikoyan MiG-29 (*izdeliye* 9.16) radar testbed

A production MiG-29 (*izdeliye* 9.13) coded '16 Blue' became a testbed for the Phazotron N010 Zhuk (Beetle) fire control radar developed for the MiG-29M. The latter had major differences in the fuselage structure, so, quite simply, the fuselage nose ahead of the windshield was severed and replaced by a new MiG-29M-style nose incorporating the experimental radar. Designated *izdeliye* 9.16, the aircraft made its first post-conversion flight on 12th January 1987; the tactical code '16 Blue' may have been derived from the manufacturer's designation. V. F. Babin was the engineer in charge of the flight tests.

The trials, which took place at GNIKI VVS in Akhtoobinsk, had to be suspended for a

The logo of GosNII AS which undertook research on digital avionics with the MiG-29 (*izdeliye* 9.21) development aircraft.

lengthy period when the experimental radar went unserviceable; finding out and eliminating the cause took a lot time. Once the problem had been fixed, the aircraft returned to Akhtoobinsk where it remained operational until approximately 1990, flying quite actively. However, the MiG-29M had entered flight tests by then and further trials of the N010 radar proceeded on the 'the real thing' at the insistence of the military.

Mikoyan MiG-29E ('aircraft 211', *izdeliye* 9.21) Skif avionics testbed
In the late 1980s the State Research Institute of Aircraft Systems (GosNII AS – *Gosudarstvennyy naoochno-issledovatel'skiy institoot aviatsionnykh sistem*) in Zhukovskiy began practical research on digital avionics suites for fourth-generation fighters. The programme was code-named *Skif* (Scythian; the Scythians were one of the ancient pagan peoples inhabiting pre-Christian Russia). The BTsK-29 suite (*bortovoy tsifrovoy kompleks* – digital avionics suite for the MiG-29) included a new speed and altitude data system (with self-contained pitot/static sensors), new navigation systems and multiplex fibre-optic data exchange channels. All of this significantly enhanced the fighter's combat potential.

To verify the new digital suite an early production MiG-29 coded '11 Blue' (f/n 1601) was converted into an avionics testbed known in-house as *izdeliye* 9.21 in late 1987. Since it was also known as 'aircraft 211', the code '211' was applied to the dielectric fin caps in small digits; the testbed also had the unofficial designation MiG-29E (*eksperimentahl'nyy* – experimental). Outwardly the MiG-29E could

be identified by numerous additional pitot heads on the nose pitot boom and on the fuselage underside immediately aft of the radome.

The first flight took place on 1st August 1986, but the actual tests under the Skif programme did not begin until April 1988. Shortly afterwards the programme was suspended. Yet the effort was not in vain; the recommendations of GosNII AS concerning the structure and hardware of the digital avionics were incorporated in later upgrades of the MiG-29.

Mikoyan MiG-29 microwave landing system testbed
In 1988 an uncoded MiG-29 referred to in LII records as 'No. 9165' – apparently MiG-29 (*izdeliye* 9.12) c/n 2960509165 – was used for testing the aforementioned Platsdarm microwave landing system.

Mil' Mi-4 missile guidance system testbed
A Mi-4 *Hound* utility helicopter was involved in the tests of the K-10S cruise missile's guidance system.

Mil' Mi-8LL laser designator testbed
A Mi-8 helicopter (exact version and identity unknown) served as a testbed for the Prozhektor laser target designator (see An-26P above). The tests were held in one of the Soviet Union's southern regions (in an area with level steppeland) under the direction of V. G. Filatov; testing took place on the ground and in flight both in daytime and at night.

The logo of the Mil' Moscow Helicopter Plant (Mil' OKB) whose helicopters served as night vision system and radar testbeds.

'11 Blue/211', the MiG-29E ('aircraft 211') testbed for the BTsK-29 digital avionics suite. Note the numerous additional pitot heads on the nose.

Mil' Mi-8 night vision systems testbed

A Soviet Air Force Mi-8 (tactical code unknown) was used by LII in 1979-1990 for testing night vision systems developed for ensuring night operations of combat helicopters and for locating people in distress at sea. The c/n has been quoted as 3115; yet, it is by no means certain this is a Kazan'-built Mi-8T *Hip-C* – the helicopter may well have been an Ulan-Ude built example (c/n 9733115). In fact, it is not even certain this was a 'first-generation' Mi-8 at all – it may also be Mi-8MT *Hip-H* c/n 93115, which likewise belonged to LII (though *this* helicopter was built in 1980!).

Mil' Mi-8T MCM systems testbed

Another Soviet Air Force Mi-8T (tactical code unknown, c/n 3611) was used by LII in 1972-1995 to test mine countermeasures (MCM) systems – not maritime but developed for the ground forces. This time the dates of operation leave no doubt this is a Kazan'-built aircraft – Ulan-Ude built Mi-8T c/n 9743611 did not exist yet, to say nothing of Mi-8MT c/n 93611.

Mil' Mi-8MT radar systems testbed

An uncoded Soviet Air Force Mi-8MT (c/n unknown) was converted by LNPO Leninets for testing new radar systems. The helicopter's appearance was really bizarre. A rectangular-section adapter occupying the entire width of the fuselage supplanted the lower row of the flight deck glazing; it carried what appears to be a phased-array radar of unknown type in a truncated, suitcase-like radome, making the machine look uncannily like a hippopotamus. Like the other LNPO Leninets testbeds, the Mi-8MT was based at Pushkin.

Mil' Mi-24V targeting system testbeds

Two Mi-24V *Hind-E* assault helicopters coded '19 Red' and '73 Red' were converted into testbeds for the PrPNK-28 targeting/flight instrumentation/navigation suite (*pritsel'nyy i pilotazhno-navigatsionnyy kompleks*) developed for the Mi-28 attack helicopter. The *Havoc*'s extreme nose with the laser rangefinder/LLLTV turret and thimble radome for the missile guidance system was grafted onto the *Hind*'s nose in lieu of the USPU-24 gun barbette and Raduga-Sh missile guidance system, giving the helicopter a really bizarre appearance. By 1998 '73 Red' was retired at the Mil' OKB's flight test facility in Panki south of Moscow.

Sukhoi Su-7BM flight data recorder testbed

The same Su-7BM c/n 5106 that had served as an IFR system testbed (see Chapter 2) was used by LII in 1968-75 for testing new flight data recorders (FDRs), including the MSRP-12 fitted to many Soviet combat aircraft, and perfecting the methods of FDR data analysis.

Sukhoi Su-7U blind landing system testbed

An unidentified Su-7U *Moujik* trainer was used in developing the **Kos***mos* (Outer space) blind landing system which was housed in an external pod.

Sukhoi T43-7 and T43-10 development aircraft

In 1960-61 two of the Su-9 interceptor prototypes – the T43-7 and T43-10 – served as avionics testbeds for the AP-28Zh-1 autopilot.

The Mi-8MT radar testbed at Pushkin. Note the gun camera above the entry door.

Reduced to a hulk, the Mi-24V used as a testbed for the Mi-28's PrPNK-28 navigation/attack suite sits on the dump at Panki.

The latter aircraft manufactured in 1960 was later damaged beyond repair in an accident, so a further aircraft (a production Su-9) was fitted with the autopilot in order to continue the trials. The autopilot later became a standard fit on production *Fishpot-As*.

Sukhoi T43-2 and T43-15 radar testbeds

In 1960-61 the T43-2 and T43-15 (the latter was converted from a production Su-9, c/n 1115310) served as testbeds for the modified TsD-30TP fire control radar. This version did not enter production and operational Su-9s were still equipped with the standard TsD-30 radar.

The logo of the NIIP avionics house whose radars were tested on several Sukhoi fighters.

Sukhoi T43-17 avionics testbed

Another Su-9 bearing the in-house designation T43-17 was used to test a new avionics suite subsequently fitted to the Su-11 interceptor. No details are known.

Sukhoi Su-9 automatic flight control system testbed

Following instructions from GKAT, in 1961 OKB-51 converted a Moscow-built Su-9 (c/n 100000603) into a testbed for the Polyot-1 automatic navigation/instrument landing system developed for new advanced fighters. The aircraft was equipped with the SAU-1I automatic flight control system. LII and GK NII VVS pilots tested the system in various flight modes, including automatic landing approach and low-level flight, until 1964 (some sources say 1968).

Sukhoi Su-15 ECM/IRCM equipment testbed

From 1969 onwards LII used a specially equipped Su-15 (identity unknown) to study the reflection of radar echoes from targets, clouds of chaff and the ground. Later this aircraft was further modified as a testbed for passive electronic countermeasures (ECM) and infra-red countermeasures (IRCM) equipment, serving to verify almost all chaff/flare dispenser types used by the Soviet/Russian Air Force.

Sukhoi SL-15R (T-58R) avionics testbed

In May 1972 the Leningrad-based NII-131 of the Ministry of Electronics Industry, aka LNPO Leninets, converted the abovementioned Su-15 coded '11' (c/n 0715311) into an avionics testbed. The Oryol-D58M fire control radar was replaced by a prototype *Rel'yef* (Profile) terrain following radar developed for the T-6 (Su-24) tactical bomber; this radar enabled automatic terrain-following flight during low-level air defence penetration. The aircraft was

designated SL-15R or T-58R, the SL standing for *samolyot-laboratoriya* (in keeping with LNPO Leninets's practice) and the R suffix referring to the Rel'yef radar.

Outwardly the SL-58R was no different from any production Su-15, since the experimental radar was housed within the standard radome.

Sukhoi Su-15 communications equipment testbed

Another Su-15 (identity unknown) was used in 1968-69 to test the new R-832M *Evkalipt-SM* (Eucalyptus-SM) communications radio.

Sukhoi Su-15TM avionics testbeds with modified radars

On 6th September 1976 the notorious Lt. Viktor I. Belenko, a pilot of the PVO's 530th Fighter Regiment stationed in the Far East, defected to Japan in a MiG-25P, landing at Hakodate airport and laying the latest state-of-the-art in Soviet interceptor design wide open for inspection by the West. In response to this, in November 1976 the Soviet Council of Ministers issued a directive requiring measures to be taken in order to minimise the damage done by Belenko's treason. The MiG-25P received the greatest attention, undergoing a complete change of the weapons control system which resulted in the advent of the MiG-25PD/MiG-25PDS. The Su-15TM, on the other hand, would make do with a modest upgrade of the radar.

In early 1977 two production Su-15TMs (c/ns 1315349 and 1415307) were set aside for modification under this programme, undergoing tests in June-October that year. It

took another 12 months to perfect the modified weapons control system, and in 1978 the modifications received approval for incorporation on operational aircraft.

Sukhoi Su-15TM IFF system testbed

In keeping with the same 'anti-Belenko' directive of November 1976 another production Su-15TM coded '61' (c/n unknown) was modified for testing the new SRO-1P Parol'-2 IFF suite. The modified aircraft underwent trials at GNIKI VVS alongside the abovementioned MiG-25P IFF system testbed and the Parol' suite was recommended for production, becoming standard on Soviet military aircraft.

Sukhoi Su-27UBK weapons control system testbed

The prototype of the Su-27UB's export version, Su-27UBK ('09 Blue'), served as a testbed for an upgraded weapons control system which allowed it to fire R-77 (RVV-AE/AA-12 *Adder*) active radar homing AAMs.

The T43-15 (c/n 1115310), an avionics testbed equipped with the TsD-30TP fire control radar. Note the camera pod under the nose and the '43-15' tail titles.

Su-27UBK '09 Blue' fires an R-77 (RVV-AE) missile during tests, which implies the radar is updated.

311

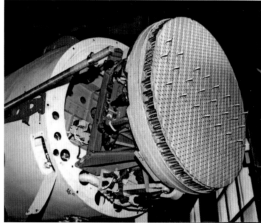

Above: and right: The T10M-11 ('711 Blue outline') seen during installation of the NIIP N011M Irbis radar.

Below and right: '503 Blue', the modified Su-30MKK prototype with an NIIP Irbis experimental radar.

Tu-4 '80 Red' was modified for testing the RBP-4 radar, which can be seen aft of the standard radome (indicated by arrows).

Sukhoi Su-27M (Su-35) radar testbed

The last-but-one pre-production Su-27M (Su-35) fighter, the T10M-11 ('711 Blue outline', c/n 79871011104, f/n 11-04), was converted into a testbed for the N011M Bars (Leopard) phased-array fire control radar developed by the Moscow-based Instrument

Design Research Institute named after Viktor V. Tikhomirov (NIIP – *Naoochno-issledovatel'skiy institoot priborostroyeniya*). This radar found use on the Su-30MKI two-seat multi-role fighter for India. Upon completion of the radar's tests '711 Blue' was used in a different programme, becoming the Su-37.

Sukhoi Su-30MKK radar testbed

The third example of the Su-30MKK – the version of the Su-30MK for China (the second K stands for *kitaïskiy*, Chinese) – became a testbed for the NIIP Irbis (Snow Leopard) active electronically scanned array (AESA) fire control radar developed for the current Su-35S (T-10BM) and the Sukhoi T-50 fifth-generation fighter. The aircraft was coded '503 Black' (c/n 79810380...01, f/n 80-01) and initially

wore an overall zinc chromate primer finish, later receiving a grey colour scheme and becoming '503 Blue'.

Tupolev Tu-4 avionics testbeds

The Tu-4 bomber was used extensively for verifying various new (mostly military) avionics systems. These included the Rym-S (Lifting lug) targeting system, a remote guidance/targeting system for torpedo boats and the PRS-1 Argon gun ranging radar for the tail gunner's station. The latter programme was especially intensive – a small batch of Tu-4s was actually built with PRS-1 radars; this model later became standard for Tupolev and Myasischchev bombers.

The Tu-4 served as the testbed for the DISS-1 Doppler speed and drift sensor (**dop**lerovskiy izme**rit**el' **sko**rosti i **sno**sa), the ARK-5 Amur ADF, the RSIU-2 and RSIU-3 communication radios, the **Booi**vol-Kod (Buffalo-Code) blind landing system, SHORAN slot antennas built into the fin, VLF trailing wire aerials (TWAs) and many other avionics items which found use on Soviet aircraft.

Another Tu-4 coded '80 Red' was a testbed for the RBP-4 Rubin navigation/bomb-aiming radar. The new radar was installed aft of the Tu-4's standard retractable radome.

Tupolev Tu-16 data link system testbed

In the late 1950s/early 1960s the Tupolev OKB modified a Tu-16 bomber to serve as a data link system testbed during the development of the '121' supersonic GLCM and its production derivative – the Tu-123 reconnaissance drone, aka DBR-1 *Ya*streb (Hawk).

Tupolev Tu-16 missile guidance system testbed

LII converted the first Kuibyshev-built Tu-16 *Badger-A* ('57 Red', later recoded '24 Red', c/n 1880101) for testing missile guidance systems. A missile's seeker head in a conical metal fairing tipped by a dielectric radome was installed at the extremity of the nose (on the navigator's station glazing); to prevent the heavy assembly from breaking loose it was firmly secured to the forward fuselage structure by four sloping twin bracing struts and a horizontal beam ahead of the flight deck windscreen. A ciné camera in an egg-shaped fairing was fitted aft of the flight deck to record the guidance system's accuracy.

Tupolev/Myasishchev 17LL-2 testbed

A Tu-16K-10(ZA) *Badger-C* IFR-capable naval missile carrier coded '57 Red' (c/n 4652042) was obtained from the Soviet Navy by the Myasishchev Experimental Machinery Factory and converted for testing the search and targeting system (STS) and gun turret developed for the M-17. (Best known as a reconnaissance and research aircraft, the M-17 was conceived for use against drifting reconnaissance balloons which were a real menace until the late 1970s.) The twin radomes of the YeS target illumination/missile guidance radar were replaced by the M-17's fuselage nose incorporating sighting windows, and the standard DT-7V dorsal cannon barbette was replaced by one developed for the high-flyer. The distinctive nose profile made the 17LL-2, as the aircraft was known, look uncannily like a saiga antelope. Live weapons trials were carried out on this aircraft, using real balloons as targets.

The first Kuibyshev-built Tu-16 ('57 Red', c/n 1880101) served as a missile guidance systems testbed. The missile's radar seeker head was installed on the nose glazing. Note the retaining braces and the ciné camera 'egg' aft of the flight deck.

The same testbed at a later date, following recoding as '24 Red'.

Close-up of the nose of Tu-16 '24 Red'; the missile's radar seeker head has been removed, leaving only the braced mounting platform.

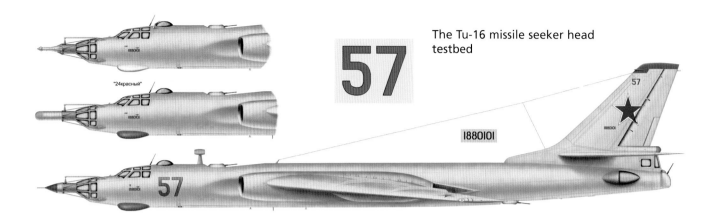

The Tu-16 missile seeker head testbed

Tupolev Tu-16 automatic landing system testbed

Another Tu-16 (identity unknown) served as a testbed for an automatic landing system. The tests took place at LII in 1958.

Tupolev Tu-22LL equipment testbed

In 1971 the Kazan' aircraft factory No. 22 converted a Tu-22R *Blinder-C* (c/n …41605…) into an equipment testbed designated Tu-22LL. Coded '20 Red' and later '05 Red', the aircraft was initially used by LII to test new aerial cameras for night recce (NA-Ya-7 and NA-Ya-8) coupled with the **Ya**vor-8 (Sycamore) high-power searchlight for target illumination. They were later replaced by various colour and spectrum (IR) photography cameras of the **Zon**a (Zone) family capable of discerning targets hidden under camouflage netting and the like. Neither of these cameras was fitted to production Tu-22Rs and Tu-22RDs.

Later the Tu-22LL became an avionics testbed for various targeting and ELINT systems. In its final configuration it sported a large trian-

gular aerial under the nose. On 18th August 1991 it took part in the Aviation Day flypast at Zhukovskiy, accompanied by a MiG-21UM *Mongol-C* chase plane. The Tu-22LL appeared to have some trouble taking off for that demo flight, as the starboard engine's afterburner would not ignite.

Tupolev Tu-104 navigation systems testbed/navigator trainer

In the 1970s a Khar'kov-built Soviet Air Force Tu-104 *Camel* medium-haul airliner coded '05 Black' (c/n 6350104), originally a VIP jet, became a testbed for the Tu-22M's flight avionics/navigation suite. Upon completion of the tests the aircraft was transferred to the Air Force's strategic bomber arm, DA (**Dahl'**nyaya avi**ah**tsiya – Long-Range Aviation), serving with the 43rd TsBP i PLS at Dyaghilevo AB near Ryazan' (central Russia). Thus it can also be regarded as a navigator trainer. However, unlike the Soviet Navy's Tu-104Sh navigator trainer conversions, it retained the standard 'glass nose' and lacked external stores pylons.

The logo of the Myasishchev EMZ operating the 17LL-2 testbed. The company also built testbeds for other OKBs.

The 17LL-2 targeting system testbed for the M-17 sits idle in the Myasishchev compound at Zhukovskiy.

Tupolev Tu-104A navigation system testbeds

Khar'kov-built Tu-104A CCCP-42394 (c/n 9350804) was one of the earliest testbeds based on the type. It was modified in the late 1950s for verifying an astroinertial navigation system developed for the Boorya interconti-nental GLCM. The system, which was a prod-uct of the Mars design bureau, was the first Soviet navigation system to use mathematical integration of inputs from sensors based on different principles (that is, gyros and star trackers). This feature made it possible to exclude noise signals and system errors.

Another Tu-104A, this time an Omsk-built example in Soviet Air Force markings (tactical code unknown, c/n 86601303), was also used for testing INS and astroinertial navigation sys-tems in 1969-80.

Tupolev Tu-104 avionics/weapons testbed with Smerch fire control radar

As part of the effort to create the Tu-128 *Fiddler* supersonic heavy interceptor forming the core of the Tu-128S-4 aerial intercept weapons system, in 1960 the Tupolev OKB's flight test facility converted the second Omsk-built Tu-104 *sans suffixe*, CCCP-42326 (c/n 66600102), into a testbed for the Smerch (Tornado) radar targeting system. The RP-S Smerch fire control radar developed for the Tu-128 under the leadership of Fyodor F. Volkov was enclosed by an ogival radome and mounted on a slightly downward-angled adapter supplanting the nose glazing. The radar control panels, test/recording equipment racks and test engineers' workstations were located in the flight deck and the cabin.

Two APU-128 missile pylons (*aviatsionnoye pooskovoye oostroystvo* – aircraft-mounted launcher) were fitted under the outer wings, enabling CCCP-42326 to carry and launch K-80 long-range air-to-air missiles. The latter had been developed for the Tu-128 by OKB-4 under Matus R. Bisnovat (later called GMKB Vympel, 'Pennant' State Machinery Design Bureau).

The long pointed nose gave the aircraft a rather weird appearance. Air Marshal Yevgeniy Ya. Savitskiy, Commander of the Air Defence Force's fighter arm, took one look at the test-bed and christened it *Booratino* (the Russian equivalent of Pinocchio). The sobriquet stuck and was later applied to other radar testbeds with similar noses.

An important feature of the first 'Pinocchio' was that some of the weapons control system's modules could be inspected and adjusted in flight if need arose, saving time by reducing the aircraft's downtime. The tests of the radar included live missile launch-es and contributed a lot to the successful test-ing of both the Tu-128 and the K-80 AAM, which became the R-4 (AA-5 *Ash*) in produc-tion form. After running out of service life CCCP-42326 ended its days on the dump at Zhukovskiy, complete with 'Pinocchio nose' and missile pylons.

Tupolev LM-104-518 avionics/weapons testbed with Zaslon fire control radar

Two more 'Pinocchio-style' avionics/weapons testbeds bearing the designation LM-104-518 were involved in the development of the Mikoyan MiG-31 heavy interceptor. The last three digits were a reference to *izdeliye* 518, the in-house code of the Ye-155MP intercep-tor which immediately preceded the MiG-31; L obviously stood for [*le**ta**yushchaya*] *labora-tor*iya, while the M may denote that the con-version was undertaken by the Myasishchev OKB. The manufacturing documents for the conversion, however, were prepared by the Mikoyan OKB.

The testbeds were meant to verify the interceptor's weapons control system based on the N007 fire control radar, aka RP-31 *Za**slon*** (Shield, or Barrier; NATO codename *Flash Dance*). Development of this phased-array radar with an antenna of 1.1 m (3 ft 7⅜ in) diameter began back in the late 1960s at the Zhukovskiy-based OKB-15 (aka KBR – *Kon**strook**torskoye byu**ro** rah**dio**stro**yen**iya*, Radio Equipment Design Bureau), which merged with NII-339 in 1969 to become NPO Fazotron. The development work lasted nearly 20 years.

The nose of the Tu-104 CCCP-42326 on the dump at Zhukovskiy. This testbed had an RP-S Smerch fire control radar replacing the nose glazing. The aircraft retained the pre-1973 liv-ery throughout its flying career.

Despite having the same designation, the two aircraft were not identical. The first LM-104-518 was converted from Khar'kov-built Tu-104 *sans suffixe* CCCP-42324 (c/n 7350201) and was purely an avionics testbed. The aircraft arrived at the Myasishchev OKB's experimental facility in Zhukovskiy in August 1969; the Mikoyan OKB, KBR and the Research & Flight Test Centre (NILITS – Na**ooch**no-iss**led**ovatel'skiy i **lyot**no-ispy-**tah**tel'nyy tsentr) assisted with the modification work. The radar was installed on a rather crude-looking cylindrical adapter and featured an ogival radome. Changes were made to the wing structure for the purpose of installing AKU-33 missile ejector racks (*aviatsi**on**noye katapool'tnoye oostroystvo*) but these were apparently never fitted. NPO Vzlyot (Take-off), which tested the Zaslon WCS, accepted the aircraft in 1970.

The other example was converted from a late Omsk-built Tu-104A (CCCP-42454, c/n 96601703). In addition to the radar installation, it did have missile pylons under the wings, and these were of a different type (AKU-410). Two ram air turbines driving generators were mounted side by side aft of the nosewheel well to cater for the experimental radar and the test equipment, with a test equipment heat exchanger next to them.

The pylons carried Vympel K-33 long-range AAMs. Tests of this weapon (known as the R-33/AA-9 *Amos* in production form) commenced in March 1975. The first two launches were in ballistic mode because the missile guidance system was still inoperative. In both cases CCCP-42454 fired the missile from the starboard pylon; a MiG-21U *Mongol-A* trainer flew chase just 20-30 m (65-100 ft) off the Tu-104's starboard wing, filming the launch.

The results proved encouraging and the K-33 was cleared for test launches against target drones in full guided mode.

NPO Vzlyot personnel participating in the tests of the Zaslon WCS on the Tu-104 included section chief A. A. Kirilin, Merited Test Pilot B. D. Mochalov and V. N. Shootov as engineer in charge. NPO Fazotron assigned the radar's programme chief A. I. Fedotchenko, laboratory chief A. V. Nesteruk, the radar's project designer I. G. Lobachov and project supervisor to the test programme. CCCP-42454 later became the Tu-104A Tsiklon weather research aircraft (see Chapter 6).

Tupolev LM-104-23S avionics/weapons testbed

Another radar testbed built by the Myasishchev OKB to Mikoyan OKB specifications was designated LM-104-23S. It served for testing the *Sap**feer**-23* (Sapphire) fire control radar developed for the MiG-23S *Flogger-A* tactical fighter, hence the designation. This time the Equipment Engineering Research Institute (*NII app**ara**tostro**yen**iya*) had a hand in shaping the specifications.

A Kazan'-built Tu-104B formerly operated by GosNII GVF (CCCP-42498, c/n 021801) arrived at Zhukovskiy for conversion in April 1968. The Sapfeer-23 (or S-23) radar with a simple conical radome was installed in the usual fashion, supplanting the navigator's station glazing. Additionally, the LM-104-23S had a TP-23 FLIR sensor (*tep**lopelen**gah**tor* – IR search & track unit), a *Lazoor'-M* (Prussian Blue) command link system forming part of the GCI system in which the MiG-23S was to operate, and wing hardpoints allowing missile pylons to be fitted. The cabin housed banks of

Tu-104A CCCP-42454 was the second of two LM-104-518 avionics/weapons testbeds under the MiG-31 interceptor's development programme, featuring an RP-31 fire control radar.

test/recording equipment and test engineers' workstations.

The modified aircraft was delivered to the customer in October 1968. The radar was tested successfully but the envisaged missile pylons were never fitted for various reasons. CCCP-42498 served on as an avionics testbed until 1981 when it was struck off charge.

Tupolev Tu-104B AFCS/navigation system testbed for Tu-144

The Tupolev OKB's own Tu-104B with the non-standard registration CCCP-06195 (c/n 021502), which was used for carrying top-ranking design staff on business trips, became an avionics testbed under the high-priority Tu-144 supersonic airliner programme. (Note: The CCCP-06xxx block was partially set aside for MAP enterprises, containing a 'mixed bag' of aircraft.) The machine was transported to Moscow where MMZ No. 156 outfitted it with the NK-144 navigation suite tailored to SST operations.

In 1974 the machine was transferred to LII. It shouldered an immense amount of flight test work when the ABSU-144 automatic flight control system (AFCS; *avtomaticheskaya bortovaya sistema oopravleniya* – automatic on-board control system) was put through its paces. A complete Tu-144 instrument panel was installed amidships in the cabin, effectively turning the aircraft into an in-flight simulator. Therefore CCCP-06195 was used a lot for training MAP and Aeroflot pilots and navigators in anticipation of the Tu-144's service entry. The aircraft was also used in state acceptance trials of the NK-144 navigation suite, thereby saving a lot of time.

CCCP-06195 ended its life as a hulk on the dump at Zhukovskiy in the late 1980s.

Tupolev Tu-104 navigation system testbed for Tu-154

In the 1970s one of LII's Tu-104s (identity unknown) was used as a testbed for the NPK-154 flight/navigation avionics suite (*navigatsionno-pilotazhnyy kompleks*) developed for the Tu-154 medium-haul airliner. This was the first such system in the Soviet Union to feature a digital processor.

Tupolev Tu-104 automatic landing systems testbed

Another Tu-104 (identity unknown) served as a testbed for automatic approach and landing systems at LII in 1961-64.

Tupolev Tu-110 Sapfeer avionics/weapons testbed

In 1971 the third production Tu-110 *Cooker* airliner (originally CCCP-Л5513) was converted into an avionics/weapons testbed designated Tu-110 *Sapfeer* (Sapphire). It served for testing the PARG-30VV semi-active radar seeker head (*poluaktivnaya rahdiolokatsionnaya golovka* [*samonavedeniya*]) developed for the K-25 air-to-air missile. This weapon was the Soviet equivalent of the Raytheon AIM-7 Sparrow, several examples of which had been captured intact in Vietnam and delivered to the Soviet Union for study.

The conversion was performed jointly by the Tupolev OKB, the Flight Test & Research Centre (NILITS – *Naoochno-issledovatel'skiy i lyotno-ispytahtel'nyy tsentr*), NKOA (?) and the MKB *Koolon* (Coulomb) weapons design bureau responsible for the K-25. It involved installing an RP-21 Sapfeer fire control radar in an ogival radome on an adapter supplanting the navigator's station glazing (hence the aircraft's designation); appropriate changes were made to the navigator's station proper. Pylons for carrying two K-25 AAMs on APU-25-21-110 launch rails were fitted under the wings, changes were made to the navigation suite, a cooling system for the radar set and the test equipment was provided, and test engineers' workstations were outfitted in the cabin.

NILITS took delivery of the converted aircraft on 31st December 1971; considering the date, the aircraft was obviously reregistered in the CCCP-3658... block by then.

Tupolev Tu-124 weapons control system testbed

After lengthy trials on the LM-104-518 the Zaslon WCS continued its trials on a suitably modified Tu-124 *Cookpot* short-haul airliner (identity unknown).

Tupolev Tu-124 automatic landing system testbed

In 1969-78 one of the initial-production long-nosed Tu-124s, CCCP-45003 (c/n 0350102?), was used as a testbed by LII to create and refine the Tu-154's ABSU-154 automatic landing system.

Tu-124
CCCP-45003
found use as an
automatic land-
ing system
testbed.

Tupolev Tu-126LL avionics testbed (*izdeliye* LL 'A' and LL '2A')

In 1969 the Soviet Council of Ministers issued a directive initiating development of a mission avionics suite for a new-generation airborne warning & control system (AWACS). The aircraft itself, designated A-50 or *izdeliye* A (NATO *Mainstay*), was to be developed by the Beriyev OKB (then headed by Chief Designer Aleksey K. Konstantinov) pursuant to a joint Council of Ministers/Communist Party Central Committee directive issued in 1973. The Shmel' (Bumblebee) mission avionics suite designed by NIIP (aka NPO Vega-M) was to have an enhanced 'look-down' capability enabling the A-50 to detect low-flying targets.

As part of the effort to create the A-50, its mission avionics had to be flight-tested. To this end the CofM Presidium's Commission on defence industry matters (VPK) adopted a special ruling on 25th August 1975 requiring a single Tu-126 *Moss* AWACS aircraft to be converted into an avionics testbed called Tu-126LL or LL 'A'; the 'A' refers to *izdeliye* A.

Accordingly the Tu-126 prototype, which had been delivered to the Soviet Air Force's 67th Independent Airborne Early Warning Squadron at Siauliai, Lithuania, upon completion of the trials, was withdrawn and ferried to

Taganrog for conversion. This involved removal of the original rotodome and all other elements of the old Liana (Creeper) mission avionics suite, as well as of the Kristall-L data link system, the equipment in the relief crew rest area, the PSN-6A life rafts and the ASO-2B-126 chaff dispensers. Instead, a new RA-L mission avionics suite was fitted (L stood for *labora**tor**nyy obra**zets*** – 'laboratory sample', that is test article) comprising a new RA-10 rotodome, avionics cooling systems (including a liquid cooling system for the search radar's antenna array), a power supply system and a test and data recording equipment suite. The power supply system included a special TG-60/GS turbine-powered generator (**toor**boghene**rah**tor) driven by engine bleed air and installed in a fairing above the wing leading edge.

The internal layout of the Nos. 2 and 3 cabin bays was changed appreciably; these featured new equipment racks and crew workstations and two extra bunks. The crew complement was increased to 13 (five flight crew, seven mission equipment operators for the RA-L suite and a test equipment operator).

The new rotodome was both larger and almost 4,000 kg (8,820 lb) heavier than the standard one. Hence limits had to be imposed

The Tu-126 pro-
totype was con-
verted into the
Tu-126LL avionics
testbed for the
Shmel' avionics
suite developed
for the A-50
AWACS.

on the maximum take-off weight, top speed and G loads of the LL 'A' testbed. Unlike the original design, which had a single dielectric segment, the new rotodome had front and rear dielectric portions of identical size, only the narrow centre portion being made of metal. Another outward identification feature of the testbed was the addition of massive cylindrical equipment pods at the wingtips. Their front and rear portions were dielectric, enclosing flat-plate antennas.

Manufacturer's flight tests of the LL 'A' lasted from 1st August to 21st October 1977. During this time the RA-L suite's power supply and cooling systems were tested and the structural strength and vibration characteristics of the new rotodome were verified. The actual first flight took place on 15th August with Vladimir P. Dem'yanovskiy in the captain's seat; V. V. Toolyakov was the engineer in charge. At this stage the aircraft logged a total of 27 hours 39 minutes in nine test flights; the crews noted that the new rotodome had virtually no effect on the machine's handling.

When the Ilyushin and Beriyev OKBs started work on the upgraded A-50U *Mainstay-B* (*izdeliye* 2A) equipped with the Shmel'-2 mission avionics suite, the Tu-126LL testbed was modified accordingly and designated LL '2A'. The modified aircraft arrived at Tret'yakovo airfield in Lookhovitsy, Moscow Region, on 13th March 1987. However, in 1990 all further work on this programme was terminated. The Tu-126LL was flown to Zhukovskiy and struck off charge. The stripped-out hulk sat on the local dump at least until August 1995, whereupon it was finally broken up – a shame, since not a single Tu-126 has been preserved.

The following description outlines the basic changes introduced on the Tu-126LL.

The crew comprises 13 persons: the usual flight crew of five, a test equipment operator and seven mission equipment operators (chief technician, technician, back-up technician, data processing operator, data registering operator, data recording operator and maintenance engineer). The radar intercept operators' consoles of the deleted Liana suite in Bay 1 are used as back-up equipment; the former radar intercept officer's station is converted into the test equipment operator's station.

Bay 2 accommodates the maintenance engineer's workstation, a rest area for two, an avionics rack and a spares box on the right; located opposite are the data registering operator's console, two more avionics racks and a second spares box. A dorsal fairing located

above this bay between frames 40-48 houses the turbine generator.

Bay 3 seats the chief technician, technician and back-up technician on the right and the data processing operator with his equipment rack on the left.

Bay 4 is entirely occupied by the modules of the RA-L suite; so is Bay 5, which houses the transceiver units. To reduce the harmful effect of the electromagnetic pulses generated by the equipment, Bays 4 and 5 are separated by a solid metal bulkhead. For the same purpose the navigator's station, flight deck and cabin windows are all provided with a wire mesh and gold plating to keep the radiation out.

Unlike the standard Tu-126, which has AV-60K eight-bladed contraprops, the Tu-126LL is fitted with AV-60N propellers of identical diameter (5.6 m; 18 ft 4$^{15}\!/_{32}$ in).

As noted earlier, the RA-10 rotodome of the RA-L system features two identical dielectric portions attached to a narrow metal centre segment; the diameter is 10.2 m (33 ft 5$^{37}\!/_{64}$ in). The dielectric portions have a honeycomb construction and feature metal attachment rims. At 6 rpm, the RA-10 rotates slower than the standard rotodome of the Tu-126.

To ensure proper operating conditions for the mission equipment, which generates a lot of heat, the Tu-126LL is fitted with several equipment cooling systems operated by the technician. These include a closed-circuit air cooling system with an air/air heat exchanger; several direct air cooling systems utilising cabin air; and two liquid cooling systems catering for the modules installed between frames 71-80. The air/air heat exchanger and the water radiator of the cooling system are installed in a ventral fairing beneath Bay 5 (frames 77-83).

The logo of NPO Vega-M which developed the radar suite tested on the Tu-126LL.

Tupolev Tu-134K (SL-134K) radar testbed

According to Tupolev OKB sources, an avionics testbed designated SL-134K (identity unknown) was converted in 1976 from a Tu-134K (the VIP version of the short Tu-134 *sans suffixe*) for testing first the Tu-22M3's PNA (NATO codename *Down Beat*) navigation/target illumination radar and later the Taïfoon (Typhoon) fire control radar. (The latter bit sounds rather improbable, as the Taïfoon was a 1960s-vintage radar developed for the Sukhoi Su-15T interceptor; see SL-18P above.) The radar was installed on a special adapter supplanting the nose glazing.

According to the Czech magazine *Létectvi +Kosmonautika* (Aviation and Spaceflight),

Tu-134K CCCP-65669 (c/n 0350916) served as a testbed for the N019 Rubin (RP-29; NATO codename *Slot Back*) fire control radar developed for the Mikoyan MiG-29 (*izdeliye* 9.12) *Fulcrum-A* tactical fighter. This pulse-Doppler radar was a product of the Moscow-based NPO Phazotron (aka NII Radio); the identical names of the RBP-4 ground mapping radar and the RP-29 are pure coincidence. However, the aircraft itself was possibly owned by LNPO Leninets and designated SL-134K, which is in line with Leninets's system of designating its avionics testbeds.

No proof of this has been found to date, but if it is true, CCCP-65669 was later reconverted to standard and transferred to the Perm' aero engine factory, an MAP enterprise.

Tupolev Tu-134AK navigation system testbeds

Sometime before 1992 the Moscow-based NPO Vzlyot converted a Tu-134AK VIP aircraft, CCCP-65604 No. 2 (c/n (93)62561, f/n 5403), into an avionics testbed for testing some sort of navigation equipment. (The registration had previously belonged to Tu-134 *sans suffixe* c/n 6350101.) Three dielectric panels of unequal size were incorporated on each side of the extreme nose; two additional slot aerials were built into the fin on each side below the usual three, as in the case of the Tu-134IK (see below). The rear fuselage underside featured tandem camera windows in flat-bottomed fairings, with a round dielectric blister further forward; the cameras were probably fitted to verify navigation accuracy by capturing landmarks as the aircraft overflew them! The white 'plug' in the first full-size window to starboard, which is one of the Tu-134AK's identification features, was missing – probably because a test engineer's workstation was installed in lieu of the forward toilet located there on the VIP version of the Tu-134A.

CCCP-65604 No. 2 was first seen at Zhukovskiy on 11th August 1992, the opening day of MosAeroShow '92. By June 1993 the aircraft had been reconverted to standard and seconded to the United Nations Peace Forces.

Another Tu-134AK owned by NPO Vzlyot, CCCP-65908 (c/n (23)63870, f/n 6307), was a navigation systems testbed at one time, featuring the same ventral tandem cameras/dielectric blister as on CCCP-65604 No. 2, plus L-shaped aerials at the wingtips. This equipment was removed before 1995.

Tupolev Tu-134AK radar testbed

In 1987 on more Tu-134AK, CCCP-65907 (c/n (33)63996, f/n 6333), became a testbed for the Phazotron N010 Zhook (Beetle) fire control radar developed for the MiG-29K *Fulcrum-D* shipboard fighter (*izdeliye* 9.31) and the MiG-29M advanced tactical fighter (*izdeliye* 9.15). The ogival radome was mounted on a slightly downward-canted adapter with angular bulges on the sides supplanting the nose glazing (the chin-mounted ROZ-1 radar was retained). As on CCCP-65604 No. 2, there was no white 'plug' in the first full-size window to starboard – possibly because the navigator's station was relocated there. Some Russian publications call this aircraft Tu-134LL – but see CCCP-65562 below!

Again, CCCP-65907 was first seen at Zhukovskiy on 11th August 1992. In January 1994 it was reconverted to standard and sold to Alrosa-Avia as RA-65907. (Incidentally, in the West there were similar radar testbeds based on aircraft sharing the Tu-134's general layout – such as the Defence Evaluation and Research Agency's BAC-111-479FU ZE433 (c/n 245) with the Ferranti Blue Fox radar developed for the BAe Harrier FA.2.)

Tupolev Tu-134A/Tu-134AK missile guidance system testbeds

Tu-134A CCCP-65738 (c/n 1351508) owned by the Siberian Aviation Research Institute (SibNIA – *Sibeerskiy naoochno-issledovatel'skiy institoot aviahtsii*) in Novosibirsk was a long-serving research aircraft used in various programmes. In particular, by February 2008 the aircraft – by then carrying the registration as 65738 without the RA- prefix, as experimental aircraft in Russia do – had been converted into an avionics testbed. A missile's radar seeker head mounted on a tapered metal adapter supplanting the navigator's station glazing; three small dielectric blisters (apparently associated with satellite navigation) were mounted dorsally above the Nos. 2, 10 and 11 cabin windows (the forward one was offset to starboard). Like all other SibNIA aircraft, 65738 wore the institute's distinctive blue/white livery with the Russian Navy flag on the engine nacelles. However, by September 2008 the radar seeker head had been removed and the nose glazing reinstated.

An unidentified Tu-134A or Tu-134AK in Aeroflot colours (possibly again CCCP-65908) owned by NPO Vzlyot served as a testbed for a cruise missile's radar seeker head. In similar

fashion to CCCP-65907 the long ogival radome (albeit of smaller diameter) was mounted on a metal adapter supplanting the nose glazing; it had three small round sensors mounted in T fashion below the radome and two more side by side above the radome. A 'pitchfork' ILS aerial was installed ahead of the weather radar's radome between the standard pitots.

Tupolev Tu-134AK ASW system testbed

In 1981 or 1982 Tu-134AK CCCP-65687 (c/n (93)62400, f/n 5302) was transferred from the Soviet Air Force to MRP (possibly to NPO Vzlyot) and converted into a testbed for a system designed for detecting submerged nuclear submarines. No details of the aircraft's appearance are available. Tragically, the aircraft crashed near Severomorsk-1 AB on 17th June 1982 during Stage B of the state acceptance trials, killing the designers of the system, which prevented it from entering service.

Tupolev Tu-134UBL radar testbeds

In 1982 or 1983 the first prototype Tu-134UBL *Crusty-B* bomber crew trainer, '11 Red' (c/n (13)64010, f/n 6401) was transferred from GNIKI VVS to LNPO Leninets. Soon afterwards the main portion of the Tu-134UBL's distinctive long pointed 'beak' was removed at the production break in line with the forward pressure bulkhead and replaced by an adapter mounting what

looked like a version of the Obzor navigation/target illumination radar fitted to the Tu-95MS *Bear-H* missile strike aircraft. The new blunt and flattened radome resembling the *Bear-H*'s 'duck bill' gave the aircraft a really weird look.

In this guise the aircraft was first seen in the summer of 1990 when a group of Western hot air balloons flew over Pushkin airfield where LNPO Leninets had its flight test facility – the wind happened to be blowing the wrong way during an air rally. After sitting idle for several years '11 Red' had become so weathered that the tactical code almost vanished, assuming a nondescript colour, and the original test registration CCCP-64010 was bleeding through the grey paint on the engine nacelles; this fooled some Western observers into thinking the aircraft was coded '100 Blue'! Eventually the aircraft was sold to Meridian Airlines in 2002 and converted to a Tu-134B-3M executive jet as RA-65945.

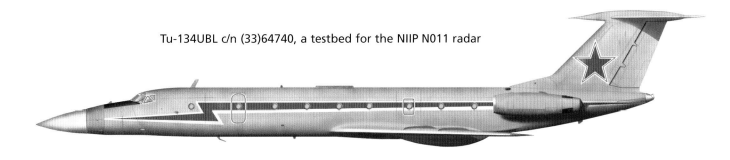

Tu-134UBL c/n (33)64740, a testbed for the NIIP N011 radar

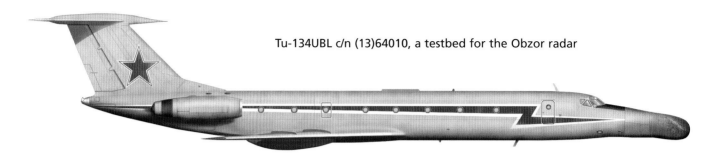

Tu-134UBL c/n (13)64010, a testbed for the Obzor radar

The nose of another converted Tu-134UBL (c/n (13)64010) fitted with an Obzor radar.

Tu-134Sh-1 '10 Red' (c/n 3350303) became an SL-134Sh testbed with a missile seeker head in the nose.

In 1984 an uncoded Tu-134UBL (reportedly ex '21 Red'; c/n (33)64740, f/n unknown) operated by LII became a testbed for the Phazotron N007 Zaslon (RP-31) fire control radar developed for the MiG-31 interceptor.

Later, the same aircraft was used to test the Phazotron N019M Topaz radar developed for the upgraded MiG-29S (*izdeliye* 9.12S/ *Fulcrum-A* and *izdeliye* 9.13S/*Fulcrum-C*), or possibly the N019ME for the MiG-29SE export version. Once again the radar was mounted on a special adapter with a hemispherical projection on top resembling the MiG-29's OEPS-29 infra-red search & track unit/laser rangefinder (**op**tiko-elek**tron**naya pri**tsel'**naya sis**tem**a – optoelectronic targeting system).

By June 1994 Tu-134UBL c/n (33)64740 was re-equipped for testing the N011 fire control radar developed for the Sukhoi Su-27M

(Su-35/T-10M) multi-role fighter by NIIP. The large white-painted radome mounted on an adapter was very obvious. The trials were completed in 1997.

Tupolev Tu-134UBL avionics testbed
Sometime before 1993 Tu-134UBL '30 Red' (ex-'42 Red'?, c/n (33)64845, f/n unknown) operated by GK NII VVS was converted into an avionics testbed of unknown purpose. Two small teardrop fairings were mounted under the forward and rear fuselage on short pylons; their front and rear portions respectively were dielectric. Wiring conduits ran along the upper rear fuselage sides from the emergency exits to the engine nacelles.

The aircraft was first seen as such at Zhukovskiy in September 1993 during the MAKS-93 airshow. Seven years later, on 21st September 2000, the freshly painted aircraft wearing red 'ROSSIYA' (Russia) titles made a formation flypast with Su-27UB *Flanker-C* '80 Red' during the air fest at Akhtoobinsk marking the 80th anniversary of GK NII VVS (929th State Flight Test Centre).

Tupolev Tu-134Sh-SL (SL-134Sh) missile guidance system testbed
Pursuant to an MAP/Air Force joint decision of 21st July 1977 several Tu-134Sh *Crusty-A* navigator trainers were converted into avionics testbeds designated Tu-134Sh-SL or SL-134Sh. Interestingly, not all of them belonged to LNPO Leninets, despite the SL designator.

The first of these was probably converted from Tu-134Sh-1 '10 Red' (c/n 3350303) which became a testbed for missile guidance systems. The nose glazing was replaced by a conical metal fairing tipped by a radome and carrying sensors. The guidance system avionics occupied all of the former navigator's station, with an access panel on the starboard side of the extreme nose. The navigator now sat on the port side opposite the service door, and a full-

The freshly converted Tu-134SL, CCCP-65098, in its original configuration with a PNA-D radar. Note the Tu-134A nose titles belying the fact that the aircraft is a converted Tu-134Sh-1.

Here the same aircraft is seen as the L17-10V (65098) with the Su-34's V004 radar. Note the Tu-134Sh nose titles and the 'KhK Leninets/ NPP Mir' titles.

size window was added ahead of the entry door to admit daylight to his workstation (just like on the later Tu-134UBL, Tu-134B-1, Tu-134SKh and Tu-134BV versions). Unusually, the c/n was actually painted on the port side of the nose.

Tupolev Tu-134Sh-1 ELINT system testbed

In the late 1980s NPO Vzlyot converted Tu-134Sh-1 '01 Red' (c/n 63550705, f/n unknown) into a testbed for the *Bagration* ELINT suite. (General Pyotr I. Bagration was a hero of the Patriotic War of 1810-12 who contributed a lot to Napoleon's defeat. The c/n was also reported in error as 3350705 and 6350705.)

The aircraft featured two fairly large bulged metal panels wrapped around the fuselage sides ahead of the wings (beneath the fourth, fifth and sixth cabin windows). Each panel was faceted and mounted eight square-shaped white antenna plates (four large ones and four smaller ones), and the whole assembly was very prominent indeed. The star tracker prisms were removed, leaving only the mounting blisters.

The standard BD-360 racks for practice bombs under the wing roots were replaced by a special frame mounting two different equipment pods. The first of these was cylindrical and unpainted, with a conical front end sprouting four probe aerials and a cut-off rear end; the aircraft was last noted with this

pod at Zhukovskiy in August 1992. The other pod was substituted sometime between August 1992 and August 1993; it was grey, looking like an outsize drop tank with horizontal fins equipped with endplates (in reality these were probably aerials). By August 1995 Tu-134Sh-1 c/n 63550705 had been withdrawn from use.

Tupolev Tu-134SL (Tu-134Sh-SL, SL-134Sh) radar testbed

In early 1978 LNPO Leninets took delivery of a Tu-134Sh-1 (c/n 73550815; f/n 0805?) which was registered CCCP-65098 and converted into the Tu-134SL avionics testbed; this aircraft, too, has been referred to as a Tu-134Sh-SL or SL-134Sh. The aircraft was painted in full 1973-standard Aeroflot livery and wore 'Tu-134A' nose titles. The BTs-63 star tracker mounting blisters on the centre fuselage were deleted; so were the AFA-BA/40 strike camera

CCCP-65098 in its second configuration with the N001 radar. Note the ventral test equipment pod.

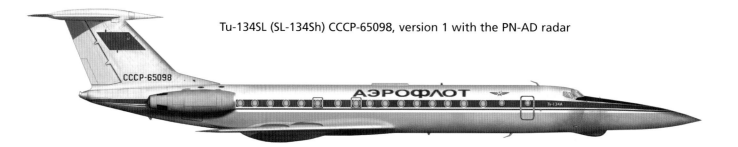

Tu-134SL (SL-134Sh) CCCP-65098, version 1 with the PN-AD radar

and the associated fairing under the rear fuselage, but the characteristic design of the APU's lower intake with two narrow grilles instead of a single large one was the giveaway that this was not a Tu-134A after all. Quite possibly the aircraft had been earmarked for conversion from the start and was never delivered to the Air Force.

CCCP-65098 had three distinct configurations. Originally the aircraft featured a Tu-22M3 nose grafted on instead of the normal nose glazing. Since this 'beak' was incompatible with the Tu-134Sh-1's deep chin radome housing an RBP-4 navigation/bomb-aiming radar, the latter was replaced by a stock Tu-134A radome housing an ROZ-1 weather radar. Hence the Tu-134SL was commonly misidentified as either the Tu-134UBL prototype or Tu-134UBK naval aircrew trainer. Since the *Backfire-C*'s PNA radar had been tested on another aircraft back in 1976, CCCP-65098 apparently carried a different version of this radar – possibly the PNA-D. A pylon was fitted under the fuselage ahead of the wings, offset to port, for carrying a cylin-

drical test equipment pod housing what appeared to be a long-range camera.

Later the *Backfire* nose was removed. Instead, CCCP-65098 was fitted with a short adapter mounting the N001 Mech (Sword) fire control radar for the Su-27 *Flanker-B* (T-10S) fighter. This is a coherent pulse-Doppler radar with an antenna diameter of 1.076 m (3 ft 6⅜ in). Interestingly, it was not an LNPO Leninets product but was developed by NIIP, which probably had no aircraft available for conversion at the time. Equally interestingly, the N001 appears to have the same NATO codename as the MiG-29's N019 radar, *Slot Back*. A smaller pod on a shorter centreline pylon was added ahead of the existing one; this appears to house a laser ranger.

In late 2000 CCCP-65098 was refitted again, receiving a navigation/attack radar developed by the Leninets Holding Co. for the Su-34 (T-10V) multi-role combat aircraft (some sources call it V004). The phased-array radar is enclosed by the Su-34's characteristic downward-angled 'duck bill' radome with sharp chines – a feature which has earned Sukhoi's

CCCP-65098 in its second version with the NIIP N001 radar

The same aircraft as the L17-10V (65098) with the Leninets V004 radar

The Tu-134LL (RA-65562) on display at the MAKS-95 airshow.

new combat jet the nickname *Ootko**nos*** (Platypus); this necessitated installation of a new radar mounting adapter.

Sporting the experimental-style registration 65098 with no prefix and 'KhK Leninets – NPP Mir' (***Khol**dingovaya kom**pa**niya **Len**inets – Na**ooch**no-proiz**vod**stvennoye predpri**ya**tiye Mir*, Leninets Holding Co. – Mir Scientific & Production Enterprise) titles instead of Aeroflot markings, the freshly overhauled and refitted Tu-134SL was unveiled at the Business Aviation-2001 airshow at Pushkin on 5-8th August. Surprisingly, it now wore 'Tu-134Sh' nose titles. Less than a week later the aircraft was displayed at the MAKS-2001 airshow, albeit on the trade days only (13-17th August). The placard in front of the aircraft identified it as the L17-10V testbed (presumably meaning '*labora**tor**iya* (testbed) No. 17 under the T-10V programme'. 65098 was displayed again on the trade days of the MAKS-2003 airshow (18th-22nd August).

Tupolev Tu-134LL (Tu-134Sh-SL, SL-134Sh) multi-role testbed

In 1991 the State Research Institute of Aircraft Systems (GosNII AS) received a second-hand Tu-134Sh-1 trainer (tactical code unknown, c/n 2350201; the c/n has also been quoted as 2350104 and 2350204) which was registered CCCP-65562 and extensively modified for research and development work. This aircraft has been called Tu-134LL in some publications; yet, again Tupolev OKB sources call it Tu-134Sh-SL or SL-134Sh.

Its original mission was to test the guidance systems of TV- and laser-guided 'smart weapons'. Later, with the onset of *konversiya* (the adaptation of defence industry assets for civilian needs), the Tu-134LL was magically transformed into a geophysical survey and environmental monitoring aircraft. Its peaceful uses

included photo mapping, thermal imaging, ecological survey and damage assessment in the wake of natural and man-made disasters.

The Tu-134LL was readily identifiable by the large square-section ventral canoe fairing stretching all the way from the third full-size window on each side to the rear emergency exits. The forward half of this fairing housing cameras and sensors was additionally secured to the fuselage by five prominent metal hoops. A small cylindrical pod with conical ends was pylon-mounted low on the starboard side of the fuselage immediately ahead of the service door. An angular fairing with three ventral sensor windows was located aft of this door, with a long rod aerial further aft. Two more rod aerials were mounted dorsally fore and aft of the wings, with an L-shaped aerial in between. The star tracker mounting blisters were deleted.

The aircraft was painted in full Aeroflot livery and wore 'Tu-134' nose titles – yes, Tu-134 *sans suffixe*, though 'Tu-134A' would have been more appropriate. Since the manufacturer's plate in the nosewheel well is only marked 0201, some people could be misled to believe this was a converted and reregistered Tu-134

Close-up of the Tu-134LL's nose with a missile seeker head and sensors.

Tu-134LL RA-65562 in its 1995 configuration

RA-65562

Ту-134

sans suffixe (CCCP-65609 No. 1 c/n 6350201)! However, the deep chin radome and the APU intake treatment revealed what it really was.

The front end of the canoe fairing accommodated a T-2 TV system used for target tracking and image recording; it was mounted in a revolving turret with a traversing angle of ±75° and a depression angle of 90° (that is, the camera could shoot anywhere from straight ahead to straight down). The TV camera's field of view could be selected between 16° and 2.8°. An AFA-BAF-40 oblique camera and an LDI-3 laser rangefinder were slaved to the TV system. The camera lenses were protected from foreign object damage by a movable shutter on take-off and landing.

The canoe fairing also accommodated five vertical cameras on two AFUS-U tilting mounts (**a**ero**fot**o'usta**nov**ka – aerial camera mount) and three GUT-3 and GUT-8 gyrostabilised mounts. Camera options included seven models: the Soviet TK-10, TAFA-10 or East German Zeiss Ikon LMK (f = 100 mm), the AFA-41/20 (f = 200 mm), the Soviet TEA-35 (f = 350 mm), the AFA-42/100 (f = 1,000 mm) and the AS-707 four-spectrum camera (f = 140 mm).

The mission equipment included an OD-4M optical seeker (f = 1,000 mm) with 80x zoom for locating and tracking the 'targets', a TV scanner with a 156x40° field of view and a laser scanner with a 120° field of view having a passive day mode and an active night mode. For thermal imaging the Tu-134LL was fitted with an IR line scanner with a 120°/60° field of view, a two-mode IR line scanner with a 120° field of view, and an IKR4-2 four-channel IR radiometer (*infra**kras**nyy rad**iom**etr*) with a 3° field of view. The experiment was controlled by a 66-MHz Compaq 486 computer. The data furnished by the various mission systems was stored by a Schlumberger 27-channel digital data recorder and a Soviet Gamma recorder. The latter captured flight and navigation parameters, as well as the time when cameras and the like were activated, and could inscribe this information into the TV image.

The aircraft featured an I-21 inertial navigation system, a GPS kit and its Soviet/Russian GLONASS counterpart, plus a programmable device enabling the jet to fly a shuttle pattern automatically. Missions were flown at up to 10,000 m (32,800 ft) and 650 km/h (403 mph); the drag generated by the various appendages reduced the top speed to 800 km/h (496 mph). Maximum range was 3,000 km (1,860 miles) and endurance 3.5 hours. The crew included ten equipment operators.

CCCP-65562 was first noted at Zhukovskiy, its home base, on 11th August 1992. During the MAKS-95 airshow the aircraft was displayed statically as RA-65562 and minus Aeroflot titles/logo. By then the pod on the starboard side of the nose had been altered to feature a TV seeker head of an air-to-surface missile; this was removed in 1997. By August 1999 the Tu-134LL had been retired.

Tupolev Tu-134IK (Tu-134A-IK) development aircraft

In 1983 a one-off custom-built aircraft designated Tu-134IK (sometimes called Tu-134A-IK) was manufactured for the Soviet Ministry of Defence. It has been suggested that the aircraft was used for testing ASW (submarine detection) systems; the letters IK may stand for *izme**rit**el'nyy **kom**pleks* (measurement suite). The Tu-134IK was developed jointly with the Khar'kov aircraft factory and LNPO Leninets.

The aircraft, which received the non-standard registration CCCP-64454 (c/n (33)66140, f/n 6341), is a unique hybrid between the Tu-134A and the Tu-134UBL, combining the former version's glazed nose and ROZ-1 radar in a chin radome with the *Crusty-B*'s centre fuselage featuring half the passenger version's number of windows and one large emergency exit on each side. (In other words, the Tu-134IK's window arrangement is almost identical to that of the Tu-134UBL, except for the absence of the port side window ahead of the entry door. Hence the unusual registration

The Tu-134IK (Tu-134A-IK) development aircraft in 2001 colours

RA-64454

is very probably explained as follows: the aircraft was built using components of an unbuilt Tu-134UBL which was to receive the c/n (23)64454.)

The aircraft bristles with non-standard antennas. Two oval fairings incorporating dielectric panels are mounted on the fuselage sides just of the entry door, with a slightly larger third fairing mounted ventrally a bit further forward. A blade aerial is fitted dorsally just of the entry door. Two additional slot aerials are built into the fin on each side below the usual three serving the RSBN-2S SHORAN and the ATC transponder; another pair of slot aerials is built into the sides of the nose immediately forward of the service door. The fin has a wide unpainted leading edge. The aircraft wore full 1973-standard Aeroflot colours and 'Tu-134A' nose titles, hence the alternative Tu-134A-IK designation.

The Tu-134IK was unveiled at the open doors day at Pushkin, LNPO Leninets's flight test facility, on 18th August 1991 during the annual Aviation Day air fest. On 15th August 1999 the aircraft was noted as RA-64454 without Aeroflot titles/logo at Chkalovskaya AB where it was probably undergoing trials at the 929th GLITs. It is now based at Levashovo AB near St. Petersburg, which is a Russian Naval Aviation/Baltic Fleet Air Arm base, and is reportedly stripped of all mission equipment for use as a transport.

Tupolev Tu-134BV automatic landing system calibrator/testbed

On 2nd December 1983 the Khar'kov aircraft factory manufactured another unique Tu-134 custom-built for LNPO Leninets. Designated Tu-134BV and registered CCCP-65931 (c/n (33)66185, f/n 6347), it was one of several assorted aircraft involved in the development of the Buran space shuttle. Its mission was to verify the **Vym**pel (Pennant) automatic approach and landing system designed for the Buran (hence the V suffix to the designation); development of this system had been under way since 1979. Thus to all intents and purposes CCCP-65931 was both an avionics testbed and a navaids calibration aircraft.

The starboard and lower antenna arrays of the Tu-134IK.

CCCP-64454, the Tu-134IK, at an open house at Pushkin in 1991. Note the window placement and 'Tu-134A' nose titles.

The Tu-134BV
testbed/flight
checker at
Pushkin in 2001
as RA-65931.

The nose of the
Tu-134BV, show-
ing the cheek
fairings with slot
antennas.

The Tu-134BV is again a hybrid – this time between the Tu-134B-1 and the Tu-134UBL, combining the 'head' of the former aircraft with the 'body' of the latter (with modifications in both cases). This accounts for the window arrangement unique to this aircraft – 1+door+1+3+1+exit+1+1 to port and service door+galley window+1+1+3+2+exit+3 to starboard. The additional windows cut in a stock *Crusty-B* centre fuselage section are positioned to provide adequate natural lighting for the many test equipment operators' stations; the extra window ahead of the entry door is for the navigator's station.

The fuselage nose features two prominent bulges immediately aft of the radome incorporating slot aerials. A third, flat-bottomed fairing under the nose houses a retractable high-powered light for photo theodolite measurements. L-shaped AShS aerials are installed above the wing leading edge, above the wing

trailing edge and atop the fin; the latter has a Tu-134SKh-style leading edge treatment.

Once again the Tu-134BV was publicly unveiled at the Aviation Day air fest at Pushkin on 18th August 1991. Reregistered RA-65931, it was displayed there on 14th June 1997 at an air event marking the 60th anniversary of Valeriy P. Chkalov's cross-Polar flight from Moscow to Vancouver. On 2nd-5th August 2001 the aircraft was in the static park at the Business Aviation-2001 airshow at Pushkin.

Tupolev Tu-142MP avionics testbed

A single Tu-142M *Bear-F* Mod anti-submarine warfare aircraft was modified for testing the new Atlantida (Atlantis) ASW suite. This one-off avionics testbed was designated Tu-142MP. Apparently the Atlantida suite did not live up to the designers' expectations and remained experimental.

Tupolev Tu-144 automatic landing system testbed

In 1976-81 a production Tu-144 supersonic airliner, CCCP-77108 (c/n 10042) was used for testing the ABSU-144 automatic landing system in 45x800 m weather minima – that is, decision altitude 45 m (150 ft), horizontal visibility 800 m (2,620 ft). Upon retirement the aircraft was relegated to the Kuibyshev Aviation Institute (KuAI; now Samara State Aviation University, SGAU) as an instructional airframe.

The Tu-134BV automatic landing system testbed/flight checker

Tupolev Tu-154A/Tu-154B-1 automatic landing system testbeds

In 1975-82 Tu-154A CCCP-85055 (ex-Egyptair SU-AXI, c/n 74A055) was used for testing the ABSU-154 automatic landing system in various weather conditions – ICAO Category I, II and III weather minima. Later, in 1980-92, Tu-154B-1 CCCP-85119 (c/n 75A119) was also used for the same purpose.

Tupolev Tu-154A visibility simulation system testbed

In the 1980s CCCP-85055 was converted into a testbed for an optoelectronic visibility simulation system (SIV – *sistema imitahtsiï vidimosti*). Outwardly the conversion was identifiable by a bulge incorporating a large optically flat circular window on the starboard side where the service door had been; a smaller sensor blister was located ahead of the forward cabin's emergency exit (the latter was faired over with sheet metal). By August 1992 the aircraft had run out of service life, sitting engineless at Zhukovskiy; it was broken up by August 1999.

Yakovlev Yak-18T navigation system testbed

At the MAKS-2003 airshow (19-24th August 2003) a Yak-18T registered ФЛА-РФ-01001, previously a propeller testbed with the registration applied as 01001ФЛА РФ, was displayed statically and in flight as an avionics testbed for testing new navigation system components. Its tasks included research of communication, navigation, surveillance/air traffic management (CNS/ATM) system elements and modes, control loop research and evaluation of the ADS-B co-operative surveillance technique.

ФЛА-РФ-01001 was fitted with GLONASS and GPS satellite navigation receivers, a repro-grammable-in-flight computer and data link using the VDB and VDL-4 data transfer protocols. The onboard suite permitted precision navigation with an error margin of less than 0.1 m ($3^{15}/_{16}$ in). Outwardly the testbed, which was operated by the Moscow-based Fine Instruments Research Institute (NIITP), differed from the standard Yak-18T only in having a small flat-topped SATNAV antenna supplanting the anti-collision light on the fin.

Two years later the same aircraft was exhibited by LII in the static park of the MAKS-2005 airshow with the registration changed to RF-051. It has since been stripped of non-standard equipment and sold to a private owner, being reregistered RA-0762G.

Yakovlev Yak-25 cruise missile simulator

At least two Yak-25s were intended to be used for 'simulating in flight the P-6 and P-35 sea-launched cruise missiles'. Apparently this means they were to be used as guidance system testbeds.

Yakovlev Yak-25 navigation system testbed

Along with the aforementioned MiG-19 and Tu-104A, in 1956-57 an unidentified Yak-25 was used for testing the navigation system of the Boorya GLCM.

Extremely weathered Tu-144 CCCP-77108 survives as an instructional airframe in Samara. This aircraft was an ALS testbed.

Yak-18T RF-051, a navigation system testbed operated by LII, at the MAKS-2005 airshow.

Seen here at Zhukovskiy following retirement, Yak-38U '24 Yellow' was used for verifying automatic landing approach procedures.

Yakovlev Yak-27 reconnaissance equipment testbed

In 1958 a production Yak-27R *Mangrove* reconnaissance aircraft was used for the development of reconnaissance equipment intended for the Yak-28R *Brewer-D* reconnaissance aircraft. The aircraft was fitted with the RBP-3 bombing sight, AP-28 autopilot and DAK-I remote-controlled astrocompass.

Yakovlev Yak-38U automatic landing system testbed

A Yak-38U *Forger-B* shipboard V/STOL attack aircraft trainer served for verifying the type's automatic landing approach mode and exploring weather minima in 1984-89. This was most probably the aircraft coded '24 Yellow' (c/n 7977764148236; f/n 0307?) which later became one of the gate guards at LII.

Yakovlev Yak-40REO avionics testbed

The Yak-40REO (***rah**dioelek**tron**noye obo**roo**dovaniye* – electronic equipment, or avionics) was one of the numerous avionics testbeds developed by NPO Vzlyot. Converted from a late-production Yak-40 *Codling* feederliner registered CCCP-88238 (c/n 9640951), it served for testing such items as inertial navigation systems, altimeters, Doppler systems and so on. The Yak-40REO was readily identi-

fiable by a huge ventral fairing ahead of the wings and a large port side observation blister with a rear fairing.

By 1992 these excrescences had been removed and CCCP-88238 had been used in a different test programme, as evidenced by the ASO-2 flare dispensers attached under the wing roots and to the underside of the integral airstairs. By 1993 the aircraft was reconverted to Yak-40 'Salon 2nd Class' executive configuration as RA-88238. It was eventually sold to the Ukraine in 1997, becoming UR-BWH.

Yakovlev Yak-40-25 avionics testbed

A production Yak-40 (identity unknown) was converted into a testbed for verifying the mission avionics of the Mikoyan Ye-155R (MiG-25R *Foxbat-B*) reconnaissance aircraft. The nose of the MiG-25RBK housing an SRS-4A Koob (Cube; *izdeliye* 30A) signals intelligence pack (***stahn**tsiya raz**ved**ki **svya**zi* – SIGINT system) was grafted onto the airliner's forward fuselage, replacing the Groza-40 weather radar.

Yakovlev Yak-42R radar testbed

This aircraft (identity unknown) was a specially converted machine which was used for flight-testing the N010 fire control radar of the Yak-41 *Freestyle* V/STOL shipboard fighter. Hence the R may refer to the radar.

The Yak-40REO, CCCP-88238, in front of a LII hangar, showing the ventral fairing and the observation blister.

5 Weapons testbeds

A number of aircraft were used for testing and perfecting new air-to-air and air-to-surface weapons. As a rule, outwardly such aircraft differed little from standard machines because the weapons were suspended externally, not requiring changes to the airframe.

Antonov An-12BL weapons testbed

Suppression of enemy air defences (SEAD) was recognised as a separate role for combat aircraft during the Vietnam War when surface-to-air missile (SAM) systems started posing a major threat. Specialised SEAD aircraft (known in US Air Force slang as 'Wild Weasels') were developed from fighter-bombers or attack aircraft. But have you ever heard of a 'Wild Weasel' transport? In 1970 the Antonov OKB converted an An-12BK *Cub* into a weapons testbed designated An-12BL ([*letayushchaya*] *laboratoriya*).

The aircraft was armed with four Kh-28 (*izdeliye* 93; NATO AS-9 *Kyle*) anti-radar missiles developed by MKB Raduga for destroying AD radars. Two of these bulky weapons were carried on pylons flanking the forward fuselage and the other two on pylons under the outer wings. A *Metel'-A* (Snowstorm-A) radar detector/guidance system was installed in a thimble radome ahead of the navigator's station glazing. Subsequently the Kh-28 entered service and was carried by various Soviet tactical aircraft along with the Metel'-A system (in podded form).

A poor but interesting shot of the An-12BL, showing two of the four Kh-28 ARMs and the thimble radome of the Metel'-A guidance system.

331

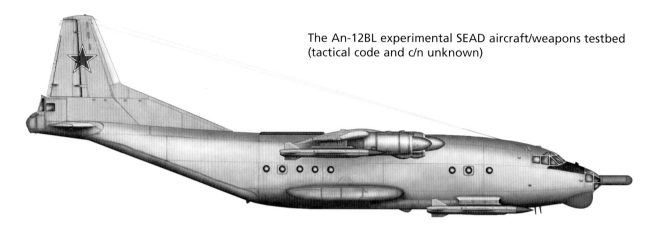

The An-12BL experimental SEAD aircraft/weapons testbed (tactical code and c/n unknown)

Mikoyan/Gurevich MiG-15 (*izdeliye* SU) development aircraft/ weapons testbed

The logo of MKB Raduga, a stylised rainbow. Its Kh-28 missile was tested on the An-12BL.

Until the advent of air-to-air missiles, a fighter pilot usually had to take aim by pointing the whole aircraft, which took considerable time. In a dogfight, this put him at a disadvantage: he had to provide target lead by aiming at a point ahead of the target on its anticipated course. If the target was more agile than his own aircraft, the pilot had no choice but to break off the attack and start anew – and the few seconds lost in so doing could prove fatal. Conversely, on a fighter with movable armament the pilot could bring his guns to bear on the target much quicker and more accurately – even when pointing the aircraft itself was impossible. Ideally, this gave him first-shot, first-kill capability.

The Mikoyan OKB began working on movable cannon armament in late 1949. Pursuant to MAP order No. 658 dated 14th September 1950 a Kuibyshev-built MiG-15 (*izdeliye* SV) *Fagot-A* serialled '935 Red' (c/n 109035) was converted to take the V-1-25-Sh-3 experimental weapons system replacing the standard armament. The system was developed by OKB-15

under Boris G. Shpital'nyy, the man behind the ShKAS machine-gun and ShVAK cannon of pre-Second World War vintage; the Sh in the designation stood for Shpital'nyy and 25 was the number of the plant where his OKB was based.

The system was built around two 23-mm Sh-3 cannons with 115 rpg in faired flexible mounts on the lower forward fuselage sides, giving the aircraft a jowly, bulldog-like appearance in a head-on view. The cannons could elevate +11°/–7°. The aircraft received the in-house designation '*izdeliye* SU' (that is, *izdeliye* S s oopravlya*yemym* [*vo'oru-zhen*iyem] – with movable armament).

The V-1-25-Sh-3 system also included a movable sight and remote-controlled electric actuators that moved the cannons. The latter were cocked by a purpose-built electropneumatic mechanism and elevation was controlled by two knobs, one on the stick and one on the throttle. Cannon and sight movement was synchronised via an electromechanical elevation-aiming synchro transmission.

The prototype was completed on 29th December 1950. Manufacturer's flight tests began on 2nd January 1951 and were completed on 27th March. On 20th June test pilot

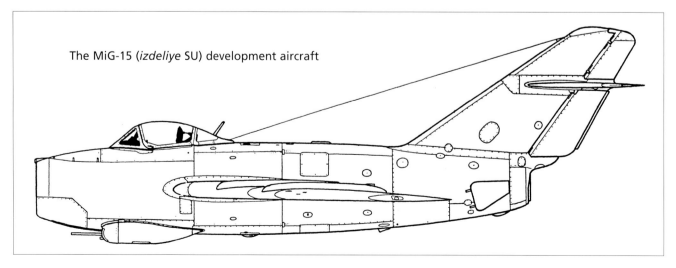

The MiG-15 (*izdeliye* SU) development aircraft

'935 Red' (c/n 109035), the *izdeliye* SU test-bed for the V-1-25-Sh-3 experimental movable weapons system.

The nose of *izdeliye* SU with the cannons level (at zero deflection) and at maximum depression.

Yuriy A. Antipov ferried the prototype to GK NII VVS for state acceptance trials, which began ten days later and continued until 10th August. GK NII VVS test pilots Viktor P. Trofimov, Vladimir N. Makhalin, Ivan M. Dzyuba, Lookin, Vasiliy S. Kotlov, Toopitsyn and Vladimir N. Filippov made 63 flights in the SU, including several flights to test the new armament. In the course of the trials, large blast panels made of heat-resistant steel were riveted to the underside of the fuselage nose and nose gear doors when it turned out that firing the cannons damaged the aluminium skin.

The State commission's report said that the movable cannon armament enhanced the MiG-15's combat potential. For instance, head-on attacks could be made without the danger of collision and the pilot could bring his guns to bear on the target for longer periods. The V-1-25-Sh-3 system could be mastered by the average pilot in 20 to 25 flights. Still, *izdeliye* SU did not enter production because the system was too complicated and the elevation angle was too small. (In 1953, Mikoyan dusted off the movable cannon idea, developing the *izdeliye* SN prototype, a heavily-modified MiG-17 with the SV-25-MiG-17 weapons system comprising three elevating 23-mm TKB-495 cannons, which likewise remained a one-off.) At transonic speeds, the movable cannons were simply not effective enough.

Mikoyan/Gurevich MiG-15*bis* development aircraft with ORO-57 rocket launchers (*izdeliye* SD-57)

In the early 1950s the Soviet military (and hence the Soviet 'fighter makers') began attaching considerable attention to giving fighters strike capability, especially the ability to carry medium- and heavy-calibre unguided rockets. Numerous models of high-velocity aircraft rockets (HVARs), folding-fin aircraft rockets (FFARs), launchers for same and bomb shackles were tested and the effect of rocket launches on drop tanks located alongside the launchers was investigated.

In June 1952 a MiG-15*bis* (*izdeliye* SD) *Fagot-B* serialled '803 Red' (apparently a Kuibyshev-built example, c/n 1...8003) was fitted with two experimental rocket pods, each featuring twelve ORO-57 launcher tubes (*odinochnoye re'aktivnoye oroodiye* – 'single jet gun', by analogy with recoilless guns) for 57-mm (2.24-in) ARS-57 *Skvorets* (Starling) FFARs. The pods were carried on D4-50 shackles at the standard drop tank hardpoints. The aircraft received the designation '*izdeliye* SD-57' (that is, *izdeliye* SD armed with 57-mm rockets). The programme was undertaken under contract with the OKB-16 weapons design bureau. The ARS-57 FFAR (designated S-5 in service) was widely used on later Soviet tactical aircraft.

'803 Red', the MiG-15*bis* (*izdeliye* SD-57) development aircraft armed with 12-round FFAR pods holding ARS-57 rockets.

Mikoyan/Gurevich MiG-15*bis* development aircraft with ORO-57 rocket launchers (*izdeliye* SD-5)

In November 1952 another MiG-15*bis* was converted at plant No. 21 in Gor'kiy. The aircraft had D3-40 shackles for carrying two FFAR pods, each with eight ORO-57 launcher tubes for ARS-57 FFARs. Firing was electrically controlled; AKS-2 gun cameras were fitted aft of the pods to record test launches. Designated *izdeliye* SD-5, the aircraft was tested but did not enter production.

Mikoyan/Gurevich MiG-15*bis* development aircraft with ORO-57 rocket launchers (*izdeliye* SD-5E)

In June 1952 MiG-15*bis* '803 Red' (the SD-57) was refitted with two enlarged FFAR pods, each with twelve ORO-57 launcher tubes; the pods were carried on D4-50 shackles. In this configuration the aircraft was known as *izdeliye* SD-5E; again, it did not enter production.

Mikoyan/Gurevich MiG-17 (*izdeliye* SI-5) experimental fighter-bomber/weapons testbed

One of the aircraft used to test heavy unguided weapons was *izdeliye* SI-5, a production MiG-17 *sans suffixe* (*Fresco-A*) adapted to carry two 190-mm (7.48-in) TRS-190 HVARs (***toor**boreak**tiv**nyy **snaryad** – spin-stabilised rocket) or two 210-mm (8.26-in) S-21 HVARs. The TRS-190 rockets were carried in PU-O-46 launch tubes (*pooskovaya oostanovka, odinochnaya* – aircraft-mounted launcher, single). Instead of the usual stabilising fins they had two angled nozzles which caused it to rotate like a bullet, thus stabilising it. The S-21s were of more conventional design and

were carried on PU-21 launch rails. The standard cannon armament was retained.

The SI-5 was equipped with a new ASP-5N gunsight linked to an SRD-1M gun ranging radar (aka Radal'-M; the name is a contraction of ***rah**diodal'**no**mer* – radio rangefinder); a PZV-5U control box was fitted when S-21 rockets were carried. The electric system included an MA-500 transformer associated with the rocket armament. Apart from the armament, the SI-5 could be identified by the non-standard canopy windshield; the bulletproof windscreen was located further forward and there was an upper glazing panel in addition to the two curved sidelights. The canopy design was similar to that of the Mikoyan/Gurevich I-1 (I-370) experimental fighter which was a sort of cross-breed between the MiG-17 and the MiG-19 *Farmer*.

Development began in August 1953; the completed aircraft was rolled out on 14th December. The gunsight and gun ranging radar were tested and refined by the Ministry of Defence's NII-2 research institute and the Central Design Bureau No. 589 (TsKB-589) of the Ministry of Defence Industry (MOP – *Mini**ster**stvo obo**ron**noy pro**mysh**lennosti*). Shortly afterwards the ASP-5N gunsight was removed for modifications by TsKB-589; the work was completed in April and the gunsight returned to Mikoyan on 26th April 1954 for re-installation. In late May the SI-5 entered state acceptance trials, completing them satisfactorily in August.

Mikoyan/Gurevich MiG-17F (*izdeliye* SI-7) experimental fighter-bomber/weapons testbed

A standard MiG-17F *Fresco-C* was modified in 1954 to test the ARS-70 ***Las**tochka* (Swallow) unguided rocket system. Two pods, each holding five 70-mm (2.75-in) ARS-70 FFARs, were carried on pylons installed between the main gear units and the regular drop tank hardpoints; two more pods could be carried on these hardpoints. Like the SI-5, this aircraft designated *izdeliye* SI-7 was equipped with an ASP-5N gunsight and an SRD-1M gun ranging radar.

Mikoyan/Gurevich MiG-17 (*izdeliye* SI-15) experimental fighter-bomber/weapons testbed

On 15th December 1951 the Council of Ministers issued a directive ordering a MiG-17 to be adapted for firing ARS-140-150 unguided rockets. The aircraft was to begin trials in the first quarter of 1953.

The advanced development project was prepared by the Mikoyan OKB branch office at the Gor'kiy aircraft factory No. 21 in 1952. Two Gor'kiy-built MiG-17s *sans suffixe* (identity unknown) were converted to test the new armament in 1953; the manufacturer's designation was *izdeliye* SI-15.

Mikoyan/Gurevich MiG-17 (*izdeliye* SI-16) experimental fighter-bomber/weapons testbed

On 27th May 1953 MAP issued an order concerning modification of a MiG-17 with two pylons for carrying two ORO-57 FFAR pods with eight ARS-57 rockets apiece; the aircraft was to begin trials in the fourth quarter of the same year. The manufacturing drawings were completed by September and conversion work began on a Gor'kiy-built MiG-17 serialled '005 Red' (c/n N54211005).

The pylons had a vertical leading edge and a raked trailing edge and were located just inboard of the drop tank hardpoints, allowing 400-litre (88 Imp gal) slipper tanks to be carried. Both FFAR pods fired simultaneously and could be jettisoned in an emergency. The aircraft featured an AP-57 computing gunsight (*avtoma**tich**eskiy pri**tsel***).

Completion was delayed until January 1954 because plant No. 598 was late in delivering the gunsight. Designated *izdeliye* SI-16, the aircraft entered flight test on 21st January; the manufacturer's flight tests consisted of two stages (airframe trials and rocket launches). On 27th March the SI-16 was delivered to GK NII VVS for state acceptance trials which were successfully completed in June and the FFAR pod installation was recommended for production.

Mikoyan/Gurevich MiG-17 (*izdeliye* SI-19) experimental fighter-bomber/weapons testbed

The same CofM directive of 15th December 1951 (see SI-15) stated that a MiG-17 was to be adapted for firing TRS-190 HVARs. The aircraft was to commence state acceptance trials in August 1952.

All manufacturing drawings had been issued by June 1952. After that, the fourth Kuibyshev-built *Fresco-A* serialled '104 Red' (c/n 1401004) was fitted with two pylons in line with the innermost wing fences for carrying ORO-190 launch tubes. The pylons were L-shaped, extending far beyond the wing leading edge. Alternatively, the rockets could be carried on special pylons installed at the regular drop tank hardpoints; these were similar to the ones fitted to the SI-16, with a vertical leading edge and a raked trailing edge. In this case the inboard pylons were removed.

'005 Red' (c/n N54211005), the SI-16 weapons testbed with two ORO-57 FFAR pods on long pylons. The pods could be carried simultaneously with 400-litre drop tanks.

335

The SI-19 weapons testbed ('104 Red', c/n 1401004) with two TRS-190 HVARs in ORO-190 launch tubes on special dog-leg pylons.

The launch tubes were suspended on D3-40 shackles; they came in two sizes and the conical noses of the rockets protruded from the shorter version. The TRS-190 rockets were fitted with EV-51 electric impact/proximity fuses (*elektrovzryvahtel'*) developed by NII-137 in the Ministry of Agricultural Machinery framework (yes, this ministry acted as a cover for ordnance development programmes – *Auth.*). An AP-2R automatic sight was fitted for aiming the rockets.

Designated *izdeliye* SI-19, the aircraft was rolled out on 13th August 1952 and delivered to GK NII VVS after manufacturer's flight tests on 9th September. Aleksey G. Solodovnikov was the institute's project test pilot.

Unbelievably, development of the TRS-190 with a high-explosive/fragmentation warhead was officially ordered *a year after the SI-19 flew* by a CofM directive of 19th September 1953 and an MOP order of 10th October 1953! This directive tasked the Mikoyan with equipping the fighter with ORO-90 launch tubes, an AP-21 (AP-2R) gunsight and a PZV-52 fuse charging device (*pribor zaryadki vzryvahteley*); the deadline for state acceptance trials was now the third quarter of 1953.

Even before that (on 12th August 1953) the SI-16 completed renewed manufacturer's tests. By 1st January 1954 it passed Stage 1 of

The logo of the Almaz-Antey concern, the successor of the KB-1 responsible for the K-5 AAM tested on several Mikoyan fighters.

the state acceptance trials. Stage 2 was held at a GK NII VVS shooting range and went smoothly. Still, like the previous aircraft armed with TRS-190 HVARs, the SI-16 did not enter service.

Mikoyan/Gurevich MiG-17 experimental fighter-bomber/weapons testbed

To meet a Soviet Navy requirement drawn up in March 1954, plant No. 81 converted a late-production Novosibirsk-built MiG-17 *sans suffixe* serialled '1628 Red' (c/n 1615328) for testing the B-374 rocket system. The system was developed for use against enemy landing ships, other small vessels and surfaced submarines. It comprised two five-tube rocket pods; each pod held fifteen 85-mm (3.34-in) TRS-85 spin-stabilised rockets, three per tube. The pods were carried on pylons installed between the main gear units and the regular drop tank hardpoints. The rockets could be fires in single salvos or in ripples; if drop tanks were carried, these had to be jettisoned first.

Manufacturer's flight tests took place in May-June 1955. After the deficiencies discovered at this stage had been corrected, the aircraft was handed over to NII-15 (the Navy's counterpart of GK NII VVS) for state acceptance trials, which took place between 13th October 1955 and 16th June 1956. The initial verdict was that using TRS-85 rockets on the *Fresco* was advisable. The ultimate decision concerning the updating of in-service aircraft, however, was delayed until the trials had been completed. Further trials, however, showed that the TRS-85 was directionally unstable when fired at high speeds; this could only be cured by a complete redesign of the weapon and the programme was abandoned.

MiG-17 '1628 Red' (c/n 1615328) with experimental five-tube rocket pods on pylons.

Mikoyan/Gurevich MiG-17P (*izdeliye* SP-6 modified) weapons testbed

In the early 1950s the Soviet government issued a number of directives concerning the development of air-to-air missiles and missile-armed interceptors. Several design bureaux were tasked with the development of AAMs, among them KB-1, an MOP division, which began development of the K-5 radar-guided beam-riding missile. (KB-1 later became the NPO *Al**maz*** (Diamond) research & production association; it is now part of the Almaz-Antey missile system concern.)

(In passing, it may be noted that at the time KB-1 was headed by Sergey L. Beria – the son of the infamous Lavrentiy P. Beria, Stalin's feared Minister of the Interior. After Stalin's death in 1953 L. P. Beria was found guilty of high treason and executed, sharing the fate of many he had sent to death. Hence Beria Jr. was removed from office and replaced by K. Patrookhin.)

One of the said directives issued on 26th November 1953 marked the birth of OKB-2 within the MAP framework; it was led by Pyotr D. Grooshin, best known for his Sh-Tandem experimental tandem-wing attack aircraft of 1937. One of its principal tasks was AAM development. By then Grooshin had already gained some experience in the design of missile systems, having contributed to the development of the S-25 Berkoot (Golden Eagle) surface-to-air missile at the Lavochkin OKB since 1951.

OKB-2 took up residence at plant No. 293 (previously allocated to the defunct Bolkhovitinov OKB) and was assigned responsibility for the K-5 AAM – that is, except for the missile's radio command guidance system, which remained the responsibility of KB-1. Two other bureaux, OKB-134 under Ivan I.

Close-up of the experimental rocket pods on the same aircraft.

Toropov and OKB-4 under Matus R. Bisnovat, were also developing AAMs. On 30th December 1954 the Council of Ministers issued directive No. 2543-1224 ordering these missile systems to be tested on interceptors. The 'missilisation' programme was a monstrous research and development effort that

The modified SP-6 was effectively the prototype of the MiG-17PFU missile-armed interceptor. These views show the four missile pylons.

required weapons, aircraft and avionics designers to work in close co-operation.

The Soviet Air Force and MAP had a hard time choosing the right missile platform. Hence all four Soviet fighter makers were ordered to adapt their production and/or experimental fighters for carrying AAMs and present them for evaluation. Specifically, the Mikoyan OKB was to equip the MiG-17P (pere**khvat**chik – interceptor) Fresco-B and the izdeliye SM-7 – the MiG-19P Farmer-B interceptor prototype – with the K-5 weapons system and test them jointly with OKB-2.

Hence the five MiG-17P (izdeliye SP-6) prototypes were suitably converted into testbeds for the K-5 (RS-1-U; NATO AA-1 Alkali) AAM. Four APU-3 launch rails were mounted on pylons with D3-40 shackles extending far beyond the wing leading edge between the inboard and centre wing fences, so that the aircraft could still carry 400-litre slipper tanks. The starboard NR-23 cannon was retained as a backup weapon, and the aircraft was equipped with an ASP-3NM gunsight.

After successfully passing manufacturer's tests and state acceptance trials the modified SP-6 was recommended for production. Actually it never entered production as such, but a missile-armed version did enter service as the MiG-17PFU Fresco-E (izdeliye SP-15) mid-life upgrade.

Mikoyan/Gurevich MiG-17PF (izdeliye SP-9) weapons testbed

Mikoyan engineers tried using unguided rockets on interceptors as well as on tactical fighters. A Council of Ministers directive to this effect was issued on 19th September 1953; the aircraft was to commence state acceptance trials in the third quarter of 1954.

On 26th November 1954 – rather later than anticipated – a production Gor'kiy-built MiG-17PF Fresco-D serialled '627 Red' (c/n N58210627) was delivered to the Mikoyan OKB's experimental shop. Conversion work continued throughout December and into January 1955; the resulting weapons testbed was designated izdeliye SP-9. The standard cannons were replaced by a quartet of ARO-57-6 Vikhr' (Whirlwind) automatic rocket launchers (avtoma**tich**eskoye re'ak**tiv**noye o**roo**diye – 'six-round 57-mm automatic jet gun'), aka 3P-6-III; these looked like outsize six-shooters with a long barrel and a revolving drum for six ARS-57 FFARs. The launchers were mounted symmetrically, two on each side, on a redesigned weapons tray with exhaust gas outlets at the back. Thus the SP-9 was the nearest Soviet equivalent of the North American F-86D Sabre with its battery of Mighty Mouse FFARs.

Outwardly the installation looked like four ordinary heavy cannons, except that there were no spent case outlets where one would expect them to be. The barrels were recessed in deep narrow troughs and there were two large bulges on each side over the ammunition drums immediately aft of the nose gear unit. Additionally, four pods with eight ARS-57 FFARs each could be carried on the regular drop tank hardpoints and two streamlined pylons inboard of these. Alternatively, the SP-9 could carry two drop tanks and two FFAR pods with five 70-mm ARS-70 rockets each. The pods were attached on D3-40 shackles. Two podded ciné cameras were installed under the wingtips to record rocket launches. The aircraft was tested in May-July 1955 but the results are not known.

Mikoyan/Gurevich MiG-17PF (izdeliye SP-10) weapons testbed

When the programme described above had been completed, MiG-17PF '627 Red' was modified again to test a new rapid-firing double-barrelled cannon. Two such cannons were mounted on a suitably modified weapons tray from a standard MiG-17. After this conversion the aircraft was redesignated izdeliye SP-10. The cannons could be fired separately or simultaneously by means of several buttons on the stick. After being tested the new cannon was deemed substandard and did not enter production.

(Some sources give a totally different story, claiming that the SP-10 was an avionics test-

Close-up of the SP-9's nose with the weapons tray lowered, showing the battery of ARO-57-6 Gatling-type rocket launchers.

bed for the Aïst radar – the one fitted to the MiG-17/*izdeliye* SN development aircraft – and was completed on 17th October 1954.)

Mikoyan/Gurevich MiG-17PF weapons testbed with K-13A (R-3S) AAMs

In keeping with a joint GKAT/Air Force order signed on 21st January 1963 the Mikoyan OKB prepared a set of drawings for installation of two launch rails for K-13A IR-homing short-range AAMs on the MiG-17PF with appropriate changes to the electric system. The K-13A was a copy of the early-model Naval Air Weapons Centre AIM-9 Sidewinder and the reverse-engineering effort was undertaken by Ivan I. Toropov's OKB-134. A production MiG-17PF was fitted experimentally with missile launch rails at plant No. 134 in the fourth quarter of 1963 and tested successfully in 1964, whereupon the missile entered production as the R-3S (NATO AA-2 *Atoll*), becoming the main weapon of the MiG-21, starting with the MiG-21F-13 *Fishbed-C* version.

(Note: In 1957 MAP lost its ministerial status and was 'demoted' to the State Committee for Aviation Hardware (GKAT – *Gosudarstvennyy komitet po aviatsionnoy tekhnike*) because of the Soviet leader Nikita S. Khrushchov's disdainful attitude to manned combat aircraft. In 1965, however, MAP regained its original name and 'rank' after Khrushchov's removal from power.)

Mikoyan/Gurevich MiG-17PF (*izdeliye* SP-16) avionics/weapons testbed

Two production MiG-17PFs were converted into testbeds for the ShM-60 radar jointly with the KB-1 design bureau. The aircraft were armed with K-5M (RS-2-U; NATO AA-1A *Alkali*) AAMs developed for the MiG-19PM *Farmer-D* and thus effectively converted into MiG-17PFUs.

The RS-2-U was an improved version of the RS-1-U with better performance (including ceiling) and kill capability. It was part of the S-2-U weapons system designed to replace the S-1-U system fitted to the MiG-17PFU. The missile had a 13-kg (28.65-lb) fragmentation warhead, a launch weight of 84 kg (185 lb), a top speed of 1,650 km/h (892 kts) and semi-active radar homing. As with the MiG-17PFU, the missiles could be fired singly, in pairs or all together with a couple of seconds in between. However, the RS-2-U had limited manoeuvrability and thus could be considered a serious threat only for sluggish heavy bombers. Besides, the missile's maximum kill range (6 km/3.72 miles) exceeded the radar's target tracking range.

Known as *izdeliye* SP-16, these aircraft successfully passed manufacturer's trials in October 1957; the objective was to test the new radar's and missiles' suitability for the new MiG-21 fighter.

Mikoyan/Gurevich MiG-19 high-speed bomb delivery research aircraft

In 1950 the Ilyushin IL-28 *Beagle* tactical bomber designed to withstand a dynamic pressure up to 3,000 kg/m^2 (614.75 lb/sq ft) attained initial operational capability with the Soviet Air Force. This brought about a decision to investigate how bombs behaved when leaving the aircraft's bomb bay at the maximum dynamic pressure. Engineer M. P. Lisichko, who had a lot of experience in conducting various experimental programmes, was tasked with this one.

Test drops of 250-kg (551-lb) bombs recorded by ciné cameras showed that the upper bombs in the IL-28's rear bomb cassette left the bomb bay at a high angle of attack. This was confirmed by numerous experiments conducted by NII-2 (an MoD research institute concerned with aircraft weapons), GK NII VVS and the Air Force Engineering Academy named after Nikolay Ye. Zhukovskiy (VVIA – *Voyenno-vozdooshnaya inzhenernaya akademiya*).

In order to study bomb separation patterns Lisichko designed special underwing pods – a sort of submunitions containers from which scaled-down bombs could be dropped at high speeds. Test results obtained with a specially-modified MiG-19 confirmed the designers' theories completely.

Mikoyan/Gurevich MiG-19S (*izdeliye* SM-9/3T) weapons testbed

The Mikoyan OKB, too, experimented with the MiG-19's armament. As mentioned above, captured AIM-9 Sidewinder AAMs were obtained via mainland China in 1958, and a reverse-engineered version soon entered production in the USSR as the K-13 (R-3S). The missile was intended both for operational fighters and for new ones then under development.

In February 1959 a Novosibirsk-built MiG-19S *Farmer-C* was converted into a weapons testbed designated *izdeliye* SM-9/3T. Unusually, the aircraft was coded '08 Yellow' (only Yakovlev aircraft wore yellow tactical codes at the time). Two K-13 AAMs were carried on APU-26 missile rails; these were attached to short pylons fitted to the standard drop tank hardpoints. The missiles could be fired singly or in a salvo; a safety system ensured that the missiles could only be fired when the landing gear was retracted. The wing cannons were removed to make room for test equipment.

The manufacturer's flight tests commenced on 11th February 1959, when Aleksandr V. Fedotov made the first flight. The rest of the programme, however, was performed by another Mikoyan OKB test pilot, Pyotr M. Ostapenko, and lasted until 3rd March. The launch of the K-13 and its performance at the initial trajectory stage were examined, as well as the effect of the missile's exhaust on engine operation. Top speed with two AAMs was Mach 1.295 at 10,800 m (35,430 ft) and 910 km/h (492 kts) at 17,600 m (57,740 ft). The missiles had no adverse effect on the aircraft's handling.

Mikoyan/Gurevich MiG-19P (*izdeliye* SM-7A) experimental interceptor/weapons testbed

When the K-5 (RS-1-U) AAM became available in the mid-1950s, the Mikoyan OKB decided to adapt it to the MiG-19P all-weather interceptor. The first missile-toting version of the *Farmer-B* armed with four RS-1-U AAMs and designated *izdeliye* SM-7A was developed in 1954. The RP-1 Izumrood-1 radar fitted to the MiG-19P was replaced by the RP-5 Izumrood-5 radar linked to an ASP-5NV optical sight. The missiles were carried on APU-3 launch rails attached to pylons extending beyond the wing leading edge with Model 369-Sh shackles. An SSh-45 gun camera was fitted to film the radar display, confirming the 'kill'; another gun camera (an AKS-3) filmed the missile's trajectory. Additionally, two S-13-300 high-speed still cameras were fitted under the wings to capture the missile separation sequence. The cannons were deleted. Seven MiG-19Ps were thus converted by the Gor'kiy aircraft factory in 1956 and tested with satisfactory results.

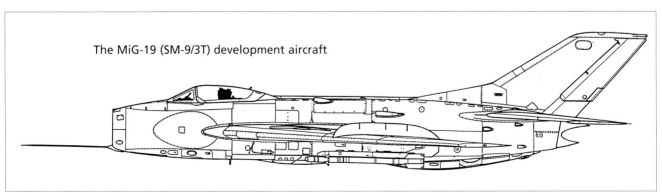

The MiG-19 (SM-9/3T) development aircraft

Mikoyan/Gurevich MiG-19P (*izdeliye* SM-7M) experimental interceptor/weapons testbed

On 7th January 1956 Artyom I. Mikoyan endorsed the project of a new version of the MiG-19P designated *izdeliye* SM-7M (*modifit-seerovannoye* – modified). This follow-on version of the SM-7A developed pursuant to the same CofM directive No. 2543-1224 of 30th December 1954 was armed with four RS-2-U (K-5M) AAMs on pylon-mounted APU-4 launch rails, with appropriate changes to the wing structure.

The RP-5 radar used on the SM-7A gave way to the RP-2-U, a modified RP-1 with an automatic tracking range of 3.5-4 km (2.17-2.48 miles); the U suffix denoted *oopravleniye* [*snaryadami*] – missile control or guidance. The radar was linked to an ASP-5N-VU sight; it detected targets in instrument meteorological conditions (IMC) and gave information on target heading and range, enabling the pilot to get within missile launch range. It also sent coded signals in the main channel for threat classification jointly with the IFF equipment. Like its predecessor, the RP-2-U had an azimuth scan of ±60° and an elevation scan of +26°/–14°. As the target came within tracking range, a reference grid and information on range, bearing and elevation relative to the fighter were automatically fed into the optical sight, allowing the pilot to take aim regardless of visibility conditions.

Two MiG-19Ps were converted to SM-7Ms at the Gor'kiy aircraft factory, using drawings supplied by the Mikoyan OKB. The fighters passed their state acceptance trials satisfactorily and the S-2-U weapons system was cleared for installation on the interceptor, paving the way for the MiG-19PM *Farmer-D*.

Mikoyan/Gurevich MiG-19 (*izdeliye* SM-2/A) development aircraft/weapons testbed

Pursuant to the abovementioned Council of Ministers directive dated 30th December 1954 the Mikoyan OKB converted a Gor'kiy-built MiG-19 *sans suffixe* serialled '420 Red' (c/n N59210420) into a ground attack aircraft designated *izdeliye* SM-2/A. The reason for this designation is unclear, since the aircraft was originally a standard SM-9.

The SM-2/A was armed with 70-mm (2.75-in) ARS-70 spin-stabilised rockets, carrying two drop tanks and two RO-70-5 five-round rock-

et pods on MiG-19PM-style pylons with Model 369-Sh shackles fitted just inboard of the main gear units. Alternatively, two more pods were carried on the 'wet' hardpoints. Aiming for cannon firing or rocket launch was done, using the ASP-5N-V2 sight linked to the Radal'-M gun ranging radar. Two additional blade aerials were located in tandem low on the forward fuselage port side. An SSh-45 gun camera was fitted. The SM-2/A was tested and found satisfactory.

The SM-2/A weapons testbed, '420 Red' (c/n 59210420), with four rocket pods; weight-equivalent dummies are fitted in this photo.

Close-up of the RO-70-5 five-round FFAR pods tested on the SM-2/A.

Mikoyan/Gurevich MiG-19 (*izdeliye* SM-2/B) development aircraft/weapons testbed

Also in 1955, another Gor'kiy-built MiG-19 ('406 Red'; c/n N59210406?) was fitted with ORO-190K launch tubes, each firing a single 190-mm TRS-190 spin-stabilised unguided rocket. The aircraft could carry either two launch tubes on Model 369-Sh shackles plus two drop tanks or four launch tubes and no drop tanks. The cannons were retained. Except for the armament, the aircraft was identical to the SM-2/A. This aircraft, known as *izdeliye* SM-2/B, was also tested successfully in 1957.

Mikoyan/Gurevich MiG-19 (*izdeliye* SM-2/V) development aircraft/weapons testbed

In 1956, SM-2/B '406 Red' was converted once again and redesignated *izdeliye* SM-2/V (V is the third letter of the Cyrillic alphabet).

The SM-2/D weapons testbed ('420 Red', c/n N59210420) with four rocket pods, each containing nine TRS-85 rockets.

manufacturer's test programme in July-August 1957 the SM-2/D underwent joint trials with NII-1 and the Air Force, starting in October.

The aircraft was armed with two 212-mm (8.34-in) ARS-212M HVARs (aka S-21) and could carry two drop tanks. The rockets were fired from APU-5 launch rails carried on Model 369-Sh shackles. The sighting equipment was unchanged. To preclude an inadvertent launch on the ground, a safety mechanism prevented the rockets from being fired when the landing gear was down. The SM-2/V successfully underwent state acceptance trials in 1957.

Mikoyan/Gurevich MiG-19 (*izdeliye* SM-2/G) development aircraft/weapons testbed

About the same time the Mikoyan OKB converted a MiG-19 for testing ARS-160 HVARs on dual launchers. In Cyrillic alphabetical order this version was designated *izdeliye* SM-2/G. Yet this aircraft was never tested, as MAP terminated the programme when the conversion was almost complete.

Mikoyan/Gurevich MiG-19 (*izdeliye* SM-2/D) development aircraft/weapons testbed

In June 1957 the SM-2/A ('420 Red') was refitted with four nine-round rocket pods firing 85-mm (3.34-in) TRS-85 spin-stabilised rockets; each pod had three launch tubes with three rockets apiece. In this guise the aircraft was designated *izdeliye* SM-2/D (in alphabetical order). After making ten flights under the

Mikoyan/Gurevich MiG-19 (*izdeliye* SM-2/I) development aircraft/weapons testbed

This is an out-of-sequence designation, since I is the ninth letter of the Cyrillic alphabet; the SM-2/Ye, SM-2/Zh and SM-2/Z never existed. Pursuant to the same CofM directive of 30th December 1954 an unserialled Gor'kiy-built MiG-19 *sans suffixe* (ex-'549 Red', c/n N59210549) was converted into a weapons testbed for the K-6 AAM. This missile was envisaged for the Sukhoi T-3 interceptor but never entered production. No target illumination radar was fitted and the wing cannons were retained. Two pods housing ciné cameras were mounted on pylons under the inner wings to record missile launches.

The SM-2/I was rolled out on 8th March 1956, completing initial tests at Zhukovskiy in April. Stage 1 of the K-6's tests took place in May-December 1956 with satisfactory results. After that, a modified version of the missile with all-movable control surfaces was tested in 1957; the aircraft made ten live launches. Upon completion of the test programme the SM-2/I was written off as time-expired.

Mikoyan/Gurevich MiG-19 (*izdeliye* SM-21) development aircraft/weapons testbed

Also in 1956, the SM-2/V ('406 Red') was converted for the third time and successfully tested with the APU-5 launch rails moved outboard to the drop tank hardpoints. Since the various configurations of the SM-2 were designated by letters in Cyrillic alphabetical order, logically this should have been the SM-2/G but the aircraft was known as *izdeliye* SM-21 – a reference to the S-21 rockets it carried.

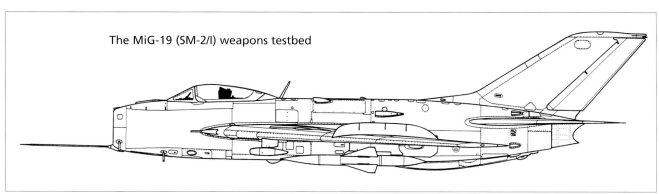

The MiG-19 (SM-2/I) weapons testbed

This modified MiG-19 *sans suffixe* (c/n N59210549) is the SM-2/I, a testbed for the K-6 AAM. Note the podded ciné camera for recording missile launches.

Mikoyan/Gurevich MiG-19 (*izdeliye* SM-2/M) development aircraft/weapons testbed

This was the last of the SM-2 weapons testbed series with yet another out-of-sequence suffix letter (the SM-2/K and SM-2/L never existed). The aircraft, the eighth Gor'kiy-built *Farmer-A* ('108 Red'?, c/n N59210108) was converted in 1955 to test the K-5M (RS-2-U) AAM developed for the MiG-19PM. Two missiles were carried on APU-5M pylons mounted near the wing cannons on Model 369-Sh shackles. A similar arrangement was later used on MiG-19Ps retrofitted with two missile pylons.

Mikoyan/Gurevich MiG-19 (*izdeliye* SM-6) development aircraft/weapons testbed

Attempting to improve the MiG-19P's armament, in 1956 the Mikoyan OKB developed a version of the MiG-19P armed with two K-6 AAMs tested earlier on the SM-2/I – again pursuant to the CofM directive of 30th December 1954. Designated *izdeliye* SM-6, the aircraft featured a new Almaz-3 target illumination radar and had the cannons deleted. The missiles were carried on APU-5M-K pylons mounted at the usual drop tank hardpoints. Two ciné cameras (Ye-13-300 and AKS-2) were carried in pylon-mounted pods under the inner wings to record missile launches. As on the SM-2 series, two downward-pointing blade aerials were mounted low on the forward fuselage port side.

Two *Farmer-Bs* (c/ns 210101 and 210102) were completed to SM-6 standard at the Gor'kiy aircraft factory in January 1957. (The c/ns show that the SM-6 was recognised as a separate sub-type at the factory (as evidenced by the batch number 01) but, surprisingly, no product code was allocated.) Next month the two uncoded aircraft embarked on a major test programme. The first prototype was used for determining target detection/lock-on range and target tracking accuracy, while the

second aircraft served for performance and reliability testing of the Almaz-3 radar. Various versions of the K-6 were launched from the first prototype to determine their performance, including stability and manoeuvrability characteristics (25 launches were made in all). The tests continued into 1958; on 2nd April MAP issued an order terminating the SM-6 programme due to the cancellation of the K-6 weapons system.

Mikoyan/Gurevich MiG-19S (*izdeliye* SM-12/3T and SM-12/4T) weapons/avionics testbeds

Upon completion of the state acceptance trials the third and fourth SM-12 prototypes were modified to carry K-13 AAMs and redesignated SM-12/3T and SM-12/4T respectively. As with the SM-9/3T, the T probably stood for *teplovaya golovka samonavedeniya* (IR seeker head), since the K-13A was a heat-seeking

Another MiG-19 used as a weapons testbed, the SM-21 ('406 Red', c/n 59210406) with an ARS-212M (S-21) heavy unguided rocket on the port outboard pylon.

missile. Originally the first prototype was also to be converted, becoming the SM-12/2T, but conversion was soon halted and the new components fitted to the SM-12/4.

The SM-12/3T was the first to be completed. The aircraft had two APU-26 missile rails on short pylons fitted to the standard drop tank hardpoints. The SRD-5 **Ba**za-6 (Base) gun ranging radar gave place to a new Kvant (Quantum) radar linked to an ASP-5NV-U1 gunsight and an SIV-52 IRST was fitted; thus the aircraft also served as an avionics testbed. The SM-12/3T had the so-called 'Version 1' of the Kvant which automatically and continuously determined true target range, showing it on a dial-type indicator, comparing it with the authorised missile launch range and showing the pilot when it was OK or too close to fire.

The changes were not limited to weapons and avionics. The BU-14MSK tailplane actuator and BU-13MK aileron actuator were replaced with more efficient BU-19D and BU-18D units respectively.

The other aircraft, SM-12/4T, was fitted with an improved Kvant-1 radar with a modified antenna linked to an ASP-5N gunsight. The Kvant-1 allowed for accurate rangefinding when the cannons and FFARs were used.

The SM-12/3 and SM-12/4T were used for performance and handling trials with K-13 AAMs and then for live missile launches in which the gun ranging radar was used. After that, both aircraft and a pre-production MiG-21F-13 (c/n N74210302) with APU-13D missile rails took part in full-scale trials of the K-13 weapons system which began on 21st October 1959.

The logo of GMKB Vympel incorporating a stylised Cyrillic V. The company created the R-27 and R-77 (RVV-AE) AAMs tested on the MiG-29 (aircraft 970 and aircraft 971).

Mikoyan/Gurevich MiG-21*bis* 'No. 908' weapons testbed

In order to provide the closest possible simulation of in-flight conditions during ground tests of various systems and weapons, the need arose to design and build ground test rigs featuring the jet engines of existing or future aircraft. Such rigs would be used for performing missile and rocket launches at the NIPAV test range (*Na**ooch**no-ispy**tah**tel'nyy poli**gon** avi-atsi**on**novo vo'oru**zhen**iya* – Aircraft Weapons Research & Test Facility) in Faustovo, Moscow Region. They would feature actual aircraft suspended on special fixtures.

This kind of simulation was considered proper because the motion of the rocket or missile with respect to the aircraft would be pretty much the same, regardless of whether the aircraft itself was flying or stationary. For the purpose of holding such tests the Mikoyan OKB teamed up with OKB-2-300 and LII to create the 21-L-11 ground test rig based on a MiG-21 airframe and convert a further MiG-21*bis Fishbed-L* into a weapons testbed. The latter was referred to in LII documents as 'MiG-21 No. 908', the figures being the last three of the aircraft's c/n (750…908) or f/n.

Mikoyan MiG-23 weapons testbeds

According to press reports, in 1976-86 a MiG-23 *Flogger* of an unspecified version referred to as 'No. 9002' (these are probably the last four digits of the c/n, not the f/n) served for testing new 'dogfight missiles' and evolving techniques of using them in a head-on engagement. The tests performed by LII also involved two MiG-21s referred to as 'No. 3217' and 'No. 2024', presumably acting as 'adversary aircraft'.

In order to verify the Vympel K-27 (R-27; NATO codename AA-10 *Alamo*) medium-range AAM and the Vympel K-73 (R-73; NATO codename AA-11 *Archer*) short-range AAM in readiness for their use on the MiG-29 and the Sukhoi Su-27, one of the MiG-23ML *Flogger-G* (*izdeliye* 23-12) prototypes coded '123 Blue' was modified for live firing trials. This aircraft was referred to in some sources by the cryptic designation '*samo**lyot** RV*' (Aircraft RV).

Mikoyan MiG-29 'aircraft 970' and 'aircraft 971' weapons testbeds

Two early-production MiG-29 *Fulcrum-As* known in-house as 'aircraft 970' and 'aircraft 971' had a modified weapons control system, serving as weapons testbeds in the interests of the GMKB Vympel design bureau. They were used to test the **semi-active radar-homing** R-27T/TE *Alamo-A*, the IR-homing R-27RE *Alamo-B* and **active radar-homing** R-77 (RVV-AE; NATO codename AA-12 *Adder*) medium-range air-to-air missiles. Both fighters originally lacked tactical codes, sporting only the prototype code (970 and 971) on the dielectric fin caps. 'Aircraft 970' (c/n 2960507687), which had first flown on 24th December 1984, made its first post-modification flight in May 1985 and later became '70 Blue/970'. The second aircraft ('71 Blue/971', c/n 2960515117) followed in January 1986.

The test programme was completed in August 1989. It included missile separation safety trials in all flight modes, verification of

The MiG-29 weapons testbed known as 'aircraft 970' had no external recognition features. Note the open airbrake and the absence of the tactical code, which was added later.

the changes to the Phazotron N019 *Slot Back* radar (as introduced on the later N019M Topaz capable of attacking two priority threats at once) and live firing trials against Lavochkin La-17M, Mikoyan M-21 (MiG-21) and Tupolev M-16 (Tu-16M) target drones.

Mil' Mi-24V weapons testbed

During the Afghan War Mi-24 assault helicopter pilots kept urging the Mil' OKB to give the *Hind* some protection for its behind. While the Mi-8MT *Hip-H* – another Afghan War workhorse – had a hatch in the port half of its clamshell cargo doors where a 7.62-mm (.30 calibre) Kalashnikov RPK machine-gun or a similar weapon could be mounted to cover the rear hemisphere, the Mi-24 had none and consequently often got shot up after making an attack. About 48% of all damage from ground

The same aircraft, now coded '70 Blue'. Note the quarter-scale gliding model of a *Fulcrum-C* in the foreground.

fire on the Mi-24 was in the rear hemisphere, compared to some 27% on the Mi-8.

Hence in 1985 a Mi-24V *Hind-E* coded '43 Red' was fitted experimentally with a 12.7-mm (.50 calibre) Nikitin/Sokolov/Volkov NSVT-12,7 *Ootyos* (Cliff) machine-gun in a bulged enclosure replacing the aft avionics bay. The gunner's station was accessed from within via a crawlway passing through the rear

The MiG-29 'aircraft 971' weapons testbed, seen here with dummy R-27Rs and R-60Ms.

fuel tank between the mainwheel wells. It was so cramped that the gunner could not be accommodated entirely and his legs stuck outside, scantily protected by rubberised fabric 'trousers' (!).

Trials promptly showed that the rear gunner's station was no good. It caused a major shift in the helicopter's centre of gravity position and was always full of engine exhaust gases, making the gunner's working conditions almost unbearable. The crunch came when the modified helicopter was demonstrated to Soviet Air Force top brass; one of the portly generals got stuck in the narrow crawlway when he wanted to check out the gunner's station and the idea was abandoned. Instead, rear view mirrors were installed on operational *Hinds* so that pilots could see they were being fired upon and take evasive action.

North American B-25 weapons testbeds

B-25 bombers were used by LII in post-war experiments associated with the flight development of bomb delivery systems.

Furthermore, two B-25G bombers found use in the trials of the **Shchoo***ka* (Pike) gliding torpedo conducted at a test range near Yevpatoria on the Crimea Peninsula in 1947-48. One aircraft was used as the torpedo-bomber, while the other served for the development of the guidance system. The spacious fuselage of the B-25G provided accommodation for the electronic suite and the operator.

Finally, in 1948 an employee of the NII-2 weapons research establishment came up with an idea of determining the shell scatter of aircraft cannons by firing at the shadow of an airborne aircraft and filming the results. A B-25 (which provided the shadow) and a MiG-15 fighter were committed to the experiment; the bomber was equipped with a ciné camera to record the gunnery results and with a radio synchroniser to switch on the camera simulta-

neously with the start of the cannon fire. After a few test flights it was decided to transfer the recording camera to the MiG-15.

Petlyakov Pe-8 weapons testbed

Six late-production examples of the Pe-8 bomber powered by Shvetsov M-82 radials were used as missile carriers in the trials of the 10Kh and 14Kh air-launched cruise missiles designed by OKB-51, which was then headed by Vladimir N. Chelomey. The 10Kh was more or less a copy of the German Fieseler Fi 103 'flying bomb', better known as the V1 (*Vergeltungswaffe Eins* – reprisal weapon No. 1), but with an indigenous D-3 pulse-jet and AP-4 autopilot.

The first prototype of the 10Kh was built before the end of 1944, and the first missile from an initial production batch of 19 was completed at Plant No. 51 on 5th February 1945. Seventeen missiles from this batch were assigned for flight testing which involved the use of the modified Pe-8 bombers; the missile was suspended under the fuselage, with a very small ground clearance. The testing took place between 20th March and 25th June 1945 in the desert areas of Central Asia. Its successive stages included the checking of release procedures and guidance systems, determining the missile's flight performance and, finally, evaluating its combat effectiveness in attacks against simulated targets at a test range.

While the speed and range of the 10Kh met the specifications, reliability left a lot to be desired. Of the 66 examples launched from the carrier aircraft during the tests, 44 entered automatically-sustained guided flight. Of the 18 machines used to check the guidance precision, only five reached the target defined as a square measuring 20x20 km (12x12 miles) at a distance of 170 km (106 miles) from the release point.

Some 300 10Kh cruise missiles were built before the end of the war. They were used for further testing with a view to enhancing the

This Pe-8 bomber powered by Shvetsov M-82 radials was one of several used as weapons testbeds after the war, carrying a 10Kh missile under the fuselage.

precision and reliability of the weapon. Close to 200 of them were used during the state acceptance trials conducted between 15th December 1947 and 20th July 1948. The machines tested in 1948 differed from the 1945 models in having the German-type wings and horizontal tail replaced by those of Soviet design and in being fitted with pulse-jets of greater thrust. The 1948 model showed a considerable improvement over the 1945 predecessor, the probability of hitting the target being raised to 88% as against the earlier 36% (in the case of the V1, according to Soviet documents, it was 70%).

Despite the generally satisfactory results, Soviet Air Force Commander-in-Chief Air Marshal Konstantin A. Vershinin voiced his opposition to the introduction of this weapon whose performance, in his opinion, fell short of the day's standards. Thus, the 10Kh was never adopted for service.

OKB-51 developed several improved versions of the 10Kh. For example, a version known as 10DD had greater fuel capacity and, accordingly, longer range (DD probably stood for **dahl'**neye **dey**stviye – long range); it could be launched either from a carrier aircraft or from a ground launcher. In parallel, work proceeded on improved types of pulse-jet engines. One of these was the D-5 delivering a thrust of 420-440 kgp (930-970 lb). As early as 1944 Chelomey started projecting the 14Kh cruise missile powered by this engine. Aerodynamically it differed from the 10Kh in having revised forward fuselage contours and tapered wings instead of rectangular ones. A variant of this model, referred to as '34', featured rectangular wooden wings fitted with ailerons. Greater thrust, coupled with aerodynamic refinement of the fuselage, was expected to endow the 14Kh with a considerably greater speed. Twenty examples of the 14Kh were manufactured in 1946; ten of them passed trials at a test range in 1948, again with the Pe-8 as a carrier. A boosted version of this missile attained a speed of 825 km/h (513 mph), a marked improvement as compared to 620 km/h (385 mph) of the 10Kh.

Sukhoi Su-7KG air-to-surface missile testbed

At the end of the 1960s the Soviet Union carried out complex research and development work involving the State Research Institute for Aviation Systems (GosNII AS), the Gheofizika (Geophysics) Central Design Bureau, the

A 16KhA (version 1) missile suspended on a special pylon beneath the Pe-8 weapons testbed.

This sequence of stills from a GK NII VVS documentary shows the Pe-8 'mother ship' carrying a 10Kh missile marked '404'. The missile's pulse-jet engine is ignited while the missile is still attached, belching tremendous flames, whereupon the weapon is released.

design bureau of the Strela (Arrow) Production Corporation and the Sukhoi OKB to determine the design theory and demonstrate the viability of a new tactical strike weapon system. This consisted of an airborne laser target designator and a missile with a semi-active seeker head able to follow the laser signal reflected from the target (or, more precisely, from the surface surrounding it). This work determined methods of developing compact target desig-

The T43-12 development aircraft based on the Su-9 was readily identifiable by the large teardrop-shaped pod offset to starboard under the nose, apparently housing a TV camera for recording missile launches. Note the red-painted inert RS-2-US missiles.

nators using lasers based on activated glass with a wavelength of 1.06 micromicrons, as well as the seeker head capable of interpreting reflections from the target at a range of up to 8 km (5 miles). The configuration of an air-to-surface missile with a fragmentation/high-explosive warhead that could be carried by the Su-7BM fighter-bomber, and the principles of damping the target designator beam which would provide satisfactorily accurate target tracking while the aircraft was diving or in level flight were also worked out.

The R&D work showed that the new type of munitions would substantially increase the effectiveness of tactical aviation compared with unguided rockets and bombs against pinpoint targets, including tactical surface-to-surface missile launch sites, radar sites, aircraft on the ground and so on. The effectiveness of laser-guided missiles was even higher in comparison with the Kh-66 and Kh-23 ASMs in service with the Soviet Air Force due to the ability to fire laser-guided missiles in a salvo during a single attack and their higher accuracy.

A single Su-7BM *Fitter-A* was modified for testing the Zvezda Kh-25 laser-guided ASM and designated Su-7KG (for *kvantovyy ghenerahtor* – quantum generator, that is, laser). Development of the missile began in the early 1970s in co-operation with the Sukhoi OKB and various other R&D establishments. It was intended for use with the Su-7 and its successor, the Su-17.

The aircraft was modified as follows. In addition to the missile itself, the Prozhektor-1 (Searchlight-1) laser designator was carried in a ventral housing. The laser beam was reflected from the target and received by the missile's 24N1 seeker head and also by a receiver in the aircraft's cockpit. After the missile was launched, the pilot had only to keep the sight

(and, consequently, the laser beam) on the target under attack and the rest was done automatically.

Simultaneous tests of the Kh-25 on the Su-7KG and the Su-17M *Fitter-C* began in the winter of 1973. After an analysis of the results, it was decided to continue the work on the newer Su-17. A high degree of accuracy was achieved on this and, as a result, the missile was accepted for service. The Su-7BM and Su-7BKL fighter-bombers still in service were not, however, fitted with laser designators.

Sukhoi T43-3, T43-4, T43-5 and T43-8 development aircraft

Upon completion of the T-3-51 aerial intercept weapons system's state acceptance trials four of the Su-9 *Fishpot-A* interceptor prototypes – the T43-3, T43-4, T43-5 and T43-8 – served as weapons testbeds for IR-homing AAMs. Also, the T43-4 was used to test a so-called pre-emptive starting system which precluded engine flameout caused by missile exhaust gas ingestion when RS-2-US missiles were fired.

Sukhoi T43-5 and T43-12 weapons testbeds

In accordance with an Air Force requirement the Sukhoi OKB modified the T43-5 and T43-12 for testing a new armament arrangement: the wingtips were cropped for the purpose of installing tip-mounted launch rails for K-13 AAMs. (At the same time the Mikoyan OKB conducted a similar test programme with the Ye-6-2 prototype, a precursor of the MiG-21F.) The new arrangement was not introduced on production aircraft, as it was deemed easier to equip the K-5MS (RS-2-US) missile with an IR seeker head.

In 1962 the T43-5 and T43-12 began a new round of trials, now armed with the K-55 IR-homing derivative of the K-5MS. Due to development problems with the missile the trials dragged on for a long time and were not completed until 1967. Eventually, as already mentioned, the R-55 missile was recommended for service along with the RS-2-US.

Sukhoi T47-5 development aircraft

During the same period (after the state acceptance trials of the T-3-8M weapons system) the fifth prototype Su-11 (the T47-5) was used as a weapons testbed for the IR-homing version of the K-8M AAM developed by Bisnovat's OKB-4. The missile subsequently entered production as the R-8 (AA-3 *Anab*) and was used on the Su-11 and Su-15.

Sukhoi T47-6 avionics/weapons testbed

The T47-6 was used in the development of the T-3A-9 aerial intercept weapons system and the ill-starred Sukhoi T-37 heavy interceptor. The T47-6 acted as a testbed for the TsP-1 fire control radar and the launch control system for the K-9 (R-38) semi-active radar homing AAMs, both of which had been created for the Sukhoi T-37 heavy interceptor. The aircraft was converted from a standard initial production T-47, featuring the T-37's fixed multi-shock intake centrebody/radome and PR-38 missile launch rails under the wings.

The aircraft entered flight test in January 1960. Several versions of the intake were tried, the final one yielding unique results with absolutely stall-free operation. By then, however, the T-47 interceptor equipped with the

Oryol (Eagle) radar and armed with K-8M missiles had entered full-scale production as the Su-11 *Fishpot-B* and it was considered inexpedient to make major design changes.

Later, when the T-37 programme had been cancelled, the T47-6 was used for testing the T-3-8M aerial intercept weapons system alongside the other prototypes in the series.

Sukhoi Su-15 cannon armament testbeds

Originally the Su-15's armament consisted solely of two R-8 air-to-air missiles; however, the military kept requesting that cannon armament be incorporated as well. At first the Sukhoi OKB intended to use a single GP-9 standardised centreline pod housing a 23-mm (.90 calibre) Gryazev/Shipoonov GSh-23 twin-barrel cannon. (GP = *gondola **push**echnaya*, which translates [and is very conveniently deciphered] as 'gun pod'.) After a series of tests this was found suitable for the Su-15, and the Novosibirsk aircraft factory No. 153 even built ten Batch 12 aircraft with appropriate attachment fittings and connectors. Eventually, however, no GP-9 pods were ever delivered to Su-15 units – a strange fact, considering that the Air Force's tactical fighter units equipped with the MiG-21S *Fishbed-J* (which, too, was originally armed only with two AAMs) received these pods.

By then, however, the military had changed their requirements; the UPK-23-250 pod (*ooniver**sahl**'nyy **push**echnyy kon**tey**ner* – versatile gun pod) containing the same cannon but with 250 rounds instead of 200 became the Su-15's standard cannon armament. In addition to the greater ammunition supply, the UPK-23-250 could be fitted and removed extremely easily (unlike the GP-9).

The T47-5 development aircraft had no radar, as revealed by the all-metal shock cone in lieu of a radome, but could still perform missile launches, as indicated by the nose markings. Here it is seen with inert K-8M AAMs.

Most importantly, two such pods could be carried on the Su-15's fuselage hardpoints instead of drop tanks, giving twice the firepower.

A production Su-15 *sans suffixe* (c/n 1115342) was set aside for cannon armament tests, passing state acceptance trials in March-September 1971. Even though aiming accuracy with the Su-15's standard K-10T sight left a lot to be desired, the cannon pod installation was recommended for use against both air and ground targets. Since production of the basic *Flagon-A* had ended by then, these aircraft were retrofitted with UPK-23-250 pods in service.

Sukhoi Su-15 missile armament testbeds

When the Su-15TM *Flagon-F* armed with R-98M (AA-3A *Anab*) missiles entered service it was decided to adapt existing Su-15s *sans suffixe* equipped with the Oryol-D58 radar for using this missile which was originally designed for use with the Taïfoon radar. To this end a late-production Su-15 (c/n 1415301) was converted in 1975, successfully passing a special test programme.

By the early 1970s the **Mol**niya (Lightning) OKB had completed development of the R-60 (AA-8 *Aphid*) 'dogfight AAM' and the Air Force selected it as the main close-in weapon for all Soviet fighters. After the R-60 had been tested on the Su-15TM it was decided to arm the Su-15 *sans suffixe* with this missile as well. A suitably modified early-production Su-15 (c/n 0615327) was tested successfully in 1978-79, the Air Force's aircraft overhaul plants started upgrading Su-15s with APU-60 launch rails for two R-60s. A configuration with four R-60s on APU-60-2 paired launchers was also tested but did not find its way into service.

Sukhoi L.10-10 weapons testbed

In 1978 Su-15T c/n 0215306, a long-time 'dogship', was modified yet again, becoming a weapons testbed as part of the T-10 *Flanker-A* fighter's development programme. Designated L.10-10 (that is, 'flying laboratory' No. 10 under the T-10 programme), the aircraft had the standard PU-1-8 missile launch rails replaced with APU-470 launch rails developed for the new K-27E advanced medium-range AAM. The actual test launches of these missiles took place in 1979, the missile eventually entering production as the R-27 (AA-10 *Alamo*).

Sukhoi Su-15TM strike armament testbed

The incorporation of bomb armament on the Su-15 for use as a tactical strike aircraft had been repeatedly delayed ever since the beginning of the state acceptance trials. Finally, by August 1974 a production Su-15TM coded '76' (c/n 1015307) was modified to feature four BD3-57M bomb racks instead of the standard missile and drop tank pylons, with appropriate changes to the electrics. The K-10T sight was equipped with a tilting mechanism enabling attacks against ground targets.

The state acceptance trials proceeded in two stages; Stage A (October 1974 through May 1975) was to verify the bomb armament, while Stage B (June-December 1975) was concerned with the cannon and rocket armament. The aircraft was flown by GNIKI VVS test pilots Yevgeniy S. Kovalenko, Vadim A. Oleynikov and Valeriy N. Moozyka. The trials were successful, the report stating that the armament fit under test rendered the aircraft suitable for use against pinpoint ground targets. As a result, late-production Su-15TMs were equipped with PU-2-8 launch rails which could be easily exchanged for BD3-57M racks.

Sukhoi Su-15TM weapons testbed with internal cannon

The Sukhoi OKB did not give up on the idea of fitting an internal cannon to the Su-15; now a production GSh-23L cannon was to be mounted on the Su-15TM's fuselage underside aft of the nosewheel well. The third production *Flagon-F* (c/n 0315304), one of the five involved in the state acceptance trials, was modified for testing this installation. The 'cannon saga' continued for almost three years, the aircraft eventually undergoing a separate state acceptance trials programme in 1973. The results were very similar to those obtained with the standard UPK-23-250 gun pods; thus the built-in cannon was recommended for production. Yet the gunnery accuracy was still rather poor because the standard K-10T sight was ill-suited for working with cannons. Since a specialised gunsight could not be installed due to lack of space, the built-in cannon never found its way to the Su-15 production line.

Sukhoi LL-OS weapons testbed

In 1989 LII and the Sukhoi OKB converted the first production Su-27UB trainer (the T10U-4) into a testbed for air-to-air missiles launched backwards as an effective protection against stern attacks; these were presumably to be guided by the rear warning radar envisaged for the Su-27M (Su-35). The testbed was designated LL-OS, the last two letters denoting *obraht*nyy *start* (backward launch).

Tupolev I-4 weapons testbeds

An I-4 fighter was fitted experimentally with a DRP recoilless gun (*di*na*more'ak*tiv*naya* **push**ka – lit. 'dynamic reaction gun') designed by Leonid V. Kurchevskiy. Work on this began in 1931.

Another I-4 was used by the Gas Dynamics Laboratory (GDL) for testing 82-mm (3.22-in) RS-82 unguided rockets in the summer of 1932. The rockets were fired from RO-82 launch rails under the lower wings.

Tupolev TB-1 weapons testbed

A single TB-1 bomber was used in trials of the 132-mm (5.19-in) RS-132 and 245-mm (9.64-in) RS-245 rocket projectiles at the GDL from 1932 onwards.

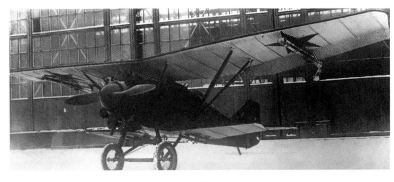

Tupolev Tu-2 weapons testbeds

In early 1947 Vladimir N. Chelomey's OKB-51 design bureau was tasked with designing new unmanned aerial vehicles; one of them was the 16Kh cruise missile – a further spinoff of the 10Kh/14Kh line – subsequently modified as the 16KhA. The original 16Kh project dated back to 1945 and featured a single D-6 pulse-jet; later it was thoroughly revised to feature two D-3 pulse-jets located on the rear fuselage sides, the jetpipes protruding aft above the tailplane fitted with endplate fins. The project programme envisaged developing initially a variant featuring autonomous guidance, to be followed by a radio-controlled version. In

A Tupolev I-4 with DRP recoilless guns under the upper wings.

This Tu-2 was converted into a weapons testbed which served for the initial tests of the 16KhA missile. Note the missile's extremely small ground clearance.

351

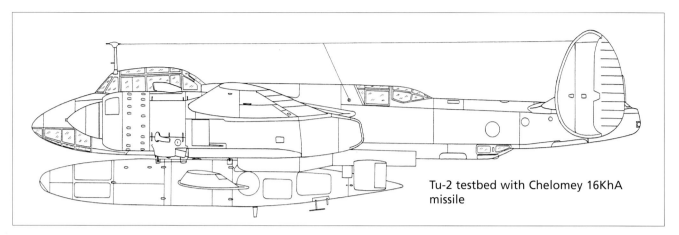
Tu-2 testbed with Chelomey 16KhA missile

the latter case, development of the associated remote control equipment proved to be an arduous task defying the originally set schedule.

By the end of 1947 five 16KhA prototypes were manufactured; a single Tu-2 bomber was converted into a carrier aircraft fitted with a special missile rack. Six 16KhAs, renamed *Pri**boy*** (Surf) by then, were subjected to pre-

An air-to-air shot of the Tu-2 testbed carrying the 16KhA missile.

Here the 16KhA is seen seconds after being released by the Tu-2 testbed.

liminary flight testing between 22nd July and 25th December 1948. Five of them were equipped with the pneumatically-actuated autonomous guidance system affording stabilisation in two planes, while the sixth machine featured an electrically-actuated autonomous guidance system affording stabilisation in three planes. In the course of the tests the D-312 pulse-jet mounted on these examples displayed a tendency to cut at flight speeds of 720-775 km/h (447-482 mph). To remedy this, more efficient D14-4 units were fitted which were capable of sustained running within the entire range of speeds between 300 and 1,000 km/h (186-620 mph). With these engines the missile attained a speed of 872 km/h (542 mph).

The Priboy missiles equipped with D14-4 pulse-jets and an autonomous guidance system successfully passed joint manufacturer's tests/state acceptance trials in 1952. However, their satisfactory performance characteristics were marred by insufficient reliability and low accuracy, which precluded service introduction of these weapons into. Further development and testing was suggested by the Air Force; however, on 19th February 1953, at the initiative of Lavrentiy P. Beriya (who supervised several important weapons development programmes at the government level), the work on Chelomey's cruise missiles was terminated and his design bureau was disbanded.

Tupolev Tu-4 weapons testbeds with guided bombs

In 1948-49 the Soviet Air Force considered the possibility of equipping the Tu-4 *Bull* bomber with cruise missiles developed by OKB-51. Also, in the late 1940s a minelayer and torpedo-bomber version of the Tu-4 for the Soviet Navy was under consideration.

Soviet development work on guided bombs, initially based on Second World War

German types such as the Fritz X, for tactical and long-range bombers began in 1950. That year the OKB-2 weapons design bureau led by Aleksandr D. Nadiradze (renamed GosNII-642 in December 1951) developed the UB-2000F and UB-5000F guided bombs (*oopravlya-yemaya **bom**ba*), which were aimed by means of an OPB-2UP optical sight and a radio command channel. During trials in 1953 a modified Tu-4 carried two UB-2000F bombs on underwing pylons.

The results proved disappointing and the 2,240-kg (4,940-lb) UB-2000F with a 1,795-kg (3,960-lb) warhead had to undergo a redesign, eventually being included into the inventory two years later as the UB-2F *Chai*ka (Seagull, aka *izdeliye* 4A22). The 5,100-kg (11,240-lb) UB-5000F with a 4,200-kg (9,260-lb) warhead tested in 1954 was of similar design but featured a TV command guidance system, a TV camera in the nose transmitting a 'bomb's eye view' to a display at the bomb-aimer's station; the bomb-aimer then manually corrected the weapon's flight path. Pretty soon, however, the Soviet Union discontinued work on guided bombs and the weapons system was never introduced on the Tu-4.

Tupolev Tu-4 weapons testbeds with AAMs

A highly unusual weapon carried by the *Bull* was the RS-2-U AAM fitted in an attempt to enhance the bomber's defensive capability in the rear hemisphere. Guidance was effected by means of the suitably modified Kobal't radar (fitted to the Tu-4 as standard); the missiles were carried on launch rails under the rear fuselage and launched by the radar operator.

A few Tu-4s modified in this fashion even saw operational service with the 25th Night Bomber Regiment. Generally, however, the system proved unsatisfactory and did not gain wide use. Target lock-on was unstable, the launch range was rather short and the missiles were expensive, not to mention the fact that they were intended for the Air Defence Force, not the bomber arm.

Tupolev Tu-4 (G-310) weapons testbed

Perhaps the most unusual weapon carried by the Tu-4 was the *izdeliye* 211 missile which evolved into the S-25 **Ber**koot (Golden Eagle) surface-to-air missile (NATO reporting name

The G-310 – a Tu-4 modified for testing the *izdeliye* 211 missile. Note the radomes on the forward fuselage. Oddly, the gear is down and the bomb bay doors open.

The missile streaks away after being launched by the G-310.

SA-2 *Guideline*). A Tu-4 was modified for air launches of this weapon in 1952; the missile minus the first stage (solid-fuel rocket booster) was carried on a special rack under the starboard wing. Twin guidance system radomes were added to the forward fuselage. The modified aircraft bore the designation G-310.

Tupolev Tu-16 with Chaika guided bomb

The Tu-16 *Badger* was perceived as a suitable delivery vehicle for the UB-5 Condor (UB-5000F) or the UB-2F Chaika (UB-2000F). Radio control or television guidance systems were used. The Tu-16 equipped to carry a guided bomb was fitted with a KRU-UB radio control transmitter, the operator either following the course of the marker flare attached to the bomb through an OPB-2UP optical sight or using images transmitted to a monitor screen from a TV camera (a 'bomb's eye view') to make course corrections through the linked

353

radio control system. The latter method proved more accurate and less reliant on weather conditions. The bombs were carried under the fuselage or under the wings.

A small number of Tu-16 bombers were equipped with the 4A-22 Chaika system which controlled the UB-2F guided bomb against small but important targets such as railway bridges, storage depots and administrative buildings from high altitude. Two UB-2F bombs were carried under the wings on special pylons and the bomber was fitted with an electric system for arming and guiding the bomb. The system could only be used in good visibility conditions when both the bomb and the target could be observed through the special OPB-2UP periscopic synchronised sight used for dropping free-fall and guided bombs in level flight. This automatically signalled proximity to bomb release, the time when the bomb bay doors were to be opened, and the moment of release. Guidance was effected through the sight by a juxtaposition of the bomb's image with an indicator registering the angles of divergence derived from calculating the bomb's trajectory and generated by a special computer.

In-flight control by the bomb-aimer was effected by a 4A-N1 data link system transmitting on three wavebands simultaneously to frustrate jamming. The bomb's receiver passed the signals to the AP-59 autopilot which activated the bomb's guidance mechanism and made the necessary corrections to keep the bomb level. The bomber's flight level dictated the required speeds for bomb release, and these varied between 550 km/h (341 mph) at 5,000 m (16,400 ft) and over 900 km/h (559 mph) above 6,500 m (21,330 ft). As the bomb's speed was lower than that of the aircraft, the pilot was obliged to reduce speed so that both the bomb and its target remained visible. The UB-2F (UB-2000F) was a general-purpose bomb (F = foo**gahs**naya – high-explosive), but the UB-2000B armour-piercing version (B = brone**boy**naya) was also envisaged.

In the late 1950s the improved Chaika-2 IR-homing bomb and the Chaika-3 with passive radar homing for use against enemy radar and ECM sites were developed, although only the 4A-22 Chaika was actually tested and accepted for operational use in December 1955. The UB-2F was the first Soviet guided bomb. Tests results showed that only two or three such bombs were needed to hit a target measuring 30 x 70 m (100 x 230 ft) as against 168 FAB-1500 conventional free-fall bombs.

Tupolev Tu-16 Condor guided bomb carrier ('order 251')

The UB-5 (UB-5000F) Condor guided bomb was intended for use against large surface ships and was essentially an enlarged Chaika with a HE warhead. Two versions were developed, one with line-of-sight radio command guidance (like the Chaika) and the other with TV guidance. The radio command system was used during tests with the Tu-4, but its accuracy proved significantly poorer than the Chaika's since the heavier Condor reached speeds in the order of Mach 1.1 and in-flight control proved more difficult. These problems were eventually overcome and satisfactory results obtained through the speed range. In August 1955 experimental Condor bombs were again tested from the Tu-4, using TV guidance, with a joystick controller for the navigator. Not only could this system be used in less favourable weather conditions, it also obviated the need for the bomber to fly over the target. Tests using an adapted Tu-16 built to 'order 251' began in March 1956. Yet, though successful, the system was not accepted for operational use. The large bomb carried externally increased drag to the detriment of speed and range. Work on guided bombs carried externally was therefore terminated.

Tupolev Tu-16 with UBV-5 guided bomb

Work on a more sophisticated guided bomb, the 5,150-kg (11,350-lb) UBV-5 fitted with a 4,200-kg warhead, began in the summer of 1956. The bomb was to have an HE or armour-piercing warhead and either TV guidance (like the Condor) or an autonomous IR homing system (like the Chaika-2). Two variants were designed: one to be carried internally and the other semi-recessed (the latter was rejected by the Air Force). The drawback of all Soviet guided bombs was that the carrier aircraft was obliged to release the bomb only a few kilometres from the target, increasing the risk of enemy interdiction. This problem could be overcome by fitting solid-fuel rocket boosters to the bombs, so that they resembled ASMs, but the advantages of guided bombs in their simplicity and low production costs as compared with ASMs began to wane. The emphasis shifted to air-launched cruise missiles able to strike enemy warships while the launch aircraft was still some distance away. This brought an end to the work on the 7,500-kg (16,530-lb) UPB rocket-powered guided bomb which was to be released at a range of 300-350 km (186-217 miles) from the target.

Tupolev Tu-16 with SNAB-3000 'Krab' homing bomb

On 14th April 1957 the Council of Ministers issued directive No. 1175-440 giving guidelines for guided weapons development until the mid-1950s. Among other things, the All-Union State Research Institute No.642 (GSNII-642) was tasked with developing a 3,000-kg (6,610-lb) homing bomb designated SNAB-3000 (*samonavodya*shchayasya *avia*bomba – homing bomb) which was also known under the codename Krab (Crab).

Development of the SNAB-3000 proceeded under the leadership of D. V. Svecharnik. The design was strongly influenced by the German Fritz X, featuring the same four large fins in a squashed-X arrangement and cruciform tailfins within a rhomboid-shaped rudder arrangement. Unlike the German prototype, however, the SNAB-3000 had swept wings. The bomb featured an AP-55 autopilot for initial guidance and an infra-red seeker head for terminal guidance. The IR seeker was activated by a timer, allowing the bomb to zero in on large targets with a high heat signature, such as factories. The warhead weighed 1,285 kg (2,830 lb).

Trials began in 1951, initially using a Tu-4 delivery vehicle and 'dumb' versions of the bomb without the homing system. The fully equipped 'smart' version entered test at GK NII VVS in Akhtoobinsk in late 1952. The bomb showed promising results at first, accurately homing in on pans with burning kerosene used as simulated targets; out of the 12 inert and live bombs dropped in 1953-54, eight fell within 47 m (154 ft) of the target.

However, the Tu-4 was slow and hopelessly outdated, so the decision was taken to use the state-of-the-art Tu-16 as the delivery vehicle. Hence a Kazan'-built Tu-16 coded '36 Black' (c/n 4200303) was fitted with special wing

pylons for carrying two SNAB-3000 bombs. It was then that problems began; the Tu-16 turned out to be too fast for the bomb which became unstable when dropped at high speeds, the accuracy dropping dramatically. The strong drag generated by the bombs reduced the aircraft's range to 3,620 km (2,250 miles) with two SNAB-3000s and 4,500 km (2,800 miles) with one bomb versus 5,430 km (3,370 miles) with a 9,000-kg (19,8440-lb) FAB-3000 free-fall bomb carried internally. To top it all, the reliability of the guidance system was all too low; of the 32 test missions flown by the Tu-16, sixteen ended in failure due to various malfunctions of the bomb. Hence on 26th August 1956 the Council of Ministers' Scientific & Technical Board convened to assess the results of the trials, cancelling all further work on the SNAB-3000.

Tupolev Tu-16 weapons testbed with a DK-20 tail turret

The DK-7 tail turret of a single Tu-16 was replaced by a DK-20 twin-cannon tail turret intended for the Tu-22 during evaluation flight tests.

Tupolev Tu-95MA development aircraft (second use of designation)

In 1983 a production Tu-95MS coded '004 Black' was converted into a testbed for new advanced weapons. Redesignated Tu-95MA, the machine had two large pylons under the wing roots in similar fashion to the Tu-95K-22 for carrying large air-to-surface missiles. The pylons proper were different, though.

Tu-16 '36' (c/n 4200303) carrying a SNAB-3000 Krab homing bomb.

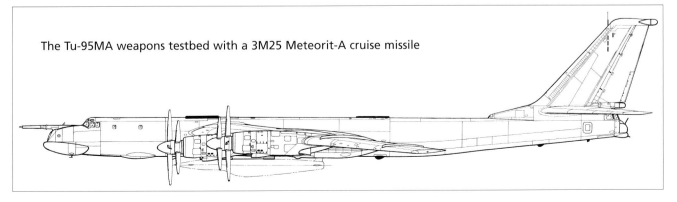

The Tu-95MA weapons testbed with a 3M25 Meteorit-A cruise missile

The Tu-95MA was operated by GNIKI VVS in Akhtoobinsk on the Volga River. The first weapon tested on it was the 3M25 Meteorit-A supersonic cruise missile, two of which were carried. Twenty launches were made, the aircraft operating from Zhukovskiy. Yet the programme ran into financial and technical difficulties and the project was put on hold.

Later, in 1992, the same aircraft was used for testing the Kh-90 Grom (Thunder) experimental cruise missile, which failed to reach production status. The designation Tu-95PA has also been quoted for the aircraft in this guise.

Tupolev Tu-95MS weapons testbed

The same Tu-95MS and another *Bear-H*, '317 Black' (c/n 19317), were used by the 929th GLITs (ex-GNIKI VVS) for testing the Raduga Kh-101 cruise missile. This weapon has a launch weight of 2,200-2,400 kg (4,850-5,290 lb) and a 400-kg (880-lb) conventional warhead; maximum range is 5,000-5,500 km (3,105-3,410 miles). The Kh-101 uses with TV terminal guidance and the circular error probable is 20 m (66 ft).

This time eight missiles were carried in pairs on four pylons (two under the wing roots and two between the engines). The tests, which began in 1999, paved the way for upgrades of the *Bear-H* (as the Tu-95MSM) and the Tu-160, which is to carry 12 Kh-101s internally.

Tupolev Tu-104 avionics/weapons testbeds

As noted in Chapter 4, the Tu-128 heavy interceptor's development programme involved the use of Tu-104 *sans suffixe* CCCP-42326 (c/n 66600102) converted into a testbed for the Smerch (Tornado) weapons control system.

'004 Black', the Tu-95MA weapons testbed, with a 3M25 Meteorit-A cruise missile on the starboard wing pylon and a camera under the wingtip. The missile's snap-action wings are folded.

The aircraft was a combined avionics/weapons testbed, with an RP-S Smerch fire control radar in the nose and two APU-128 pylons under the outer wings. The latter feature allowed CCCP-42326 to carry K-80 air-to-air missiles specially developed for the Tu-128 by OKB-4 under Matus R. Bisnovat (now called GMKB Vympel, 'Pennant' State-owned Machinery Design Bureau). The tests of the weapons control system included live missile launches and contributed a lot to the successful testing of both the Tu-128 and the K-80 AAM, which became the R-4 (AA-5 *Ash*) in production form.

One more combined avionics/weapons testbed based on the Tu-104 was the second of two LM-104-518 testbeds (see Chapter 4). It was meant to verify the Mikoyan MiG-31 heavy interceptor's WCS based on the RP-31 Zaslon phased-array radar. The aircraft, a converted Tu-104A (CCCP-42454, c/n 96601703), featured both the RP-31 radar and AKU-410 missile pylons under the wings, carrying Vympel K-33 long-range AAMs. Tests of this weapon, known as the R-33 (AA-9 *Amos*) in production form, commenced in March 1975. The first two launches were in ballistic mode because the missile guidance system was still inoperative. In both cases CCCP-42454 fired the missile from the starboard pylon; a MiG-21U trainer flew chase just 20-30 m (65-100 ft) off the Tu-104's starboard wing, filming the launch. The results proved encouraging and the K-33 was cleared for test launches against target drones in full guided mode.

The K-33 was a massive weapon, with a launch weight of 490 kg (1,080 lb). Hence the Tu-104's mass and aerodynamic balance was disrupted after an asymmetric launch and the pilot had to make a climbing turn in the opposite direction immediately after the missile left the pylon.

After lengthy trials of the Zaslon WCS, the LM-104-518 testbed was converted to the Tu-104A Tsiklon weather research aircraft.

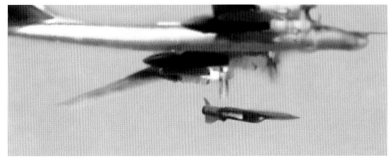

Tupolev Tu-110 Sapfeer avionics/weapons testbed

Again, as noted in Chapter 4, in 1971 the third production Tu-110 (originally CCCP-Л5513) was converted into an avionics/weapons testbed designated Tu-110 Sapfeer. It served for testing the PARG-30VV semi-active radar seeker head developed for the K-25 AAM. An RP-21 Sapfeer fire control radar on a special adapter supplanted the navigator's station glazing, and two pylons with APU-25-21-110 launch rails for carrying K-25 AAMs were fitted under the wings. The testbed was commissioned on 31st December 1971; regarding the date· the aircraft was reregistered in the CCCP-3658* block by then.

The Tu-95MA on the hardstand of the 929th GLITs in Akhtoobinsk.

The Meteorit-A seen during a test launch by the Tu-95MA. The wings and tail surfaces are deployed.

Modified Tu-95MS '317 Red' comes in to land, carrying a full load of eight Kh-101 missiles. The same arrangement is envisaged for the Tu-95MSM.

The APU-128 missile pylon under the port wing of Tu-104A CCCP-42326.

A K-33 long-range AAM on an AKU-410 missile rail attached to a BD-352 pylon under the wing of Tu-104A CCCP-42454 (the LM-104-518).

Yakovlev Yak-25M (SM-6) weapons testbeds

Two production Yak-25Ms were to be used for testing the K-6 air-to-air missile; these testbeds were designated SM-6. However, the K-6 programme was eventually cancelled, and the two aircraft were modified for other uses.

Yakovlev Yak-25K-75 interceptor prototype/weapons testbed

Pursuant to Council of Ministers directive No. 2543-1224 one of the Yak-25Ks (c/n 1608) was modified in early 1956 for testing the K-75 missile system developed by the Toropov OKB. The conversion was performed by the OKB's experimental shop (plant No. 134). Four K-75 (*izdeliye* 129) AAMs were carried on the existing wing pylons; appropriate changes were made to the Izumrood radar and an ASP-3NM automatic gunsight was installed. AUPT fuel flow limiters (*avtomaht ooprav-leniya podachey topliva* – automatic fuel feed controller) were incorporated into the fuel system; these devices throttled the

engines back briefly when the missiles were fired to preclude surging caused by missile exhaust gas ingestion.

Two AKS-2 high-speed cameras were mounted in angular fairings low on the aft fuselage sides (just aft of the airbrakes), augmented by two more cameras (possibly S-13s) under the outer wings. Two small cylindrical objects were mounted on short pylons on both sides of the tailcone; these were probably tracer flares designating the aircraft's position for a 'camera ship' during night test flights.

Appropriately designated Yak-25K-75, the uncoded aircraft underwent tests between 5th March and 10th July 1956 at the hands of project test pilot Fyodor L. Abramov, test pilot Ye. N. Pryanishnikov and weapons systems operators G. N. Oolyokhin and V. N. Makarov. During this period the aircraft logged a total of 16 hrs 34 min in 23 flights.

The Yak-25K-75's AUW was 8,830 kg (19,470 lb), including 324 kg (714 lb) for the four missiles and 2,650 kg (5,840 lb) of fuel. At 11,000 m (36,090 ft) the aircraft cracked the sound barrier, attaining Mach 1.07; top speed was 1000 km/h (621 mph) at 5,000 m (16,400 ft) and 950 km/h (590 mph) at 10,000 m (32,810 ft). As compared to the standard Yak-25M the service ceiling decreased to 13,600 m (44,619 ft) and the time required to reach 5,000 m and 10,000 m increased to 2.5 and 6.0 minutes respectively, but maneuvrability and handling remained unaffected.

Most flights were made with dummy missiles. On three occasions the Yak-25K-75 made single launches of live K-75s in level flight at 5,000 m and 12,000 m (39,370 ft). The radar could detect a medium bomber such as the IL-28 (or English Electric Canberra) flying at 5,000-8,000 m (16,400-26,250 ft) at a range of 7-7.5 km (4.34-4.65 miles), getting a good target lock-on at 4-4.5 km (2.48-2.79 miles). Generally the test results were deemed satisfactory; still, the K-75 weapons system did not enter production.

Yakovlev Yak-25K-7L weapons testbed

Another uncoded Yak-25K (c/n 0109) became a testbed for the Toropov K-7L AAM. The new missile was larger and heavier than the K-5 (RS-1-U) and K-75 (*izdeliye* 129), so the complement had to be limited to two missiles. This aircraft also had the same camera fairings on the aft fuselage sides and under the wings and

the tracer flares on the tailcone. Designated Yak-25K-7L, the aircraft was tested but once again the K-7L weapons system did not enter production.

Yakovlev Yak-25K-8 weapons testbed

By the end of 1956 the Bisnovat OKB was ready to test the K-8 missile system. Hence four interceptors, including two Yak-25Ks, were converted into weapons testbeds for this AAM. Unlike the missile testbeds described earlier, these Yak-25s were equipped with the Sokol-2K radar, a version of the RP-6 specially modified to work with the Bisnovat missile.

The K-8 was a big and heavy weapon, which meant only two of these could be carried by the *Flashlight-A*. The long trapezoidal pylons – they had to be long in order to give adequate clearance for the missile's large cruciform fins – were located under the inner

wings and were different on the two testbeds. The first aircraft serialled '110 Red' (c/n 1110) had pylons with a forward-swept leading edge, while on the other aircraft, which wore no tactical code (c/n 0119), the pylons had a sweptback leading edge. The reason for this was the need to determine the best location of the missiles' attachment points with respect to the wings. This time only two AKS-2 cameras were installed on the fuselage.

The aircraft were designated Yak-25K-8. Some documents, however, refer to '110 Red' as the Yak-25S K-8, the S standing for *sereeynyy* (production, used attributively); thus it was meant to be the pattern aircraft for production Yak-25Ks equipped with the K-8 weapons system. Yet this never happened.

Between them the two Yak-25K-8s made 111 test flights, including 72 with the missiles attached, and performed 42 launches, including two live ones. The K-8 had passive IR homing with an S1-U seeker head (not to be con-

The Yak-25K-75 testbed (c/n 1608) with four K-75 missiles. Note the non-standard pitot on the fin leading edge.

The Yak-25K-7L testbed (c/n 0109) with two K-7 missiles. Again a pitot is installed on the fin leading edge. Note the camera pod under the port outer wing.

359

One of the two Yak-25K-8 test-beds (c/n 0119). This view illustrates the size of the K-8 missile. The shape of the pylons makes an interesting comparison with the preceding version.

fused with the similarly designated weapons system developed under Pyotr D. Grooshin!) and a high-explosive/fragmentation warhead.

The K-8 (AA-3 *Anab*) was intended for several Sukhoi types, the Mikoyan I-75 and the Yak-27 *Flashlight-C* – all of them supersonic interceptors. Since the Yak-27 was already in the middle of its flight test programme, there was no point in fielding the Yak-25K-8 weapons system based on an outdated subsonic aircraft. The missile itself, however, turned out to be more successful than the AAMs tested earlier and was put into production after passing its state acceptance trials; it was later developed into several improved versions.

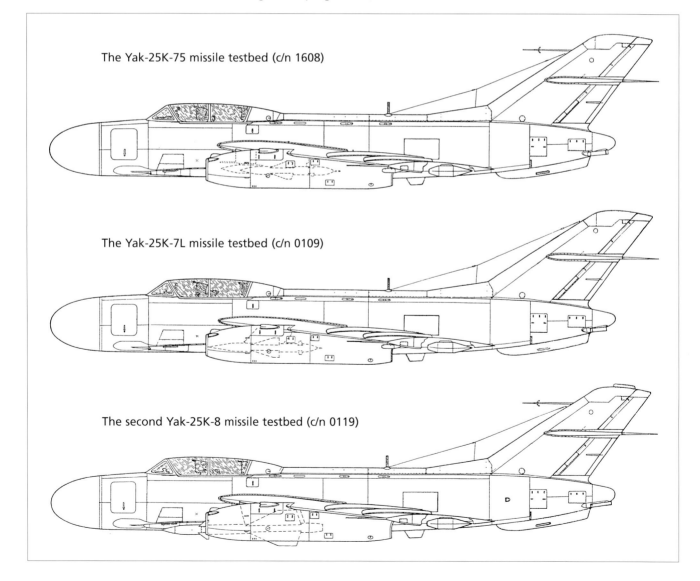

The Yak-25K-75 missile testbed (c/n 1608)

The Yak-25K-7L missile testbed (c/n 0109)

The second Yak-25K-8 missile testbed (c/n 0119)

A large number of aircraft were outfitted for meteorological research and weather control experiments. Even though some of them existed in more than one copy, no two were absolutely identical.

Antonov 'izdeliye K' experimental weather reconnaissance aircraft

On 23rd December 1946 MAP issued an order requiring the design bureau headed by Oleg K. Antonov (then based in Novosibirsk and called OKB-153) to develop and build a specialised weather reconnaissance aircraft based on the izdeliye T agricultural biplane (the SKh-1 – that is, the An-2 Colt prototype). Bearing the in-house designation izdeliye K, the aircraft was built by plant No. 153, making its first flight from Novosibirsk-Yel'tsovka (the factory airfield of plant No. 153) on 21st March 1948. Outwardly it differed from the basic utility/agricultural version in having an observer's cockpit in the rear fuselage; the observer sat under a fighter-type canopy immediately ahead of the fin. The cockpit offered a good field of view ahead and to the sides over the wings. Meteorological instruments, including small airfoil sections for icing visualisation, were mounted on the sides of this cockpit.

The trials of izdeliye K showed that, with a 700-hp Shvetsov ASh-21 radial, the aircraft was underpowered and had poor altitude performance making it unsuitable for its intended mission; a powerplant with better altitude performance was required. However, the prototype never received a suitable engine; in October 1948 it was damaged beyond repair in a hard landing when the engine throttle controls became disconnected in flight.

Antonov An-6 Meteo (An-2ZA) experimental weather reconnaissance aircraft

In 1950 OKB-153 was again tasked with developing a high-altitude weather reconnaissance aircraft and geophysical research aircraft, as well as an aircraft optimised for operations in mountainous areas with rarefied air. The engineers chose to proceed from the abovementioned izdeliye K, fitting it with a new powerplant – a Shvetsov ASh-62IR/TK radial featuring a TK-19 two-speed supercharger developed by S. A. Treskin (TK = **toor-**

The sole the *izdeliye* K weather reconnaissance aircraft based on the An-2.

The unmarked *izdeliye* K in flight with an observer in the rear cockpit.

*bokom**pres**sor*) with an RTK-1 supercharger governor (*regoo**lya**tor **toor**bokom**pres**sora*). The ASh-62IR/TK maintained a rating of 850 hp up to an altitude of 9,500 m (31,170 ft). The engine drove a V-509A-D7 four-blade wooden propeller with scimitar-shaped blades.

The aircraft, which was initially known as the An-2ZA (*zondi**rov**shchik atmos**fe**ry* – atmosphere sampler), wore a natural metal finish with Soviet Air Force insignia but originally no serial; later it was serialled '04 Red'. In December 1951 it underwent state acceptance trials at GK NII VVS. Later the aircraft was placed on the civil register as CCCP-Л1208 (that is, SSSR-L1208) and gained a new designation, An-6 Meteo.

On 9th June 1954 Antonov OKB test pilot Vladimir A. Kalinin and flight engineer V. I. Baklaykin established a Class C-1e-1 world altitude record in the An-6 Meteo, reaching 11,248 m (36,902 ft). Class C-1e-1 includes fixed-wing aircraft with a gross weight of 3,000-6,000 kg (6,610-13,230 lb). This record was officially recognised by the FAI and still stands as of this writing.

The An-6 Meteo had its peculiarities. On one occasion an attempt to explore the aircraft's spinning characteristics almost ended in disaster when the machine proved extremely reluctant to recover from the spin. It turned out that the observer's cockpit affected the airflow around the tail unit to such an extent as to make spin recovery very difficult indeed.

Unfortunately the An-6 Meteo shared the fate of its precursor; in September 1958 CCCP-Л1208 crashed on take-off from Kiev-Svyatoshino, the airfield of the Kiev aircraft factory No. 473 which became the Antonov OKB's new home. The extensively damaged aircraft was declared a write-off. No more were built.

Antonov An-2V weather reconnaissance aircraft (second use of designation)

In 1957-58 the Kiev aircraft factory No. 473 manufactured a small batch of six weather reconnaissance aircraft. Confusingly, they were designated An-2V, just like the floatplane version of the *Colt*; in this case V stood for *vy**sot**nyy* (high-altitude), not **vod**nyy (waterborne). Again, the An-2Vs were powered by supercharged ASh-62IR/TK engines, differing from the An-6 Meteo only in lacking the observer's cockpit at the rear.

The An-2ZA (An-6 Meteo) weather recce aircraft in military markings during trials at GK NII VVS in December 1951. Note the angular housing of the ASh-62IR/TK engine's intercooler; this makes an interesting comparison with the smooth cowling of the *izdeliye* K.

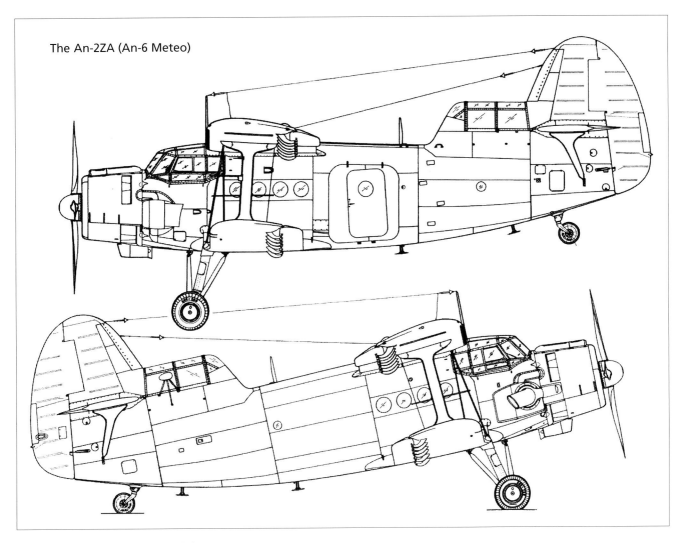

The An-2ZA (An-6 Meteo)

Antonov An-12BPTs Tsiklon
weather research aircraft

By the mid-1970s radar technologies of weather research and atmospheric research had gained wide use. The use of radar made it possible to study the distribution of clouds and precipitation over wide areas, follow the devel-

opment of storm nuclei. Therefore a special equipment suite called *Tsiklon* (Cyclone) was developed to meet an order placed by the Central Aerologic Observatory (TsAO – *Tsentrahl'naya aerologicheskaya observatoriya*), a division of the Soviet Union's State Committee for Hydrometeorology and

Another view of the An-2ZA during state acceptance trials. The starboard side view shows to advantage the exhaust-driven TK-19 supercharger.

363

Close-up of the observer's cockpit and the strut-mounted airfoil section intended for detecting icing conditions.

The An-2ZA (An-6 Meteo) at a later stage of the trials.

A post-1955 photo of the An-2ZA with the tactical code '04 Red'. The propeller spinner has been removed.

The same aircraft as the An-6 Meteo with the registration CCCP-Л1208 (that is, SSSR-L1208) and Aeroflot titles but no Soviet flag. The fourth cabin window is closed by a plug carrying a sensor; note also the sensor above the CCCP- prefix.

Unicorns'R'Us. CCCP-11531, one of the two An-12BP Tsiklon weather research aircraft, on a grass airfield. Note the rarely seen rainmaking agent dispenser pods on pylons fore and aft of the main gear fairings.

Environmental Control (**Gos**komghidro**met**). In 1976-79 two Tashkent-built An-12BPs, CCCP-11530 and CCCP-11531 (c/ns 6344503 and 6344506), were converted into An-12BPTs Tsiklon weather research aircraft (often called An-12BP Tsiklon or simply An-12 Tsiklon – and sometimes erroneously called An-12BKTs).

The technologies employed by the Tsiklon system allowed stand-off research of clouds –

CCCP-11530, the other An-12BP Tsiklon, in flight, showing the Groza-26 radar replacing the tail turret.

One of the An-12BP Tsiklons on the snow-covered GosNII GA apron at Moscow/Sheremet'yevo-1.

365

An-12BP Tsiklon CCCP-11531

6344506

An-12BP Tsiklon CCCP-11531 rests between missions at Sheremet'yevo with runway 07L in the background.

that is, without actually entering them, which could change their shape and structure (and could be dangerous for the aircraft itself, since severe turbulence is often encountered in clouds, not to mention lightning!). Apart from studying the principal thermodynamic and electric parameters of the atmosphere and cloud formations, the aircraft were designed to perform cloud-seeding missions for the purpose of making rain – for example, in order to prevent an impending hailstorm which could destroy crops, or to scatter rain clouds which could ruin a public holiday.

Outwardly the An-12BPTs Tsiklon bore a strong resemblance to the Royal Air Force's unique Lockheed C-130K-140-LM Hercules W.2 (XV208) operated by the Weather Research Flight and nicknamed 'Snoopy'. The nose terminated in a long pointed boom tipped with sensors, leaving the navigator's station with just one small window on each side; the boom was painted in black and white zebra stripes to avoid damage by ground vehicles. A small rounded radome at the base of the boom housed a specialised BMR-1 weather research radar (the standard ROZ-1 Lotsiya

chin-mounted navigation/weather radar was retained). More sensors were located dorsally and ventrally on the forward fuselage centre-line.

The DB-65U tail turret gave place to a fairly large rounded radome enclosing a Groza-26 (Thunderstorm) weather radar borrowed from the An-26 transport. The nose-mounted instrumentation boom and rear radome increased the overall length from 33.1 m (108 ft 7⁵⁄₃₂ in) to 38 m (124 ft 8 in).

Two large pylons were installed on each side fore and aft of the main gear fairings for carrying cloud-seeding chemical pods. Finally, three observation blisters were provided on each side for filming purposes. The aircraft wore the 1973-standard Aeroflot livery but the cheatline was broken beneath the flight deck by the eye-catching Tsiklon emblem.

Two research equipment consoles were located in the pressurised forward cabin, with two more in the unpressurised freight hold. The mission equipment comprised a measurement suite, data recording/processing equipment and cloud-seeding equipment. The measurement suite included, first of all, a ther-

An-12BP Tsiklon CCCP-11530

modynamic measurement system comprising a TsSV-3M-1KM central air data system, an EM TsAO electric meteorograph developed by the Central Aerologic Observatory, a PK G-load measuring kit (pere**groo**zochnyy kom**plekt**), an SAMB-70 airborne automatic weather research module, an ASTA-74 airborne automatic thermoanemometer (avtoma**tich**eskiy samo**lyot**nyy **ter**moane**mom**etr), an RV-18Zh radio altimeter and an SG-1 airborne humidity meter (samo**lyot**nyy ghi**grom**etr). It recorded the outside air temperature and its fluctua-

Two more views of CCCP-11531, showing the port pylons. Note the observation blisters.

tions, the aircraft's speed and heading, the wind speed and direction, airflow pulsations, static and dynamic air pressure, barometric and true altitude, vertical gusts and G loads acting on the aircraft. The second major component was the cloud and precipitation microstructure measurement system comprising an IRCh water/ice particle size meter (*izmeritel' razmera chastits*) and an SEIV-3 airborne electric cloud water content meter (*samolyotnyy elektricheskiy izmeritel' vodnosti*). The third component was the meteorological radar system – the aforementioned BMR-1 in the nose for vertical scanning of the atmosphere and the Groza-26 radar in the tail. Finally, there was a PNP meter for measuring electric fields and the aircraft's electric charge.

The data recording/processing equipment consisted of a K60-42 magnetic recorder, an AKS-2 ciné camera and an SYeO time synchronisation system (*sistema yedinovo otshchota*). The K60-42 automatically recorded signals generated by the thermodynamic and cloud measurement systems for future computer analysis. The ciné and photo cameras were used to film the outside conditions, using the observation blisters.

The cloud-seeding equipment designed to generate rain from cumulus and stratus clouds included seven KDS-155 dispensers (*kassetnyy derzhatel' spetsiahl'nyy* – special cassette-

The cargo cabin of the An-12BP Tsiklon where part of the experimental and data recording equipment was installed. The roller conveyor on the left probably served for dumping rain-making chemicals overboard.

type rack) and four ASO-2I chaff/flare dispensers (*avtomaht sbrosa otrazhateley* – automatic chaff dispenser) mounted on the cargo doors, an ITU solid carbon dioxide atomiser (*isparitel' tvyordoy ooglekislotyy*) and, as an option, four pods for dispensing powdered rainmaking agents. The most widely used among the latter is silver iodide. The KDS-155 and ASO-2I were adapted from stock units used on Soviet military aircraft for passive ECM and infrared countermeasures (IRCM). Instead of bundles of chaff or aluminium-coated glass needles the KDS-155 was loaded with granulated carbon dioxide (known colloquially as 'dry ice'); the ASO-2I fired 26-mm (1-in) PV-26 flares with silver iodide instead of PPI-26 magnesium flares (*peeropatron infrakrahsnyy* – infra-red [countermeasures] cartridge). A local discharge of a large amount of carbon dioxide into the atmosphere caused the temperature to fall, and microscopic particles of silver iodide triggered the formation of ice crystals. The latter became too heavy to be supported by the air currents inside the cloud and started falling as hailstones; however, these melted and turned into rain before reaching the ground. One gram of silver iodide produced on average one billion raindrops.

Since the An-12BPTs was to operate over vast stretches of water if necessary when chasing storms, it was provided with appropriate LORAN equipment.

The aircraft had a flight crew of five plus a 14-man team of researchers. In fully equipped configuration with four external pods the TOW amounted to 61 tons (134,480 lb); top speed was 550-600 km/h (340-372 mph) and the service ceiling was 8,700 m (28,540 ft). The An-12BPTs could stay airborne for up to eight hours, with a maximum range of 5,000 km (3,105 miles).

The two weather research *Cubs* were operated by the State Civil Aviation Research Institute (GosNII GA) and home-based at Moscow/Sheremet'yevo-1 but they were seldom seen there, travelling far and wide in pursuit of their mission. Among other things, they deployed to Cuba from time to time because the Caribbean, with its frequent typhoons, offered plenty of material for research. The aircraft were quite efficient economically due to the high accuracy of measurements, the ability to track changes in the weather processes in real time, influence them quickly and check the results immediately.

Regrettably, in the general chaos that followed the demise of the Soviet Union the

Tsiklon weather research aircraft found themselves unwanted. Both An-12BP Tsiklons were stripped of their special equipment and used as cargo aircraft. Interestingly, the navigator's station glazing was never fully reinstated (there was a solid panel replacing the two upper windows in the second row) and the tail radomes on both aircraft remained, though they were empty now. The aircraft served on with GosNII GA as RA-11530 and RA-11531 for a while; the latter aircraft was then sold to Angola as D2-FVG No. 1.

Antonov An-26B Tsiklon weather research aircraft

In 1987 an An-26B commercial transport in the red/white Polar version of Aeroflot's 1973-standard livery (identity unknown) was converted into the sole An-26B Tsiklon for the Central Aerologic Observatory by the manufacturer, the Kiev Aircraft Production Association (KiAPO). The c/n has been quoted in error as 14208 but An-26 c/n 14208 was exported as East German Air Force '373 Black'.

Actually, unlike the other aircraft in the Tsiklon series, the An-26B Tsiklon was a 'sky cleaner' intended for treating clouds with special chemicals for the purpose of artificially increasing the rainfall and protecting designated areas from rain. The machine was identifiable by two slab-sided **Vey**er (Fan) pods mounted on BD3-34 pylons and by the eye-catching Tsiklon emblem applied aft of the flight deck. The pods, originally designed to be fitted to military An-26s for passive ECM and

IRCM, housed six ASO-2I-E7R flare dispensers each; the latter held PV-26 cartridges with silver iodide. The cargo hold housed a heat-insulated container for disseminating granulated carbon dioxide.

Antonov An-26 Pogoda weather research aircraft

For tackling the same weather control tasks one more unidentified aircraft was modified in Kiev in 1988 to become the An-26 *Pogod*a (Weather) 'sky cleaner'. By comparison with the An-26 Tsiklon it featured a simpler set of mission equipment.

Antonov An-26 Sfera aerological research aircraft

In 1991 a single production An-26 (identity unknown) was outfitted for studying the physical properties of the atmosphere in response to an order from the Ukrainian Academy of Sciences. The machine was designated An-26 **Sfe**ra (Sphere), the name obviously deriving from the word 'atmosphere'.

Antonov An-30M (An-30 Meteozashchita) meteorological protection aircraft

In the mid-1980s several Aeroflot An-30A photo mapping aircraft equipped with the Groza-M30 radar were converted into a version designated An-30 **Met**eoza**shchit**a (meteorological protection) or An-30M. The

A fine landing study of An-12BP Tsiklon CCCP-11530.

369

No. 2 camera window was turned into an outlet for a granulated carbon dioxide dispenser; the 'dry ice' was housed in eight heat-insulated containers in the cabin, each holding 130 kg (290 lb). Additionally, two Veyer pods loaded with PV-26 silver iodide cartridges were carried on BD3-34 racks on the centre fuselage sides.

The aircraft's mission was to protect a given area from rainfall by causing rains in the adjoining areas, thereby preventing the rain clouds from entering the protected zone, or to avert an impending hailstorm which could do great damage. The An-30M 'sky cleaners' were primarily intended for ensuring good weather when especially urgent and important tests of aircraft were conducted; actually, however, they were mostly used in support of various public events, including the MAKS airshows in Zhukovskiy.

The An-30M had a take-off weight of 22,100 kg (48,730 lb) and an operational speed of 300-350 km/h (186-218 mph); maximum endurance was 4 hours 30 minutes at the altitude of 3,000 m (9,840 ft) and 5 hours 20 minutes at 6,000 m (19,680 ft).

Ilyushin IL-12 weather research aircraft

At least one IL-12 aircraft (identity unknown) was a weather research aircraft, featuring sensors and particle traps protruding from the cabin windows.

A poor-quality but interesting picture showing an IL-14P used in an atmospheric turbulence research programme. The nose-mounted sensor boom is clearly visible.

IL-14M CCCP-52056 retrofitted with an RPSN-2 Emblema radar and observation blisters was one of several examples modified for cloud-seeding.

Ilyushin IL-14 weather research versions

Control of atmospheric processes through active measures designed to influence these processes presupposes a thorough knowledge of what is happening in the atmosphere (of the laws governing the atmospheric phenomena and their causes). Physical research in clouds and the free atmosphere, as well as treating clouds with special chemicals for the purpose of dispersing them and artificially provoking rainfall, is conducted with the help of specially equipped research aircraft. These included a number of IL-14s

All research and survey versions of the IL-14 were developed and converted from standard machines by designers and production engineers of Aircraft Repair Plant (ARZ – *aviaremontnyy zavod*) No. 407 at Minsk-1 (Loshitsa) airport. The peculiarities of the tasks performed by these aircraft necessitated appropriate design changes, as well as installation of additional systems and equipment.

The designation '*Il-chetyrnadtsat' Meteo-laboratoriya*' (IL-14 weather research aircraft) covered a series of versions produced by ARZ No. 407; every single aircraft in this series was intended for its own specific tasks and featured an equipment suite unique to it. Conversion of each particular machine was made in accordance with manufacturing drawings and other documents applicable only to the given aircraft. For example, Moscow-built IL-14P CCCP-52008 (c/n 146001121) of the Ukrainian Civil Aviation Directorate/ Simferopol' United Air Detachment/370th Flight had a Groza-40 weather radar, observation blisters on both sides immediately ahead of the wings and various external sensors and particle traps. One more IL-14P (identity unknown) was used for atmospheric turbulence research in the 1970s; it had a long sensor boom under the nose carrying anemometric sensors.

Depending on the range of tasks to be fulfilled, the individual research aircraft featured their own internal layout and were fitted with additional instruments, such as electrometeographs, hygrometers, thermohygrometers, humidity pulsation sensors, ice formation sensors, zenith ozone meters, electric field intensity sensors, passive radar antennas and so on.

The IL-14 weather research aircraft tackled the following tasks: studying the atmosphere and clouds over a wide range of altitudes, using direct contact and remote sensing methods for measuring the investigated parame-

ters; subjecting the clouds to treatment for the purpose of studying the mechanism of forming artificially induced rainfall; development of cloud-seeding methods; systematic study of the sea surface pollution, of the condition of the country's waterways; development of a system and methods for interpreting and decoding data obtained from space vehicles; studying the characteristics of electric zones in the atmosphere that might pose a danger for air traffic; studying the laws of radio wave propagation in various meteorological conditions; monitoring the work of other aircraft; monitoring the environment for the purpose of discovering polluted zones in industrial areas; assessing the condition of young crops with the help of spectre-photometers for the purpose of forecasting the harvest.

Ilyushin IL-14GGO weather research/geophysical survey aircraft

Research in the field of cloud and precipitation physics and development of new prospective methods for treatment of the clouds are strongly connected with the use of specially equipped research aircraft. As far back as the 1950s, aircraft were widely used in many points on the territory of the Soviet Union for the vertical probing of the atmosphere with the purpose of obtaining data about the profiles of basic meteorological factors, for studying clouds and precipitation, and the atmospheric electricity.

To back up the research in the field of cloud physics and active influence, IL-14M CCCP-52029 (c/n 7343210) operated by the Leningrad CAD/2nd Leningrad UAD/74th Flight/4th Squadron was converted into a research aircraft for the Main Geophysical Observatory named after Aleksandr I. Voyeikov (GGO – *Glavnaya gheofizicheskaya observatoriya*). Hence it is known as the IL-14GGO.

Thanks to its fairly low speed (180-300 km/h; 112-186 mph), relatively low operating costs and unpressurised cabin the IL-14 came to be used on a very wide scale for research work in the USSR. However, the rather small payload and limited electric power supply posed stringent requirements concerning the energy consumption and weight of the mission equipment and limited the number of mission equipment operators to 5-7 persons.

Outwardly the IL-14GGO was identifiable by the nose radome housing a Groza-40 weather radar (instead of the standard solid

metal nose), the two observation blisters and the multiple sensors in the cabin windows and elsewhere on the fuselage; the configuration of these sensors changed in the course of time. The blisters had been installed immediately aft of the flightdeck; in the process of conversion, however, they were moved aft to the second pair of windows, the circular apertures at their original locations being closed by crude plugs with sensor ports.

The mission suite comprised 11 sets of equipment for meteorological probing, aerophysical measurements, the study of clouds and precipitation, and for artificially provoking rainfall; aerial cameras were also provided. The aircraft was equipped with three IT-3 infrared radiometers, two of which were mounted laterally aft of the flight deck (behind the said plugs) and the third ventrally. They allowed the thermal structure of a cloud to be captured accurately; the accuracy was not compromised by icing or humidity in clouds, as might be the case with external sensors. Copper, nickel or platinum thermometers were used to measure the outside air temperature, augmented by wire or semi-conductor temperature fluctuation sensors and a TUZ-1 ultrasonic thermometer (*termometr ool'trazvookovoy*).

Cloud water content was measured by means of a LIVO laser radar (*lazernyy izmeritel' vodnosti oblakov*). DIVO-3 (*dahtchik izmereniya vodnosti oblakov*) and SEIV-3 electric cloud water content meters using the

CCCP-52029, the IL-14GGO research/survey aircraft, runs up its engines at Leningrad-Rzhevka.

Above: The forward fuselage of the IL-14GGO, showing some of the sensors on the port side. The plugged blister aft of the flight deck (carrying sensors) reveals the aircraft's former identity as an IL-14FKM.

Test equipment placement on the IL-14GGO:

1. Groza-40 weather radar.

2. Weather radar imagery video recording device.

3. K4-51 meteorograph.

4, 5. Equipment racks.

6. IT-3 radiometer No. 2.

7. IT-3 radiometer No. 1.

8. LIVO laser radar (cloud water content measurement device).

9. Flare dispenser for IT-26 silver iodide cartridges.

10. IT-3 radiometer No. 3.

11. Dispenser for granulated rainmaking agents.

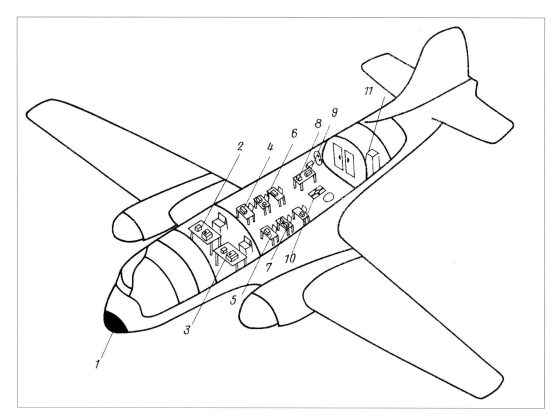

calorimetric method were also fitted. Rain density was measured using an inclined plane with water trapping gutters. The cloud microstructure was explored by means of particle traps and a microscope. Water content was measured with the help of a *Volna-1M* (Wave) piezoelectric quartz hygrometer and an STG-3 high-precision dew point thermohygrometer (*samolyotnyy termoghigrometr*).

Vertical gusts were measured in several ways, using G-force meters, barometric anemometers, thermoanemometers and cabin pressure sensors. The cloud-seeding equipment designed to generate rain from cumulus and stratus clouds included ASO-2I flare dispensers firing PV-26 flares and a hopper for dispensing powdered rainmaking agents.

The IL-14GGO was capable of conducting the probing for extended periods at altitudes up to 5-6 km (16,400-19,680 ft).

Ilyushin IL-14 cloud-seeding version

A number of IL-14s, including Tashkent-built IL-14M CCCP-52056 (c/n 7342804), IL-14P CCCP-61678 and IL-14M CCCP-91556, were converted into 'sky cleaner' aircraft for seeding clouds with rainmaking chemicals (silver iodide and granulated carbon dioxide). These aircraft had a parabolic nose radome apparently housing an RPSN-2 Emblema radar (*rahdiolokatsionnyy pribor slepoy navigahtsii* – blind navigation radar device) borrowed from the IL-18 airliner, necessitating relocation of the taxi

The IL-14GGO research/survey aircraft

lights; an observation blister was installed in the No. 2 cabin window to starboard, a ventral blade aerial was located aft of the wings and a braced probe projected aft from the tailcone.

Apart from keeping the rain away from cities during public holidays (which might otherwise be 'watered down'), such aircraft were used for provoking rainfall as a way of fighting forest fires. CCCP-52056 was operated by the Leningrad CAD/2nd Leningrad UAD/74th Flight; it crashed near Bagdarin, Buryat SSR, on 9th August 1975.

Ilyushin IL-28ZA weather reconnaissance aircraft

On 23rd February 1959 the State Committee on Aviation Hardware (GKAT) issued an order concerning the development of the IL-28ZA weather reconnaissance aircraft for civil aviation needs (again, ZA stood for *zondirov-shchik atmosfery*). A few *Beagles* were converted to this configuration. They wore civil registrations and had a sensor on the port side of the navigator's station.

Ilyushin IL-18V 'Meteor' weather research aircraft

One of the first *Coots* to be converted for research purposes was IL-18V CCCP-75716 (c/n 180001902), a former 235th Independent Air Detachment IL-18V 'Salon' VIP machine which was the 'presidential' aircraft of Nikita S. Khrushchov. By 1963 the aircraft had been transferred to Aeroflot's Polar Directorate, giving way to a newer example.

In 1963 CCCP-75716 was fitted out as a weather research aircraft by the manufacturer, MMZ No. 30 '***Znamya trooda***' (Banner of Labour) at Moscow-Khodynka, for the Central Aerologic Observatory (TsAO). Part of the mission equipment was installed in a large canoe fairing under the forward fuselage (outwardly identical to that of IL-18V CCCP-75431, see below); another external recognition feature was a satellite communications antenna in a small dielectric dome above the wing centre section. The total weight of the equipment suite was about 4 tons (8,820 lb).

At first the aircraft's mission was to corroborate the data supplied by weather research satellites. These were only beginning to gain wide use and had yet to prove their worth. Later, when the weather research satellites had earned their credentials, TsAO started using CCCP-75716 for other tasks; the first of these

was monitoring the launches of weather research rockets in the High North. Such flights sometimes involved participation of foreign scientists who were quick to recognise both the considerable capabilities of the aircraft itself and the skill of its crew (high flying skill was a requirement, since the flight profiles were rather unusual). They were in a position to form an opinion, as prior experience they had gained with other weather research aircraft was far less encouraging. The aircraft was also used for investigating jetstreams, turbulence (the clear-air turbulence phenomenon was already recognised as a flight safety threat in those days), storm fronts and the upper reaches of the atmosphere, including the ozone layer.

As the research missions changed, so did the equipment suite. The refitting jobs were performed at various locations in the Moscow Region – the LII airfield in Zhukovskiy, MMZ No. 30's flight test facility at Lookhovitsy-Tret'yakovo airfield and Chkalovskaya airbase. Some missions took CCCP-75716 over some of the Soviet Union's top secret military installations and sensitive areas where any other aircraft would have been shot down at once. Makes you wonder if these missions were not exactly of a peaceful nature…

In 1968 IL-18V CCCP-75716 was formally reassigned to Aeroflot's Central Directorate of International Services/63rd Flight based at Moscow-Sheremet'yevo. Its actual owner and operator, however, was GosNII GA and the aircraft was still used for weather research. The canoe pod under the forward fuselage was removed in March 1974, since the programme it was associated with had ended. Since CCCP-75716 could work in conjunction with Meteor weather research satellites, it was unofficially designated IL-18V 'Meteor'. Experience gained with this aircraft allowed the best ways of placing and using various meteorological equipment to be determined and later put to good use on other aircraft.

Ilyushin IL-18E 'Meteor' weather research aircraft

In 1976 the IL-18V 'Meteor', which was manufactured in 1960, was approaching the limit of its 20-year designated service life and a replacement aircraft had to be procured. It was another year before GosNII GA was able to provide the Central Aerologic Observatory with a newer *Coot*, IL-18E CCCP-75598 (c/n 186008802), which also had longer

IL-18E
CCCP-75598 in its
days as the IL-18E
'Meteor' weather
research aircraft.
Note the many
sensors and air
probes on the
fuselage, the
observation blis-
ter and the ven-
tral fairing of the
BMR-1 weather
research radar.

The logo of the
Main Geophysical
Observatory
named after
A. I. Voyeikov
(GGO) which
operated the
IL-14GGO and
IL-18GGO survey
aircraft.

range. Interestingly, the aircraft was released by the factory on 18th July 1967, despite the third digit in the c/n being a 6 (that is, 1966).

Not wishing to be snared in miles of red tape while obtaining all the necessary clearances for outfitting a new weather research aircraft from scratch, TsAO chose the way of the least resistance, requesting permission to simply transfer the existing equipment from CCCP-75716 to the new aircraft. Permission was quickly granted and the work went ahead. On 24th October 1977 the former IL-18V 'Meteor' was transferred to Ul'yanovsk Higher Flying School, the Soviet Union's top-notch civil aviation flying college, where it served as a trainer until finally retired in 1980.

Designated IL-18E 'Meteor' by analogy with its predecessor, CCCP-75598 featured a specialised BMR-1 weather research radar in a rather large round flat-bottomed radome under the wing centre section. The radar set was located in the forward baggage compartment and accessible via several maintenance hatches from the cabin allowing the radar to

be promptly fixed in flight, should it fail. Numerous air sampling traps and sensors were mounted on short struts on the forward and centre fuselage and under the wingtips, including sensors in the foremost and rear-most cabin windows to starboard and air traps in the forward emergency exit (7th window) and 13th window to starboard. Observation blisters were provided in the second and 13th window to port and the third window to starboard. Vertical cameras were installed in the front and rear cabins, while the toilets were transformed into photo processing labs, allowing the films to be developed and photos printed on board.

The conversion job, which was performed by the Ilyushin OKB's experimental facility (MMZ No. 240) at Moscow-Khodynka, was completed in June 1977. For more than 20 years that followed, CCCP-75598 served faithfully in the weather research role, taking part in numerous scientific experiments held both in the Soviet Union and abroad. By 1991, however, it became clear that the IL-18E

The IL-18E 'Meteor' weather research aircraft

'Meteor' was not being used to the full in its intended capacity, so GosNII GA proposed using the aircraft for cargo charter flights in order to generate revenue. Then, faced with the decline in the number of research programmes and skyrocketing aircraft leasing charges, TsAO had to give up using the IL-18E 'Meteor' and the aircraft was refitted for geophysical research (see Chapter 7).

Ilyushin IL-18GGO (IL-18V CCCP-75431) weather research/ geophysical survey aircraft

In late 1972 or early 1973 an IL-18V with the out-of-sequence registration CCCP-75431 (c/n 180002003) was transferred to the Main Geophysical Observatory named after Aleksandr I. Voyeikov (GGO) and converted to a geophysical survey aircraft. (This aircraft should have been CCCP-75718, the latter registration being assigned to IL-18V c/n 180002005 instead. A possible explanation is that the aircraft was originally operated by the

Polar Aviation Directorate and registered in the CCCP-04xxx block.)

CCCP-75431 featured a large canoe fairing under the forward fuselage with four ventral apertures for sensors and cameras; the lower portion could swing open to starboard for maintenance. A large teardrop metal fairing incorporating three dielectric panels was installed ventrally just aft of the wings. The first, third and tenth cabin windows to starboard were blanked off with metal plugs mounting small sensors; another sensor was installed on a short strut above the wing trailing edge. The rear emergency exit to starboard had a non-standard windowless cover.

The aircraft participated in assorted research programmes held in the interests of various ministries. For instance, on 3rd April 1975 CCCP-75431 took off from its home base of Leningrad-Pulkovo, heading for Karaganda. The mission was to survey arable lands in the Karaganda, Tselinograd and Kokchetav Regions of Kazakhstan in preparation for the wheat sowing campaign, deter-

IL-18V CCCP-75431 flies over the North Sea during the POLEX-Sever-76 experiment. The picture was taken from IL-18V 'Meteor' CCCP-75716 shortly before the latter was reconverted to standard configuration. Here CCCP-75431 is seen in pre-1973 colours.

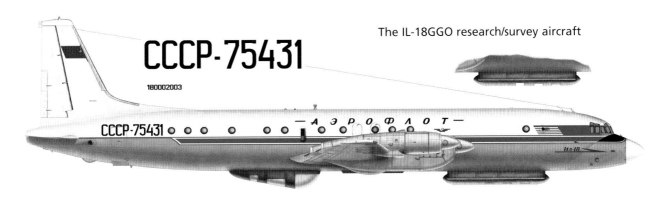

The IL-18GGO research/survey aircraft

CCCP-75431

180002003

CCCP-75431

mining how much water was needed for irrigation. The IL-18 worked in conjunction with a specially modified Antonov An-24 *Coke* airliner and an An-30 photo survey aircraft. Similar land survey missions were flown in the spring of 1977 in such far-apart areas of the Soviet Union as the Krasnoyarsk Region of West Siberia, the Stavropol' Region of southern Russia and the southern regions of the Ukraine – all major wheat-growing areas.

A year earlier the aircraft was involved in the POLEX-Sever-76 ('Polar Experiment – North 1976') together with the IL-18V 'Meteor' weather research aircraft, CCCP-75716 (which is described above), operating from Amderma and working together with the research vessel M/V *Professor Wiese*. That same year it participated in the SAMEX (*So**vet**sko-ameri**kahn**skiy mikrovol**novyy** eksperi**ment*** – Soviet/US Microwave Experiment), measuring the condition of the sea and atmosphere in two designated areas of the Pacific near the Kamchatka Peninsula and the Kurile Islands. The experiment also involved a modified Convair CV 990 Coronado, the Soviet research vessel M/V *Akademik Korolyov* and the Nimbus-5 and Nimbus-6 satellites.

In due course the aircraft's pre-1973 IL-18 colours gave way to a variation on Aeroflot's 1973-standard livery. The cheatline was red but the red colour on the tail surfaces and outer wings (as required by the Polar version of the livery) was missing, as were the Aeroflot titles and logo. By 1992 CCCP-75431 had been transferred to LII and all survey equipment had been removed, thus turning the aircraft into an IL-18Gr package freighter.

Ilyushin IL-18D Tsiklon (IL-18DTs) weather research aircraft

One of the seven assorted weather research aircraft in the Tsiklon series commissioned by TsAO in the mid-1970s was based on the long-range IL-18D – specifically, CCCP-75442 (c/n 187009702) – and designated IL-18D Tsiklon, though some sources call it IL-18DTs. Outwardly it was something of a spotter's delight, sprouting all manner of 'bumps and bulges' so characteristic of research aircraft. A pointed boom 4 m (13 ft 1½ in) long tipped with sensors was installed on the port side of the nose ahead of the flight deck glazing, increasing overall length to 40 m (131 ft 2¾ in); it was painted in black and white zebra stripes to avoid damage by ground vehicles. A specialised K-11M weather research radar in a

deep teardrop radome with flattened sides was installed immediately aft of the nose gear unit (frames 5-11), requiring the nosewheels to be fitted with a mudguard and the nose gear doors to be bulged accordingly.

In similar fashion to the SL-18D radar testbed (see Chapter 4), a FLIR or laser equipment housing was mounted on the flight deck roof and two shallow dielectric canoe fairings were fitted above and below the fuselage in line with the wing leading edge. Two elongated square-section fairings with apertures for laser measurement systems took the place of the two rearmost cabin windows on each side. Numerous sensors were mounted on struts just aft of the flight deck. The lower fuselage incorporated two optically flat camera windows protected by sliding doors for take-off and landing. An extra wire aerial ran from the fin leading edge to a strut above the forward entry door. Finally, two observation blisters were provided in the third and 13th cabin windows on each side, plus two dorsal observation blisters in line with the rear pair of emergency exits and ahead of the rear entry door. The Tsiklon emblem was painted on the nose to clarify the aircraft's 'storm chaser' role.

The mission equipment was largely the same as on the An-12BP Tsiklon, comprising a measurement suite, data recording/processing suite and cloud-seeding equipment. The measurement suite included, first of all, a thermodynamic measurement system comprising a TsSV-3M-1KM central air data system, an EM TsAO electrometeorograph, a PK G-load measuring kit, an SAMB-70 airborne automatic weather research module, an ASTA-74 airborne automatic thermoanemometer, a DISS-013-134 Doppler speed/drift sensor system, an RV-18Zh radio altimeter, an ISVP airflow structure meter (*izme**rit**el' strook**too**ry voz**doosh**novo potok**a***) and an SG-1 airborne humidity meter. It recorded the outside air temperature and its fluctuations, the aircraft's speed and heading, the wind speed and direction, airflow pulsations, static and dynamic air pressure, barometric and true altitude, vertical gusts and G loads acting on the aircraft.

The second major component was the cloud and precipitation microstructure measurement system comprising an RP-73 translucency recorder, an IRCh water/ice particle size meter, an SALYa atmospheric ice particle counter (***shchot**chik atmos**fer**nykh ledya**nykh ya**der*), SEIV-3 and IVO airborne electric cloud water content meters and an AFSO cloud phase analyser. The third component was the meteorological radar and laser system – the aforementioned K-11M search radar, a BMR-1A radar for vertical scanning of the atmosphere and an LR-3P laser polarisation meter. The fourth component was the radiometric system comprising the RAK radiometric/actinometric complex and the TETA radiometric module. Finally, there was a PNP meter for measuring electric fields and the aircraft's electric charge.

The BARS-1 data recording/processing suite ('bars' means 'leopard', but in this case it is an acronym for *bortovaya avtomatich-eskaya reghistreeruyuschchaya sistema* – on-board automatic recording system) consisted of a K60-42 magnetic recorder, a PTU-31-1-7 video recording system, a forward-looking AKS-2 ciné camera on the starboard side of the nose, an AFA-BAF-21S photo camera and an SYeO time synchronisation system. A photo/ciné film processing lab was provided in the rear cabin.

The cloud-seeding equipment designed to generate rain from cumulus and stratus clouds included three KDS-155 solid carbon dioxide dispensers and three ASO-2I flare dispensers mounted on the centre fuselage underside (frames 17-23). The aircraft had a flight crew of five and a 34-man team of researchers who sat behind 19 single and dual equipment consoles installed in all three cabins.

The first post-conversion flight took place on 4th April 1980. The maximum take-off weight was 64 tons (134,480 lb), including a 9.4-ton (20,720-lb) payload; the various external outgrowths reduced the cruising speed to 625-650 km/h (388-403 mph) and the service ceiling to 8,650 m (28,380 ft). The aircraft could stay airborne for up to eight hours, with a maximum range of 4,270 km (2,650 miles).

The IL-18D Tsiklon had quite an active service career, flying at home and abroad (among other things, it periodically deployed to Cuba to chase tropical hurricanes). After the demise of the Soviet Union it was reregistered RA-75442. Unfortunately, like some other research aircraft, the IL-18D Tsiklon fell victim to unscrupulous businessmen to whom profits were more important than science – in 1997 it was stripped of all mission equipment, leased to a succession of airlines and finally sold.

Lisunov Li-2 Meteo weather research aircraft

The Li-2 Meteo weather research aircraft was created at ARZ No. 407 in Minsk after the war. The aircraft was fitted with the necessary set of equipment; workstations for aerologists were outfitted in the cabin. The aircraft was intended for:

• exploration of the atmosphere and clouds within a wide range of altitudes, using the methods of contact and remote measurements;

This aspect of the IL-18D 'Tsiklon' gives a good detail view of the instrumented nose probe, the K-11M weather research radar aft of the nose gear, the ventral ASO-2I rainmaking chemical dispensers immediately aft and the plethora of sensors on the forward fuselage.

The IL-18D 'Tsiklon' in flight, showing off its sensor array on the centre/aft fuselage and wings.

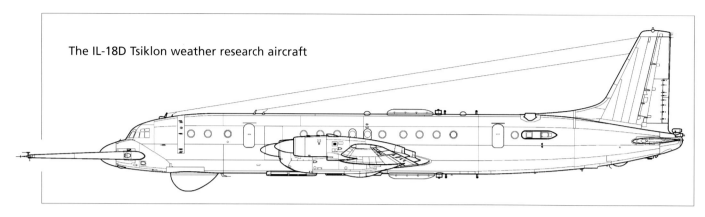

The IL-18D Tsiklon weather research aircraft

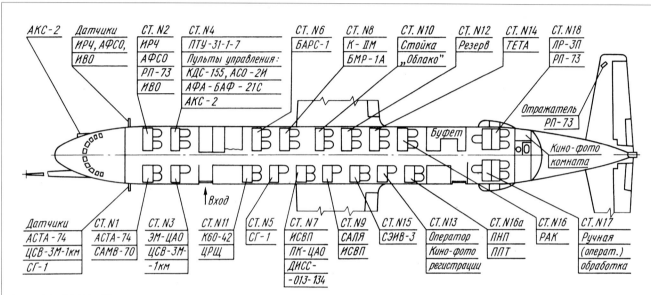

АКС-2	Датчики ИРЧ, АФСО, ИВО	СТ. N2 ИРЧ АФСО РП-73 ИВО	СТ. N4 ПТУ-31-1-7 Пульты управления: КДС-155, АСО-2И АФА-БАФ-21С АКС-2	СТ. N6 БАРС-1	СТ. N8 К-ДМ БМР-1А	СТ. N10 Стойка „Облако"	СТ. N12 Резерв	СТ. N14 ТЕТА		СТ. N18 ЛР-3П РП-73

Отражатель РП-73

Буфет

Кино-фото комната

↑Вход

| Датчики АСТА-74 ЦСВ-3М-1км СГ-1 | СТ. N1 АСТА-74 САМВ-70 | СТ. N3 ЭМ-ЦАО ЦСВ-3М- -1км | СТ. N11 К60-42 ЦРЩ | СТ. N5 СГ-1 | СТ. N7 ИСВП ПК-ЦАО ДИСС- -013-134 | СТ. N9 САЛЯ ИСВП | СТ. N15 СЭИВ-3 | СТ. N13 Оператор Кино-фото регистрации | СТ. N16а ПНП ППТ | СТ. N16 РАК | СТ. N17 Ручная (операт.) обработка |

The cabin layout of the IL-18D 'Tsiklon', showing the researchers' workstations.

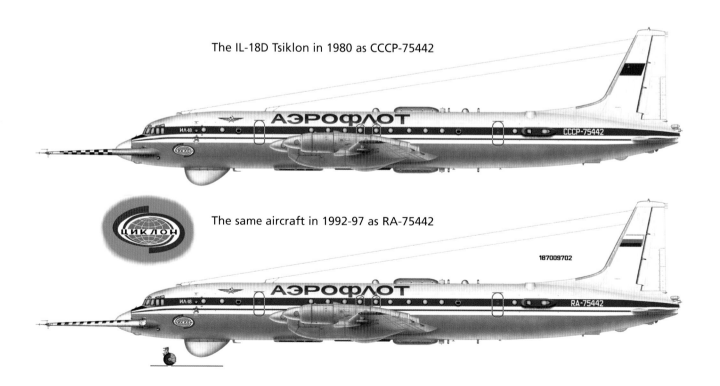

The IL-18D Tsiklon in 1980 as CCCP-75442

The same aircraft in 1992-97 as RA-75442

• studying the characteristics of electric zones in the atmosphere that could be dangerous for air traffic;

• exploration of the laws of radio wave propagation in different meteorological conditions;

• monitoring the operation of other aircraft;

• monitoring the environment with a view to detecting pollution zones.

One of the first *Cabs* to be converted to a Li-2 Meteo research aircraft was СССР-Л4890 (that is, SSSR-L4890). Subsequently several other examples, including CCCP-54909 No. 1, CCCP-83962 No. 1 and CCCP-84713 No. 1, were modified to take up this role, featuring minor differences in equipment. (The registrations were later reused for the An-12BK transport prototype (CCCP-83962) and two An-2R crop-sprayers.)

Mikoyan MiG-25MR weather reconnaissance aircraft

A handful of MiG-25s was built in a specialised weather reconnaissance version designated MiG-25MR (***met**eoraz**ved**chik*). Based on the MiG-25RB, the aircraft had the cameras and SRS-4 SIGINT pack replaced with other mission equipment appropriate for its new role.

Mil' Mi-8MT weather research helicopters

In 1990 twelve Mi-8MT *Hip-H* helicopters were converted for weather research. Unfortunately none have been identified yet.

Sukhoi Su-15UT research aircraft for exploring weather minima

In 1973 LII used a suitably modified Su-15UT *Flagon-C* trainer (c/n 0815315) to conduct research for the purpose of determining the weather minima in which aircraft could take off and land safely. For example, the Su-15's weather minima were 250 x 2,000 m (that is, cloudbase 250 m/820 ft and horizontal visibility 2,000 m/1.24 miles) in the daytime and 350 x 3,000 m (1,150 ft x 1.86 miles) at night; for the Su-15TM interceptor equipped with the RSBN-5S short-range radio navigation system (and described later in this chapter) the minima were 60 x 800 m (200 ft x 0.5 mile).

Tragically, one of the test missions on 29th October 1975 aimed at reducing the type's weather minima ended in a crash in which LII test pilot Ivan V. Makedon and 148th Combat Training & Aircrew Conversion Centre pilot

Rudol'f N. Zoobov lost their lives. Landing with the blind flying hood closed and a simulated control system actuator failure, Zoobov (who was the pilot in command) inadvertently increased the sink rate from the required 5-6 m/sec (980-1,180 ft/min) to 10 m/sec (1,970 ft/min) and then to 15 m/sec (2,950 ft/min). Despite corrective action at the last moment the aircraft landed hard between the outer and inner marker beacons, undershooting by 1,400 m (4,590 ft). Next moment, it collided with the elevated edge of a concrete road and caught fire, both pilots being thrown clear of the cockpit before the machine came to a halt. The crew was rushed to a hospital but Zoobov was dead on arrival; Makedon died in hospital on 23rd November.

CCCP-54909, a Li-2 Meteo weather research aircraft with wing-mounted sensors.

Close-up of the sensor on the starboard wing of a similar aircraft.

Li-2 Meteo CCCP-83962 No. 1 featured nose- and wing-mounted sensor probes. The registration later passed to the An-12BK prototype.

A Tu-16 in Aeroflot livery certainly was a strange sight. CCCP-42484, the second Tu-16N Tsiklon, had dorsal sensor fairings (missing on the first aircraft). The chemical dispenser pods carried on the pylons are probably modified KMGU-1 submunitions dispensers used by tactical strike aircraft and helicopters.

Tupolev Tu-16 Tsiklon-N weather research aircraft ('order 386')

On 4th April 1976 MAP issued order No.176 envisaging the conversion of several Tu-16s into Tu-16 Tsiklon-N weather research aircraft to meet the order placed by Goskom-ghidromet and TsAO. The N referred to the basic aircraft (the Tu-16's product code at the Kazan' aircraft factory No. 22 was *izdeliye* N). The Tupolev OKB collaborated with the Naval Aviation to procure two Kazan'-built Tu-16K-26 *Badger-G Mod* (or Tu-16KSR-2-5)

missile carriers manufactured in 1956, c/ns 6203203 and 6203208, for modification at the Soviet Navy's ARZ No. 20 at Pushkin in 1977.

The Tu-16 Tsiklon-N was intended for studying the principal thermodynamic and electric parameters of the atmosphere and cloud formations, as well as for cloud-seeding in order to make rain . Accordingly all armament and military equipment (except for the missile pylons) was removed and replaced by R-802V and R-802GM radios, a Rubin-1M wide-scan weather radar with a revolving antenna in a large teardrop radome under the centre fuselage and other special equipment. The former weapons bay and BD-187 wing pylons were used for carrying special bombs filled with rainmaking chemicals or special pods housing research equipment or chemical dispensers.

Trials by GNIKI VVS and TsAO lasted from November 1978 to April 1980, whereupon the aircraft were placed on the civil register in the Tu-104 series as CCCP-42355 (c/n 6203203) and CCCP-42484 (c/n 6203208). (The registration CCCP-42355 was simply taken out of the Omsk-built Tu-104A sequence, while CCCP-42484 should have gone to Kazan'-built Tu-104B c/n 021502, which received the non-standard registration CCCP-06195 instead – see Chapter 4.) In keeping with their new mission the *Badgers* gained the blue/white 1973-standard Aeroflot livery – save that the type was marked on the nose simply as 'Tu'. Like the other research aircraft in the Tsiklon series, the Tu-16s wore the

distinctive Tsiklon emblem on the nose to clarify their 'storm chaser' role.

The two Tu-16 Tsiklon-Ns were based at Chkalovskaya AB. For more than ten years they served in a variety of scenarios over central Russia and the Ukraine, including 'sky cleaning' missions during the XXII Summer Olympic Games held in Moscow in 1980 and damage control in the wake of the 1986 Chernobyl' nuclear disaster.

Tupolev Tu-16 Tsiklon-NM weather research aircraft

On 19th November 1986 the Soviet Council of Ministers ordered that both Tu-16 Tsiklon-N aircraft should be re-equipped; this would enable them to participate in international weather research programmes. Eventually only CCCP-42355 No.1 was so modified; it began its tests in 1991 but these were interrupted by the disintegration of the USSR later that year.

Close-up of the nose of Tu-16NM Tsiklon CCCP-42355 No. 1, showing the characteristic logo carried by all weather research aircraft in the Tsiklon series. Curiously, the type is marked on the nose only as 'Tu'.

Tu-16NM Tsiklon CCCP-42355 No. 1 sits in a remote corner of Chkalovskaya AB following its retirement in the late 1990s. The aircraft's origins as a Tu-16KSR-2-5 with a Rubin-1M radar are obvious, the large teardrop radome and missile pylons being clearly visible; all armament has been removed.

The Tu-16 Tsiklon-N with chemical dispenser pods

АЭРОФЛОТ

The Tu-16 Tsiklon-N and Tu-16 Tsiklon-NM were to have been used for the last time during the First Chechen War (1994-96) to facilitate operations for the federal troops, as the North Caucasus was often plagued by foul weather hampering Air Force and Army Aviation operations). However, they did not actually take part in the conflict and were retired soon afterwards. The ultimate fate of CCCP-42484 is unknown, while CCCP-42355 No. 1 was last noted in reasonably good condition at Chkalovskaya AB in August 1999.

Curiously, back in late 1987 the registration CCCP-42355 was reused for a Yakovlev Yak-42 airliner (c/n 4520424711399, f/n 1309) delivered to Aeroflot's Lithuanian CAD. Thus, contrary to all rules there were two aircraft with the same registration operational at the same time!

Tupolev Tu-16N for spraying liquid carbon dioxide

Two Tu-16N *Badger-A* tankers were used for spraying liquid CO_2 in the late 1970s as part of the Tsiklon research programme. They were based at Chkalovskaya AB.

Tupolev Tu-104A Tsiklon weather research aircraft

Upon completion of the Zaslon weapons control system's trials Tu-104A CCCP-42454 (the second LM-104-518 radar testbed) had the components of the WCS removed and was further modified for weather research, becoming yet another weather research aircraft in the Tsiklon series – the Tu-104A Tsiklon. It sported rather fewer 'bumps and bulges' than most of the other aircraft in the series. The pointed nose radome was retained, but now it was attached straight to the forward fuselage structure, not via a cylindrical adapter; thus overall length was somewhat reduced. The AKU-410 missile ejector racks gave place to BD-357 pylons, which were used for carrying pods with rainmaking agents. The ram air turbines and heat exchanger under the forward fuselage were deleted. The aircraft was repainted in 1973-standard Aeroflot colours and the Tsiklon emblem was added on the nose below the cheatline.

The research missions, which were code-named *Groza* (Thunderstorm) and flown under the auspices of the Central Aerologic

CCCP-42454 in its ultimate guise as the Tu-104A 'Tsiklon' weather research aircraft.

The Tu-16 Tsiklon-NM

Observatory, were typically flown in the southern regions of the Soviet Union. Among other things, grade 800 and 1,000 cement was used as a rainmaking agent along with granulated 'dry ice' and silver iodide. The experiments continued until the mid-1980s.

Yakovlev Yak-40 Akva weather research aircraft

In 1986 a Yak-40 *Codling* feederliner (CCCP-87937, c/n 9740856) was converted into a weather research aircraft called Yak-40 Akva. The job was performed by ARZ No. 407, which was the principal refurbisher of the Yak-40. In addition to atmospheric studies, the machine was intended for provoking rainfall by disseminating chemical agents in clouds and for ensuring protection of crops from hail in a similar way, as well as for monitoring air pollution. A special feature of the Yak-40 Akva was the ability to carry underwing SPVG-40 pods with PV-26 silver iodide flares.

A model of this aircraft was shown at an exhibition in Moscow in 1985. The model displayed a nose probe and various sensors, as well as a scabbed-on fuselage dispenser.

The Yak-40 Akva served for several years with the Ukranian Research Institute of Hydrometeorology the Goskomghidromet framework. Later it was reconverted to airline configuration, serving with Armenian Airlines as EK-87937. On 17th May 2001 the aircraft crashed in Iran while operated by the local charter carrier Queshm Air as EP-TQP.

Yavkovlev Yak-40 Meteo weather research aircraft

The Yak-40 Meteo (CCCP-87537, c/n 9520242) was a multi-role weather research aircraft equipped with various sensors, data recording and processing equipment which enabled the researchers on board to study, among other

things, the atmospheric turbulence and the effects of solar radiation on the atmosphere. The equipment also included tandem ASO-2I flare dispensers low on the forward fuselage sides for causing artificial rainfall. Again, the Yak-40 Meteo was eventually stripped of all mission equipment, serving with Kazakhstan Airlines and subsequently with Taraz Wings as UN 87537.

Yavkovlev Yak-40 Liros weather research aircraft

One more weather research aircraft based on the Yak-40, the Yak-40 *Liros* (CCCP-87536, c/n 9522041; the meaning of the word 'Liros' is unknown) was intended for studying atmospheric conditions and clouds in a broad range of altitudes and for testing the various methods of causing artificial rainfall. The Yak-40 Liros was operated by GosNII GA and based at Moscow/Sheremet'yevo-1. This aircraft, too, was reconverted to airline configuration and became EK-87536 with Armenian Airlines.

The nose of the Yak-40 Shtorm research aircraft, CCCP-87992. Note the non-standard probe and the flare dispensers for cloud-seeding.

Yavkovlev Yak-40 Shtorm weather research aircraft

The Yak-40 Shtorm (Sea Storm) was developed jointly by the High-Mountain Geophysics Institute and ARZ No. 407, which did the conversion job. Registered CCCP-87992 (c/n 9541644), the aircraft was intended for weather research and was equipped with strut-mounted sensors and probes and data recording equipment, as well as tandem ASO-2I flare dispensers low on the forward fuselage sides. In the early 1990s the Yak-40 Shtorm was reconverted to airline configuration and sold to Stavropol' Avia as RA-87992.

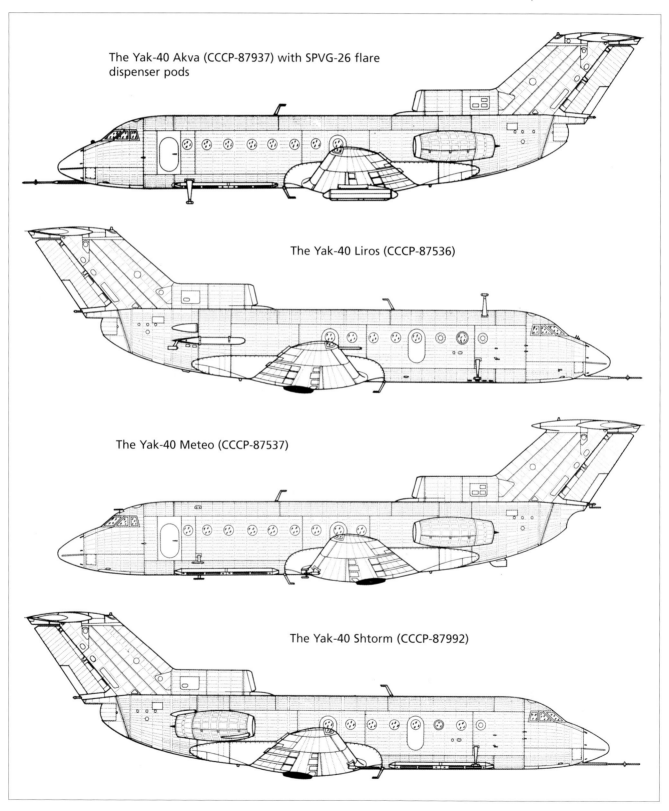

The Yak-40 Akva (CCCP-87937) with SPVG-26 flare dispenser pods

The Yak-40 Liros (CCCP-87536)

The Yak-40 Meteo (CCCP-87537)

The Yak-40 Shtorm (CCCP-87992)

A number of aircraft were modified for geophysical survey and research work, including ore prospecting.

Antonov An-2 geophysical survey versions

Several An-2 *Colt* utility biplanes were modified for geophysical survey and research. The equipment fit and hence the exterior of these aircraft varied. Some were reportedly equipped with vertical and oblique cameras for continuous strip photography while following the designated route. They were also equipped with celestial compasses and could carry teams of prospectors which they could land in areas of interest.

One aircraft (the registration is hard to read in the only available photo but may be CCCP-15454) had a towed magnetic anomaly detector (MAD) sensor – commonly known as 'bird' – for iron ore prospecting and an antenna array atop the fin. The MAD 'bird' looked like a bomb with an annular stabiliser and was stowed in a special cradle under the fuselage when not in use, being completely exposed. Another aircraft (identity unknown) had a dif-

ferently shaped towed MAD 'bird' with elongated cruciform stabilisers.

One An-2TP in 1973-standard blue/white Aeroflot livery (CCCP-96232, c/n 1G 7216) was equipped with a SKAT-77 instrumentation suite comprising a gamma spectrometer, an MAD and imaging. Outwardly this was revealed by a sensor array mounted well aft of the rudder on long booms connecting it to the aft fuselage underside and the horizontal tail (in order to prevent the massive metal fuselage from affecting the sensor). The array had a streamlined centrebody and a loop aerial around it, looking oddly similar to the cooling fans commonly found in offices!

Another An-2 in Aeroflot's 1973-standard red/white Polar colours (identity unknown) had a similarly positioned MMS-214 MAD array aft of the tail unit, a GSA gamma spectrometer and an MIR-3 radiogeodetical positioning system. Data obtained by means of these systems were stored in a magnetic data recorder for later computerised processing.

An-2R CCCP-33374 (c/n 1G 22633) belonging to the Leningrad CAD/Murmansk UAD again featured a sensor array mounted on converging booms aft of the rudder. This

385

An-2R
CCCP-33374 con-
verted for geo-
physical survey is
pictured at
Leningrad-
Rzhevka on 12th
May 1991; note
the tail-mounted
sensor on a
braced outrigger
and cable aerials.

An-2 CCCP-96232
was another geo-
physical survey
aircraft equipped
with the SKAT-77
remote sensing
suite.

was augmented by a towed MAD 'bird' which
was semi-recessed in the fuselage when not in
use and by cables stretched from the upper
wing tips to the stabilisers to form a huge loop
aerial. The aircraft wore the beige colour
scheme typical of Aeroflot's An-2R crop-
dusters but with a red/striped fin and a red
stripe down the fuselage top.

Antonov An-3T geophysical survey aircraft

An example of the An-2's turboprop conversion, the An-3T, was similarly modified for
geophysical survey. The aircraft, RA-05869 (c/n

Close-up of the
tail-mounted
sensor array on
CCCP-96232.

1G 20334-2005), started life as an agricultural An-2R (CCCP-17773, later RA-17773, c/n 1G 20334) in June 1983 before being converted to the fifth production An-3T by the 'Polyot' Omsk Aircraft Production Association in 2000. It is operated by KrasAvia (a regional airline based in Tura, Krasnoyarsk Region, central Siberia) and carries an MAD sensor on a strut-braced outrigger aft of the rudder.

Antonov An-12TP-2 geophysical survey aircraft

A single An-12B with the non-standard Polar Aviation registration CCCP 04366 (the registration was painted on with no hyphen; c/n unknown) was custom-built for long-range transport and geophysical survey duties in the Antarctic. The aircraft had a non-standard radar in a much longer and deeper radome extending

Left: The cabin of CCCP-96232, showing the equipment racks of the SKAT-77 mission equipment suite.

Above: The cabin of the An-2 pictured on this page, showing one of the mission equipment operators at work.

This unidentified An-2 in 1973-standard red/white Polar colours was converted into a very similar geophysical research aircraft (note the MMS-214 magnetometer aft of the rudder).

Anonymous-looking KrasAvia An-3T RA-05869 carries a sensor array similar to that of An-2 CCCP-33374.

all the way aft to the nosewheel well (the ventral flightdeck escape hatch was thus blocked). There was no tail gunner's station, but a long slender magnetic anomaly detector (MAD) boom protruded from the 'civil' tailcone. The pre-1973 red/white livery worn by Aeroflot *Cubs* (patterned on that of the An-10 *Cat* airliner) was modified: the belly was white instead of grey, the vertical tail and propeller spinners were painted orange, and the tops of the outer wings and stabilisers were red for high definition against white backgrounds. To make its mission clear the aircraft wore additional '*Polyarnaya aviahtsiya*' titles and penguin tail art.

Designated An-12TP-2, the aircraft took part in an expedition to the Antarctic in December 1961. On arrival in the Antarctic CCCP 04366 was refitted with a non-retractable ski undercarriage. Interestingly, the *Cub* was referred to as an An-10 in Soviet press reports of the expedition because the designation An-12 was still classified at the time; this misinformation found its way into Western publications as well.

Antonov An-12AP Magnitometr/ Relikt geophysical survey aircraft

Irkutsk-built An-12AP CCCP-12186 (c/n 1901807) was another geophysical survey aircraft developed in 1982 for the Leningrad branch of the Earth Magnetism Institute (a division of the Soviet Academy of Sciences). The design and conversion work was performed by the Soviet Navy's ARZ No. 20 in Pushkin, Leningrad Region.

The aircraft, which wore full 1973-standard Aeroflot blue/white livery, was fitted with an 8-m (26 ft 2⁶¹⁄₆₄ in) tapered MAD boom supplanting the DB-65U tail turret. A small 'superstructure' housing an L-14MA astro-inertial navigation system was mounted on top of the boom near the front, making it look like a miniature submarine. A camera fairing was mounted immediately aft of the rear cargo door segment. There were three researchers' workstations in the pressure cabin, with most of the equipment being installed in the unpressurised freight hold.

Originally the aircraft bore the codename *Magnitometr* (magnetometer, or MAD). In 1990 CCCP-12186 was upgraded by installing new scientific instrumentation and the codename was changed to *Relikt* (relic). The aircraft was used for studying the structure of the Earth's magnetic field and for making gravimetric measurements as requested by various government agencies and ministries.

Antonov An-24RV Nit' (An-24LR Nit') multi-role survey aircraft

In 1978 a single An-24RV airliner – actually an upgraded Kiev-built An-24B – with the out-of-sequence registration CCCP-47195 (c/n 07306202) was converted into a multi-mission survey aircraft. (Logically this aircraft manufactured in 1970 should have been CCCP-46389, but it is not known if the aircraft had been reregistered before 1978 or this registration had been allocated but not taken up. Theoretically the registration CCCP-47195

should have been worn by an Ulan-Ude built An-24B with a c/n something like 99902102.)

The mission equipment suite of CCCP-47195 was built around a Nit'-S1 (Thread-S1) SLAR which gave the aircraft a rather bizarre appearance: the SLAR fairings were huge, drooping below the lower fuselage contour, and were angled slightly nose-down. 'Pitchfork' arrays of triple rod aerials were

mounted in tandem under each SLAR fairing. A vertical camera of unknown type was housed in the rear fuselage, the lens being closed by a prominent teardrop fairing with clamshell doors located between the ventral fins. Observation blisters were provided in some of the cabin windows. The cabin accommodated the researchers' workstations, as well as data processing and recording equipment.

The An-12TP-2 seen during a refuelling stop en route to Antarctica. Red-tailed Polar Aviation IL-18V CCCP-75743 is just visible beyond.

The An-12TP-2 geophysical survey aircraft on wheels and on skis, 1961-62

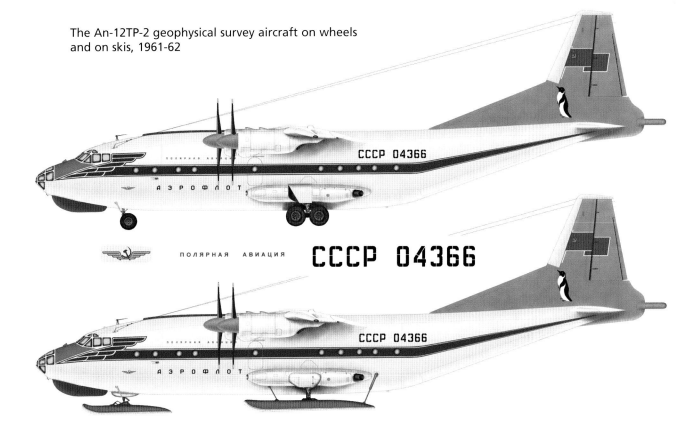

Another view of CCCP 04366, the custom-made An-12TP-2 geophysical research aircraft. The 'commercial' tailcone housed a telescopic MAD boom.

Close-up of the An-12TP-2's nose, showing the non-standard large radome. This aircraft and IL-18V CCCP-75743 made a trip to Antarctica in December 1961/ February 1962.

CCCP 04366 taxies at Ice Station Molodyozhnaya, one of the Soviet research stations in Antarctica. During its sojourn on the ice continent the aircraft was temporarily fitted with skis.

Above and left: An-12AP CCCP-12186 was converted for geophysical survey in 1982, being used by the Earth Magnetism Institute in the Magnitometr and Relikt research programmes. It is seen here at Pushkin.

Close-up of the MAD boom fitted to An-12AP CCCP-12186. Note the camera fairing aft of the cargo door.

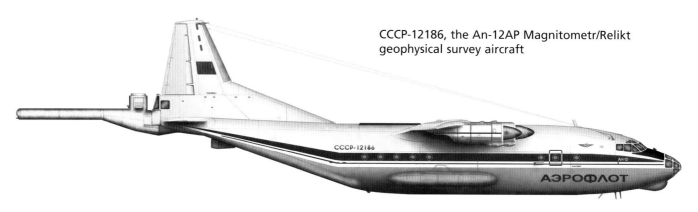

CCCP-12186, the An-12AP Magnitometr/Relikt geophysical survey aircraft

The sole An-24RV Nit' (aka An-24LR Nit'), RA-47195, parked at Pushkin on 8th August 2001. The size of the SLAR fairings is clearly visible. Note the Polar Aviation badge on the baggage door.

Known as the An-24RV Nit', the aircraft was used for prospecting natural resources, including those of the World Ocean. However, the powerful SLAR could be used with equal success for measuring the thickness of ice-fields; hence CCCP-47195 was also referred to as An-24LR Nit' (LR = *ledovyy razvedchik* – ice reconnaissance aircraft). Incidentally, the same Nit'-S1 SLAR was fitted to the Ilyushin IL-24N long-range ice reconnaissance aircraft based on the IL-18D four-turboprop long-haul airliner. Since it was to operate in the Polar regions, the aircraft wore Aeroflot's red/white Polar colour scheme (also used by regular transport aircraft flying in the High North) for high definition against white backgrounds; the Polar Aviation badge was painted on the nose.

The An-24RV Nit' (An-24LR Nit') participated in several major scientific programmes,

working in concert with several dozen other aircraft, the smallest of which was an An-2; Soviet cosmonauts aboard the Salyut-6 space station also made their contribution to the programme. Unlike most other Soviet ice reconnaissance aircraft, the An-24LR Nit' remained in service after the demise of the Soviet Union, gaining the Russian registration RA-47195.

Antonov An-30 geophysical survey aircraft

According to a press report which appeared in 1983, an Aeroflot An-30 photo survey aircraft (presumably CCCP-30067, c/n 1208) was fitted with infra-red scanners for prospecting natural resources. It was used with the participation of the *Aerogheologiya* (Airborne

Geology) Industrial Association for studying the geological properties of rocks.

Antonov An-30D (modified) geological prospecting aircraft

In 1994 a single An-30D, RA-30053 (c/n 1008) – the version characterised by strap-on extra fuel tanks on the centre fuselage sides (hence D for *dahl'niy*, long-range) – was leased for off-shore oil prospecting in Norway by Conoco (Continental Oil Co.). Oddly enough, the external tanks were removed; special equipment was installed in the cabin, with a slender boom tipped with a sensor array protruding aft from the tailcone. The aircraft was based at Kristiansund for the duration.

Ilyushin IL-12 geophysical survey aircraft

An IL-12 *Coach* airliner (identity unknown) was fitted out for aeromagnetic survey with a towed MAD 'bird' and used for geological prospecting.

Ilyushin IL-14FKM geophysical survey conversion

At least two IL-14FKM photo survey aircraft operated by the Central Regions CAD/ Myachkovo UAD/229th Flight – CCCP-91480

and CCCP-91483 (c/n 147001232) – were converted for geophysical research. Their main external identification feature was a towed MAD 'bird' housed under the fuselage aft of the camera ports when not in use.

Ilyushin IL-14FKM thermal mapping conversion

In the mid-1980s IL-14FKM CCCP-61685 operated by the same flight was outfitted with a *Voolkahn* (Volcano) thermal imager. This aircraft was used for geophysical research and for making thermal maps of cities and other areas, allowing leaks in hot water mains to be detected, among other things.

Close-up of the port SLAR fairing on RA-47195. Note the tandem pitchfork-like antenna arrays below the fairings.

An-30D RA-30053 was used for oil prospecting for a while.

Ilyushin IL-14T geophysical survey conversion

An IL-14T transport operated by the Leningrad CAD/2nd Leningrad UAD/74th Flight and registered CCCP-61611 was converted for geophysical research. It was equipped with various external sensors (including a towed MAD 'bird') and a Groza-40 weather radar.

Ilyushin IL-18 geophysical survey aircraft

IL-18D-GAL (IL-18D Antarktida)

By 1987 IL-18D CCCP-74267 (c/n 188011105), which by then had been in use for several years for supporting Soviet Antarctic research

The nose of IL-14T CCCP-61611 outfitted for geophysical survey work.

The rear fuselage of IL-14T CCCP-61611, showing the ventral sensors.

Mission equipment racks in the cabin of IL-14T CCCP-61611.

The MAD 'bird' a and ventral sensor fairings of CCCP-61611.

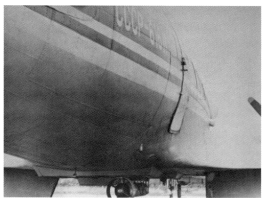

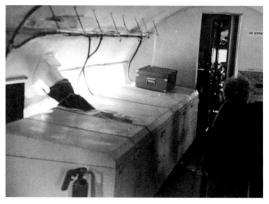

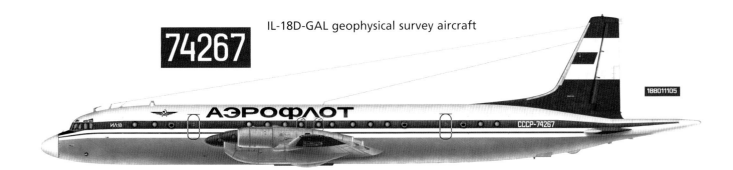

IL-18D-GAL geophysical survey aircraft

74267

188011105

АЭРОФЛОТ

ИЛ-18

CCCP-74267

stations, was modified for geophysical survey in Antarctica. To this end a tapered boom with an APM-73 MAD borrowed from the IL-38 was fitted instead of the usual tailcone; the mission equipment and operators' workstations were installed in the cabin. In some documents the modified aircraft was called IL-18D-GAL (*gheofizicheskaya aerolaboratoriya* – geophysical flying laboratory), while a newspaper publication referred to it as the IL-18D *Antarktida* (Antarctica).

The 'stinger tail' IL-18D took part in the 32nd Soviet Antarctic expedition which started off from Moscow on 26th September 1987. Taking off from Moscow-Sheremet'yevo, the aircraft captained by V. Ya. Shapkin made a short hop to Leningrad-Pulkovo to pick up the team of polar researchers, then flew to Ice Station Molodyozhnaya via Odessa, Cairo, Aden and Maputo. Upon arrival CCCP-74267 made a series of survey flights over the South Pole and the so-called Pole of Inaccessibility – a part of the ice continent which is truly hard to reach.

In 1989 the aircraft was contracted by the Norwegian company Amarok to undertake magnetic monitoring of the sea shelf north of Spitsbergen (Svalbard) Island. The work was commissioned by three petroleum companies – Conoco, Statoil (both Norway) and Elf-Aquitaine (France). The work was performed by a team from the Soviet specialist organisation **Sev***morgheo***log***iya* (North Sea Geology Trust). Operating from Murmansk-Murmashi and Longyear (the airport of Spitsbergen), CCCP-74267 flew 14 sorties between 20th April and 11th May, inspecting an area of 50,000 km² (19,305 sq miles). The flights proceeded at 600-900 m (1,970-2,950 ft) along 23 routes.

In post-Soviet days the IL-18D-GAL was stripped of mission equipment and sold to the Domodedovo Civil Aviation Production Association as RA-74267.

IL-18E CCCP-75598

In the autumn of 1991 the IL-18E 'Meteor' weather research aircraft, CCCP-75598 (see above), was stripped of its mission equipment and converted into a geophysical survey aircraft with an *izdeliye* A-723 SLAR developed by the Electronics Institute of the Ukrainian SSR's Academy of Sciences. The SLAR was installed in a large slab-sided fairing aft of the port wing and a ventral fairing in line with it.

Upon completion of this programme in 1995 GosNII GA reconverted the aircraft to standard configuration with a 72-seat tourist-class layout. By then the aircraft had gained the Russian prefix, becoming RA-75598.

Sometime before 1987 IL-18D CCCP-74267 was converted into the IL-18D-GAL geophysical survey aircraft equipped with an IL-38 style MAD boom. It is seen here at Leningrad-Pulkovo airport.

АЭРОФЛОТ

CCCP-74267

Seen on final approach to Sheremet'yevo before 1995, IL-18E RA-75598 shows its geophysical survey configuration with an A-723 SLAR.

Ilyushin IL-76MD geophysical survey aircraft

As mentioned in Chapter 4, IL-76MD 'Falsie' RA-76753, which was primarily a navigation system testbed, could be used for geophysical survey 'odd jobs'.

Mil' Mi-8 Makfar geophysical survey helicopters

In 1981 two Mi-8Ts were specially modified for geophysical survey work in the Yakutian Autonomous SSR. Known as Mi-8 Makfar, the helicopters were fitted with the Makfar-11 special measurement suite (the meaning of the name is unknown).

There is a strong possibility that the helicopters in question are actually Mi-8MTs CCCP-22936 and CCCP-22937 operated by GosNII GA and that they operated as a pair, the mission suite being too extensive to be carried by a single Mi-8. Both helicopters featured outriggers on the fuselage sides (similar to those mounting the weapons pylons on the armed versions of the *Hip*) which carried large equipment boxes, and the latter were different on the two aircraft. The c/n of CCCP-22937 has been reported as 94822 but this is doubtful, as the same c/n has been quoted for a Russian Federal Security Service/Border Guards Mi-8MTV-2 '29 Yellow' (now placed on the government aircraft register as RF-23147).

Mil' Mi-8RF radiometric survey helicopter

This aircraft converted from a Mi-8T in 1989 featured a radiometric/thermal imaging system designated RF (hence the helicopter bore the designation Mi-8RF). The system created IR-waveband images of the earth's surface.

Mil' Mi-10UPL versatile field lab

In 1966 the Mil' OKB produced a version of the Mi-10 *Harke* heavy-lift helicopter designated Mi-10UPL; it was intended to carry a mobile versatile field laboratory for ore analysis (*ooniversahl'naya polevaya laboratoriya*, hence the UPL). The helicopter – the original second prototype, CCCP-04102 (c/n 04102) – carried a heavy-duty two-axle van trailer outfitted with special equipment; on arrival at the airfield nearest to the geological prospecting site it would be disengaged and towed to the actual location by a lorry.

Mil' Mi-26L235 geological survey helicopter

In 1987 a one-off example of the Mi-26 *Halo* heavy-lift helicopter designated Mi-26L235 was built at the Mil' OKB's experimental plant in Panki. Described as 'flying laboratory', it was intended for geological prospecting work. No further details have been released.

IL-18E RA-75598 geophysical survey aircraft, 1994

Tupolev Tu-104 airborne solar observatory

In early 1961 Tu-104B CCCP-42498 (c/n 021801) belonging to what was then GosNII GVF was outfitted as a flying solar observatory for monitoring the solar eclipse that took place on 15th February 1961. Some of the cabin windows to starboard were fitted with special high-quality optically flat glass panes to avoid distorting the image captured by the cameras, spectrometers and other equipment installed in the cabin.

Taking off from Rostov-on-Don, the aircraft cruised westward at 10,000 m (32,810 ft), chasing the lunar shadow that blocked out the sun. The aircraft's high speed gave the astronomers an invaluable 80 seconds of extra time, increasing the observable eclipse time from 2 minutes 45 seconds to 4 minutes 5 seconds.

Tupolev Tu-134AK geophysical survey aircraft with IMARK suite

In 1991 NIIP (NPO Vega-M) teamed up with the All-Union Research Institute of Cosmo-aerological Methods and LII to convert Tu-134AK CCCP-65906 (c/n (33)66175, f/n 6345) into a geophysical survey aircraft fitted with the IMARK all-weather high-resolution polarimetric scanning suite (*izmeritel'nyy mnogochastotnyy aviatsionnyy rahdiolokatsionnyy kompleks* – multi-frequency airborne radar measuring suite).

The range of tasks this aircraft performed was truly vast. In geology it could be used for surface and subsurface mapping of geological formations, oil/natural gas and ore prospecting, searching for ground waters (this included tracing ground water migration which can carry pollutants and locating swamped areas). In oceanology it was detection of surface pollution and tracing the movement of icefields in Polar regions. In agriculture and forestry it was measurement of soil humidity, detection of deforestation and the like, and specifying the land evaluation cadastre.

Tu-134AK RA-65906, the IMARK survey aircraft, at the MAKS-95 airshow. The photo on the left shows the ventral SLAR canoe fairing.

Once again the aircraft was readily identifiable by its unique combination of tell-tale 'bumps and bulges'. Originally CCCP-65906 sported a natural metal fairing wrapped around the starboard side of the fuselage; it ran from the service door to the third full-size window, incorporating two yellow cruciform antennas of a synthetic aperture SLAR one above the other, and the top was level with the door. A large boxy fairing housing more SLAR antennas was located beneath the wing centre section; most of it was dielectric. Two forward-pointing probes ('roach antennae') were located on the sides of the extreme nose just aft of the navigator's glazing. The rear fuselage underside featured tandem fairings enclosing vertical cameras.

In 1993 the aircraft was modified: the lateral SAR fairing became longer and wider, terminating ahead of the seventh window; it now incorporated two pairs of 'crosses', with a large square dielectric panel consisting of small squares further aft. The enlarged fairing

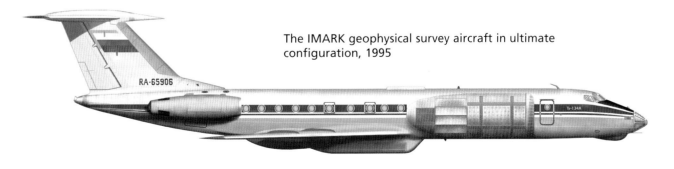

The IMARK geophysical survey aircraft in ultimate configuration, 1995

RA-65906

obstructed the Aeroflot titles, and these were painted out untidily. A blade aerial was added at the top of the fin. A further upgrade in 1995 saw the addition of a smaller square dielectric panel aft of the large one.

The SLAR worked in four wavebands – 3.9 cm (1½ in), 23 cm (9 in), 68 cm (26¾ in) and 2.54 m (100 in), with a resolution of 4-6 m (13-20 ft), 8-10 m (26-33 ft), 10-15 m (33-50 ft) and 15-25 m (50-80 ft) respectively. The radar 'looked' both sides in 3.9 cm wavelength mode and to starboard only in all other modes, scanning a strip 24 km (15 miles) wide. Scanning was performed with the aircraft flying at 500-5,000 m (1,640-16,400 ft) and 500-600 km/h (310-370 mph). The mission equipment also included IR radiometers and a microwave radiometer. The aircraft was provided with a GPS kit.

Reregistered RA-65906, the IMARK survey aircraft was used to monitor the ecology of the Volga River estuary near Astrakhan' in the summer of 1993. That same year it was on show at the MAKS-93; two years later it was

displayed again at the MAKS-95 airshow. In late 1997 RA-65906 was stripped of all mission equipment, refurbished, outfitted as an 80-seater and sold to Nefteyugansk Airlines.

Yakovlev Yak-42F research aircraft

A production Yak-42D *Clobber* short-haul airliner with the out-of-sequence registration CCCP-42644 (c/n 4520424914090, f/n 0410) was converted into a special research aircraft for earth resource/geophysical survey, as well as for environmental monitoring and atmospheric research. High-definition electro-optical sensors were carried in the passenger cabin and in large cylindrical pods under each wing just outboard of the undercarriage units, the front portions rotating to bring the terrain below into the field of view.

Designated Yak-42F (the meaning of the suffix is unknown) and wearing the red/white Polar version of Aeroflot's 1973-standard livery, the aircraft was shown at the 39th Paris Air Show in June 1991. Incidentally, this was the sole Yak-42 to wear this colour scheme.

Regrettably, the Yak-42F found itself unwanted. In the late 1990s CCCP-42644 was withdrawn from use at the Yakovlev OKB's flight test facility in Zhukovskiy. After sitting there for several years the aircraft was stripped of mission equipment and flown to the BASCO (Bykovo Air Services Co.) aircraft repair plant at the nearby Moscow-Bykovo airport in the hope of refurbishing and selling it for commercial use. Evidently the attempts to find a buyer failed and, after sitting engineless at Bykovo for years, in early 2010 the aircraft was broken up.

8 Miscellaneous applications

This chapter contains descriptions of research and test aircraft that do no fit into the previously listed categories and are too few to merit separate chapters.

Environmental research aircraft

Antonov An-2LL ecological survey aircraft

In the early 1990s the State Research Institute of Aircraft Systems (GosNII AS) in Zhukovskiy modified an An-2R crop-duster, RA-70547 (c/n 1G 14523), for ecological research and survey. Outwardly this aircraft, known as the An-2LL, was identifiable by an observation blister, which has caused it to be mistaken for an An-2PF photo survey aircraft.

Antonov An-26LL ecological survey aircraft

An ecological research and survey version of the An-26 tactical transport designated An-26LL was created in 1995 within the framework of an environment rehabilitation programme at former Soviet military bases.

The aircraft (identity unknown) was fitted with equipment for aerial photography and multi-spectrum sounding of the Earth's surface.

Antonov An-30 research aircraft (Institute of Cosmic Studies of Natural Resources)

A photo circulated by the TASS news agency in July 1981 depicted an Aeroflot An-30 registered CCCP-30067 (c/n 1208) which was described as a 'flying laboratory' (that is, research aircraft) belonging to the Institute of Cosmic Studies of Natural Resources, a division of the USSR Academy of Sciences. According to TASS, this institute was the main participant of an Industrial-Scientific Association for Cosmic Studies established under the auspices of the Academy of Sciences of Azerbaijan (then a Soviet Republic).

Presumably it was this aircraft that later took part in two so-called aerocosmic experiments involving airborne and space-based means of sampling and survey – the Black Sea experiment of 1983 and the *Gyunesh* (Azerbaijani for 'sun') experiment of 1984. These experiments also involved three other aircraft – an An-2, an Ilyushin IL-14 and a Mil'

399

Mi-8 helicopter, as well as the Salyut-7 space station. They were based on correlation of data obtained simultaneously by satellites and airborne sensors. The modified An-30 used in the Gyunesh experiment was equipped with a thermal imaging device, an infra-red spectrometric suite and a VHF radiometric suite, including an East German Zeiss Ikon MKF-6M multi-spectrum aerial camera.

Antonov An-30 research aircraft (Institute of Atmospheric Optics)

At the end of the 1990s an unidentified An-30 was converted into a research aircraft by the Institute of Atmospheric Optics of the Russian Academy of Sciences' Siberian Division. The institute was developing optical methods of analysing environment conditions. One of these methods involved the use of a laser radar, or 'lidar' by the Russian researchers. Such a laser locator was installed on the An-30. Among other things, it could be used for spotting fish shoals for the benefit of the fishing industry. The laser locator was also

capable of measuring the degree of transparency of the upper layers of the sea amenable to quickly changing their optical properties. Exactly such measurements were conducted by the Institute with the help of the An-30 in the North Sea areas off the coasts of Scotland at the invitation of the Defence Evaluation and Research Agency (DERA) of Great Britain. A published photo of this aircraft shows it has strap-on fuselage tanks – presumably this is a suitably modified An-30D.

Aviacomplex Sigma-4 – Poisk-E survey aircraft

An unregistered example of the Aviacomplex Sigma-4 two-seat ultralight aircraft (c/n 01?) was converted into a survey aircraft designated *Poisk-E* (Search-E, or Quest-E) by LII. Its missions included local environmental monitoring, damage assessment in the wake of natural or man-made disasters, searching for thermal anomalies and concealed fires, gas pipeline surveillance and searching for leaks in urban water and gas mains. To this end the aircraft was outfitted with a Chizh (Siskin) IR line scan system, a Canon D100 digital camera, an A39 film camera, a thermal imager developed by IRTIS (the Russian market leader in thermal imaging equipment), an MS-08 spectrum radiometer and a GPS receiver. The mission equipment control panel was installed instead of the left-hand seat (on the Sigma-4 the pilot sits on the right). Performance included an operational altitude range of 100-3,000 m (330-9,840 ft), a working speed of 120-200 km/h (74.5-124 mph) and an endurance of four hours.

The aircraft was in the static park at the MAKS-2005 airshow. Two years later it was displayed again at the MAKS-2007.

Ilyushin IL-14FKM aerochemical survey conversion

At the concluding stage of its service career a Moscow-built IL-14FKM photo survey aircraft, CCCP-91483 (c/n 147001232), was converted to an aerochemical survey aircraft. In this configuration it featured particle traps for air sampling built into two of the cabin windows on each side. The aircraft was operated by the Central Regions CAD/Myachkovo UAD/229th Flight based at Moscow-Myachkovo. In 1987 it took part in the ANEX-87 ('ANthropogenic EXperiment') atmospheric research programme and was retired in late 1988.

The Poisk-E survey aircraft based on the Sigma-4 light aircraft at the MAKS-2007.

The cockpit of the Poisk-E, with a mission equipment console replacing the passenger seat.

IL-14FKM CCCP-91483 converted for aero-chemical research is seen here at Prague-Ruzyne during the ANEX-87 ecological research programme.

Ilyushin IL-18V ecological survey aircraft

In the 1980s an IL-18V with the out-of-sequence registration CCCP-75423 (c/n 182005601) was extensively modified for environmental monitoring and research tasks by LII and MIREA. (Logically this aircraft should have been registered CCCP-75855; this registration was probably allocated but not taken up.) Its main external identification feature was a large cylindrical pod under the forward fuselage housing a SLAR based on the Mech system (the pod was very similar to the one on the SL-18I); this and the unusual red/white colour scheme identical to the one worn by

CCCP-75431 until 1995 caused some people to mistake it for an IL-24N. At an altitude of 6,000 m (19,685 ft) the radar could scan a strip of land 20 km (12.4 miles) wide.

A flat-bottomed ventral 'bathtub' with a large dielectric panel similar to that of IL-18V CCCP-75894 (see Chapter 4) was installed aft of the wings. Finally, four angular fairings incorporating optically flat glass windows for optical sensors or spectrometers were provided on the port side. One such window was located dorsally just aft of the forward entry door, looking up at about 45° to the direction of flight; a second window placed just a bit further aft was directed about 45° upwards,

IL-18V CCCP-75423 in the static park at MosAeroShow'92.

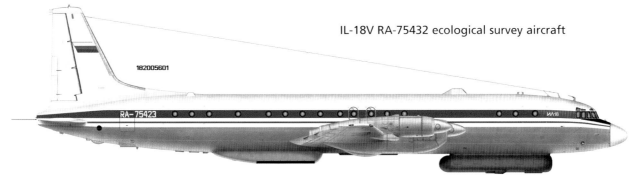

IL-18V RA-75432 ecological survey aircraft

RA-70880, the Mi-17LL research aircraft (formerly the Mi-17LIZA ecological survey aircraft) at the MAKS-95 air-show. Note the crude-looking Perspex blister, the blade aerial added under the truss-type nose boom and the white LII logo above the emergency exit.

while the other two windows located fore and aft of the rear entry door looked 45° down.

CCCP-75423 was in the static park at MosAeroShow-92, Russia's first real international airshow, at LII's airfield on 11-16th August 1992. The official exhibitors catalogue at the MAKS-93 airshow referred to this aircraft as the IL-18 ZhLIIP (*Zhookovskoye lyotno-ispytahtel'noye issledovatel'skoye predpriyatiye* – Zhukovskiy Flight Test & Research Enterprise), though this is hardly its proper designation.

In 1993 the aircraft received the Russian registration prefix; by August 1997 it was withdrawn from use. Finally, in early 1999 RA-75423 was restored to airworthy status for ferrying to Pushkin; there in the course of a complete refurbishment at the Russian Navy's ARZ No. 20 it was stripped of all non-standard appendages and converted to combi (passenger/cargo) configuration for its new owner, IRS-Aero.

Mil' Mi-8T ecological survey aircraft

A Kazan'-built Mi-8T in Soviet Air Force markings (tactical code unknown, c/n 3611) that was previously a testbed for mine countermeasures systems (see Chapter 5) was refitted for ecological monitoring tasks. The helicopter served with LII in 1972-95 but it is not known when it switched to its new role.

Mil' Mi-17LIZA ecological survey aircraft

The fleet of assorted test and development aircraft operated by LII included a Mi-17 *Hip-H* from an undelivered foreign order. Painted in orange/blue Aeroflot colours and registered CCCP-70880 (c/n 212M144), it was fitted out for environmental survey as the Mi-17LIZA, this curious suffix standing for *laboratoriya dlya izmereniya zagryazneniya atmosfery* – atmospheric pollution measurement laboratory.

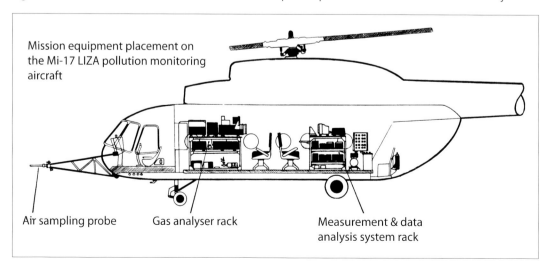

Mission equipment placement on the Mi-17 LIZA pollution monitoring aircraft

Air sampling probe   Gas analyser rack   Measurement & data analysis system rack

Outwardly 'Pretty Lisa' could be identified by the lattice-like boom on the nose glazing (the same kind as fitted to several other Mi-8s operated by the institute) carrying an air sampling probe. The cabin accommodated test equipment racks along the starboard side and an auxiliary fuel tank to port. The Mi-17LIZA was in the static park at MosAeroShow-92.

Mil' Mi-17LL research aircraft

Duly reregistered RA-70880 after 1992, the former Mi-17LIZA found use in several other test and research programmes in the 1990s; some of them had unspecified military applications. This brought about a change of designation to Mi-17LL and resulted in the addition of a crude cylindrical Perspex blister for some kind of sensor in the second window to starboard, plus a blade aerial on the underside of the nose boom. The Mi-17LL has been a regular participant of Moscow airshows, and the equipment fit has been updated periodically.

Mil' Mi-17 ecological survey aircraft

A Mi-17 – apparently not the Mi-17LIZA – was modified for determining the range and quantity of pollutants in industrial areas. The mission suite included spectrozonal mapping equipment, radiation metering equipment and a system for remote sampling of the atmosphere and the soil, using different wavebands. Additionally, the helicopter could be configured for photo mapping (vertical and strip photography) which was performed at altitudes of 50 to 6,000 m (164-19,685 ft) and speeds up to 250 km/h (155 mph).

Mil' Mi-24V ecological survey aircraft

In the late 1980s in accordance with the *konversiya* programme NPO Polyot ('Flight' Scientific and Production Association) converted a Mi-24V *Hind-E* assault helicopter (identity unknown) into an ecological survey aircraft for detecting oil spills, monitoring air pollution and the like. The USPU-24 machine-gun barbette in the nose was replaced by a flat fairing containing sensors, making the helicopter appear to be rudely sticking out its tongue. The low light level TV/FLIR pod of the Raduga-Sh targeting system under the nose and the mounting frames for 9M114 Shtoorm-V (AT-6 *Spiral*) anti-tank guided missiles at the ends of the stub wings were deleted, but the missile guidance antenna 'egg' was retained. Test instrumentation was carried in large slab-sided pods on the outer wing pylons, the inner ones being occupied by fuel tanks; the test engineer (operator) had the cabin all to himself.

The demilitarised *Hind-E* was one of the exhibits of the annual industry fair in Nizhniy Novgorod in September 1991.

Su-24 *Fencer-A* '15 White' (c/n 1515301) configured as an ecological survey aircraft with a centreline sensor pod in the static park of the MAKS-99 airshow.

The Tu-134A-1510 (RA-65740) at the MAKS-95. The port ventral pod is a development engine for a cruise missile, but the dorsal sensor and the ones in the cabin windows are for aerothermophysical and ecological research.

Sukhoi Su-24LL ecological survey aircraft

At the MAKS-99 airshow in Zhukovskiy an early Su-24 *Fencer-A* tactical bomber coded '15 White' (c/n 1515301) was exhibited as being available for contract work in monitoring ecological conditions. To this end a large sensor pod converted from the Shpil'-2M laser line-scan pod (carried by the Su-24MR, which see) could be carried on the fuselage centre-line.

Tupolev Tu-134A-1510 multi-role testbed/research aircraft

As mentioned in Chapter 3, apart from engine tests, LII's Tu-134A-1510 multi-role testbed/research aircraft (CCCP-65740, c/n 2351510) served for aerothermophysical and environmental research, such as measuring the heat signature and air pollution levels in the wake of other aircraft. This is where the heat sensors in the dorsal fairing came in. Additionally, two IR spectrometers and a gamma spectrometer were installed at the second, third and fifth

cabin windows to starboard whose glazing was replaced by special filters. The glazing of the galley window to starboard gave place to a metal plug mounting a temperature probe, an aerosol particle counter and an ozone sensor. The mission equipment included gas analysers for measuring hydrocarbon (CH), ozone, nitrogen oxide (NOx), carbon oxide and sulphuric dioxide (SO_2) concentrations. Measurement results were processed in real time by a computerised system.

Flying the Tu-134A-1510 had a few quirks. Firstly, wake turbulence during measurement missions could be quite severe, the aircraft vibrating violently and tending to roll. For instance, it was impossible to get closer than 700 m (2,300 ft) to a Su-24 tactical bomber – the heavy Tu-134 was kicked out of the wake with up to 60° bank, and normal flight could only be restored after the airliner had lost some 500 m (1,640 ft) of altitude. Secondly, the mission equipment 'acted up', and some modules which functioned beautifully on the ground simply refused to work in flight!

The development engine and test equipment pod of the Tu-134A-1510 in early configuration (CCCP-65740)

Agricultural survey aircraft

Tupolev Tu-134SKh agricultural survey aircraft

A highly specialised derivative of the Tu-134A-3 was developed for the Soviet Ministry of Agriculture in keeping with Council of Ministers directive No. 127-39 dated 28th January 1981. Provisionally designated Tu-134I (*issledovatel'skiy* – research, used attributively) at the ADP stage, the aircraft emerged as the Tu-134SKh (*sel'skokhoziaystvennyy* – agricultural). Its mission was to survey agricultural land, measuring soil humidity, detecting pest attacks and the like, thus allowing crop yields to be estimated and corrective measures taken if necessary.

The Tu-134SKh was easy to identify because the airframe incorporated major structural changes. Firstly, the overwing emergency exits and most of the cabin windows were omitted, resulting in a very characteristic window arrangement – 1+door+3+1+1+1+1 to port and service door+3+1+1 to starboard. Actually, most of the cabin was occupied by mission equipment, and the two doors were enough for evacuating the flight crew and ten equipment operators.

Secondly, four camera/sensor ports were provided on the fuselage underside; the first three were covered by an optically flat glass panel forming part of the pressure cabin, while the rearmost port was an unpressurised bay. A sliding cover forming a flat-bottomed bulge ahead of the wings closed the camera/sensor ports for take-off and landing; this required the lower anti-collision beacon to be relocated under the rear fuselage.

Thirdly, the aircraft had a 2-cm waveband Nit' S-1SKh (Thread) SLAR; the antennas were housed in cigar-shaped pods 8 m (26 ft 3 in) long carried under the wing roots on pylons. The first Tu-134SKh surveyors were built with no SLAR but retrofitted in due course.

Other identification features included a dorsal Type AShS L-shaped aerial (identical to those fitted as standard above and below the forward fuselage) further aft and an exceptionally wide unpainted fin leading edge, as on late-production Tu-134AKs. Finally, the Tu-134SKh wore characteristic markings – an ear of wheat was painted on the fuselage aft of the Aeroflot titles and a blue circle outlined in yellow with the white letters 'CX' (SKh in Cyrillic characters) was located on the nose below the 'Tu-134A-3' nose titles.

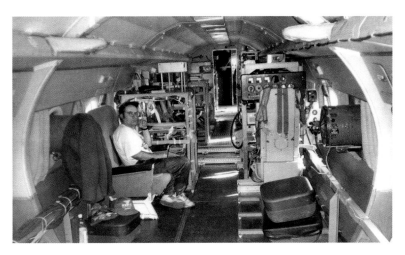

The mission equipment included three cameras – East German Zeiss Ikon MKF-6M and MSK-4 multi-spectrum cameras on fixed mounts (alternatively, indigenous AFA-41/10 and AFA-TE/35 cameras could be installed) and a TAFA-10 topographic camera on an AFUS-U tilting mount. The Soviet cameras' focal length was 100 mm for the AFA-41/10 and TAFA-10 or 350 mm for the AFA-TE/35. A processing lab was located at the rear of the cabin, allowing films to be developed and photos printed on board.

The cabin of the Tu-134A-1510, looking towards the nose.

CCCP-65917, the Tu-134SKh prototype, and its cabin; note the camera port.

to two screens, one showing the actual radar map and the other displaying the data fed into the storage system. Unlike aerial photography, which in many regions is possible only 25 to 35 days a year, radar imaging is possible day and night, in any weather and season. Radar imagery shows dry, saline and swampy areas (such as rice paddies); the SLAR can even see through a layer of snow, making it possible to check up on winter crops.

A special Mak (Poppy) navigation suite allowed the Tu-134SKh to move in a shuttle pattern during photography/remote sensing, follow a preset route and automatically return along the same route.

Empty weight was 34.4 tons (75,840 lb) with SLAR or 32.4 tons (71,430 lb) without SLAR, and the MTOW was 47.6 tons (104,940 lb). The fuel supply was 18,000 litres (3,960 Imp gal), providing a range of 3,600 km (2,240 miles) and a maximum endurance of 4.5 hours.

The prototype, CCCP-65917 (c/n (33)63991, f/n 6329), entered flight test in April 1983 and was retained by the OKB. Nine production aircraft followed:

CCCP-65721 (c/n (43)66130, f/n 6339)
CCCP-65722 (c/n (63)66420, f/n 6368)
CCCP-65723 (c/n (63)66440, f/n 6369)
CCCP-65724 (c/n (63)66445, f/n 6370)
CCCP-65725 (c/n (63)66472, f/n 6371)

Tu-134SKh CCCP-65917 on display at MosAeroShow'92.

The SLAR pods and the protective doors closing the Tu-134SKh's camera ports.

A six-channel IR scanner developed by the French company Matra fed images to a magnetic data storage system. In the USSR this scanner was known as **Yash**ma (Jasper).

The SLAR offered a field of view 15 or 37.5 km (9.31 or 23.29 miles) wide, generating a radar map to 1/100,000th or 1/200,000th scale in the former case and 1/250,000th or 1/500,000th scale in the latter case. A 'blind spot' 11 km (6.83 miles) wide remained directly below the aircraft. Radar mapping was performed at 3,000-6,500 m (9,840-21,325 ft) and 350-750 km/h (217-465 mph); an area of up to 10,000 km² (3,860 sq miles) could be surveyed in a single sortie. Radar data was fed

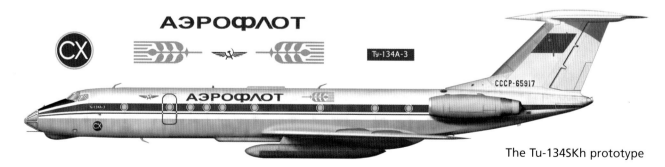

The Tu-134SKh prototype

406

CCCP-65918 (c/n (43)63995, f/n 6332)
CCCP-65928 (c/n (63)66491, f/n 6372)
CCCP-65929 (c/n (63)66495, f/n 6373)
CCCP-65930 (c/n (63)66500, f/n 6374)

All Tu-134SKh surveyors were purpose-built aircraft, not converted Tu-134As. Production was slow because the Khar'kov aircraft factory was switching to the Antonov An-72 *Coaler-A* STOL transport; the final example left the factory in the late summer of 1989.

The Tu-134SKh entered service on 28th September 1984 pursuant to MAP order No. 335, Ministry of Civil Aviation order No. 210 and Ministry of Agriculture order No. 230. All nine production aircraft were delivered to the Central Regions CAD/Voronezh UAD/243rd Flight but three were later transferred to the Ivanovo UAD/176th Flight. (Both Voronezh and Ivanovo are located in central Russia's bread belt.) Still, trials at GosNII GA continued until March 1986, with V. Vvedenskiy as project test pilot.

The first major 'operation' in which the type made its mark was the Kursk-85 experiment aimed at studying the condition of crops and developing crop yield prediction methods. Apart from the Tu-134SKh, it involved An-30s, modified An-2 biplanes, Meteor and Kosmos satellites and the Salyut-7 space station from which the Soviet cosmonauts Vladimir Djanibekov and Aleksey Savinykh took pictures of the Earth.

The jet proved much more cost-effective for aerial photography than the An-30 or the Mi-8 helicopter, the cost per square kilometre being 2.7 and 4.6 times lower respectively. In mid-1986 an all-Union research centre called Agricultural Resources Automated Data & Control System (AIUS 'Agroresoorsy') was established within the ministry, making use of several more-or-less specialised aircraft, including the Tu-134SKh.

The aircraft was progressively upgraded. In May 1987 the Nit' S-1SKh SLAR was finally cleared for full-scale operation after rigorous testing. In 1990 the Tu-134SKh received an A-723 LORAN system (identified by a new starboard strake aerial atop the fuselage which was twice taller than usual), a VEM-72F electromechanical altimeter, Kama-S/Kama-F six-channel gamma spectrometers and a *Voolkahn* (Volcano) thermal imager.

On 11-16th August 1992 the Tu-134SKh made its public debut when CCCP-65917 was displayed at MosAeroShow-92. The prototype was displayed again at the MAKS-93 airshow (31st August – 5th September 1993).

Unfortunately, Tu-134SKh operations were beset by organisational problems and all kinds of bureaucratic snags which kept the aircraft on the ground most of the time. As a result, in the early 1990s the new airlines operating them (Voronezhavia and IGAP) decided to sell them for conversion. By August 2002 nine out of ten had been converted into Tu-134A-3M biz-jets, leaving RA-65929 as the 'last man standing' (the only reason why it was not converted is that the aircraft was damaged beyond repair on 24th June 2003).

Parachute system testbeds

Antonov An-12 parachute system testbeds

An-12A c/n 9900902 was used for performing test drops of the spherical re-entry capsule of the first Soviet manned spacecraft, *Vostok* (East), in 1960 and the re-entry capsule of the *Voskhod* (Sunrise) and *Soyooz* (Union) manned spacecraft in 1967-72. The capsule (painted black and white for phototheodolite measurements) was suspended on a special bearer in the open cargo hatch.

A quasi-civil An-12BP (identity unknown, c/n 6344701) with grid-type photo calibration markings on the rear fuselage was used by the MKPK *Ooniversahl* company for testing new parachute systems. Later the aircraft gained Air Force insignia and was coded '42 Red'.

At an early stage of its test career, An-12A c/n 9900902 was used for testing the parachute recovery system of Soviet spacecraft re-entry capsules: the Vostok (below) and Soyuz (bottom).

This B-25J was used by LII for testing the crew escape system of the DFS 346 research aircraft whose cockpit section is suspended in the bomb bay.

North American B-25J parachute system testbeds

A B-25J-5-NC Mitchell bomber (43-28112, the c/n is quoted variously as 108-35125 or 108-32625) was used for post-war tests of new personnel parachutes.

In 1948 a B-25J was used by LII for testing the detachable cockpit capsule of the '346' experimental rocket-powered aircraft (the German DFS 346 that underwent further development and testing in the USSR). The cockpit capsule was dropped in flight, whereupon the prone pilot's berth was extracted from the cockpit and his parachute opened.

Mikoyan MiG-23BN stealth technology testbed

In 1983-88 the Mikoyan OKB and LII conducted experiments with a MiG-23BN *Flogger-H* fighter-bomber, 'No. 9680' (apparently this means c/n 0393219680) with a view to reducing its radar signature. By applying a coat of radar-absorbing material (RAM) – a special graphite compound – to certain parts of the airframe it proved possible to reduce the radar cross-section (RCS) tenfold.

Mikoyan MiG-29 stealth technology testbed

Concurrently the Mikoyan OKB experimented with RAM coatings on a standard production MiG-29 coded '22' (the tactical code was later removed). Again, trials showed that the coating significantly reduced the fighter's RCS, reducing hostile radar detection range, the accuracy of missiles' target lock-on and proximity fuse detonation. However, the experiment was not taken further.

Here the DFS 346 cockpit is dropped by the B-25, whereupon the pilot's berth is extracted from the cockpit and his parachute opens.

This MiG-29 (*izdeliye* 9.13) seen at Zhukovskiy during a Mikoyan presentation in February 2000 is a development aircraft for testing a radar-absorbing material coating.

Sukhoi Su-25 stealth technology testbeds

The Sukhoi OKB likewise undertook such experiments with the Su-25 *Frogfoot* attack aircraft. One of the development aircraft known at the OKB as the T8-11 ('66 Blue', c/n 25508101056) received a coat of RAM, these trials coming under the overall project name of 'Astra'. Eventually the aircraft was donated to the Central Russian Air Force Museum in Monino, where it is still on display in non-authentic colours as '66 Red'.

Later, the T8-12 ('12 Red') was used for similar 'stealth' trials, where in addition to RAM the aircraft was given a special camouflage scheme to reduce its detectability in the visible spectrum. Due to someone's negligence this very special aircraft was displayed for the general public at Moscow-Khodynka in 1989 until a visiting official blew the whistle and ordered its removal.

Air accident investigation laboratories

Antonov An-12BP (LIAT) air accident investigation laboratory

In 1972, at least seven Voronezh-built Soviet Air Force *Cubs* with no tail gunner's station – An-12A CCCP-11131 (c/n 2400702), An-12Bs CCCP-11652 (c/n 402702), CCCP-11653 (c/n 402703), CCCP-11654 (c/n 402602), CCCP-11791 (c/n 402611), CCCP-11792 (c/n 402701) and CCCP-11992 (c/n 402604) – were converted into a mobile air accident investigation laboratory designated An-12BP (LIAT) which could travel to the airbase where a crash had taken place. The LIAT suffix stood for *labora**tori**ya is**sled**ovaniya*

*aviatsi**onn**oy **tekh**niki* – aviation hardware examination laboratory. The freight hold housed special equipment for analysing the flight data recorder and cockpit voice recorder readouts of the crashed aircraft; it also featured a crew rest area, a galley and toilet facilities. The only external recognition feature was an additional TA-6 APU buried in the rear fuselage immediately ahead of the tail unit, with the air intake and exhaust on the port side. The APU provided power for the mission equipment. These aircraft later saw service with the Russian Air Force.

Antonov An-24LL flying metrological laboratory

A single An-24 airliner (identity unknown) was outfitted with a measurement suite for checking the conformity of other aircraft to current airworthiness standards. This aircraft has been referred to as the An-24LL metrological laboratory. The An-24LL was used in the early stages of a large-scale flight test programme managed by M. Kotik, a notable researcher at LII.

Mil' Mi-8TL air accident investigation laboratory (second use of designation)

At least one Soviet Air Force Mi-8T coded '45 Yellow' (c/n 4155) was converted into a mobile air accident investigation laboratory designated Mi-8TL which could travel to the airbase where a crash had taken place. As distinct from the civil Mi-8TL water bomber used for firefighting (*lesopo**zhar**nyy*), the L suffix stood for *labora**tori**ya*. The cargo cabin housed special equipment for deciphering the 'black boxes' (FDR and CVR) of the crashed aircraft.

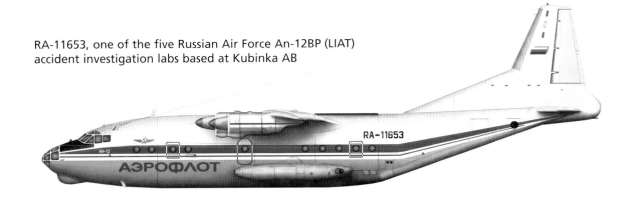

RA-11653, one of the five Russian Air Force An-12BP (LIAT) accident investigation labs based at Kubinka AB

The MiG-25PU-
SOTN chase air-
craft ('22 Blue')
on approach to
Zhukovskiy.

The MiG-25PU-
SOTN chase air-
craft ('22 Blue')
on approach to
Zhukovskiy.

The logo of NPO
Molniya which
developed vari-
ous space vehi-
cles, including
the Buran space
shuttle. A large
number of air-
craft served as
testbeds in sup-
port of the Buran
programme.

The MiG-25PU-
SOTN, overtakes
the Buran space
shuttle which has
just touched
down at
Baikonur-
Yoobileynyy air-
field after its sole
orbital flight on
15th November
1988.

Testbeds in the Buran space shuttle programme

Mikoyan MiG-25RBK testbed

A MiG-25RBK *Foxbat-B* coded '02 Blue' (c/n
N02029210) was modified for the Buran test
programme. The reconnaissance and bombing
system was replaced by additional communications gear, data link and other specialised
equipment. Samples of the Buran's ceramic
heat insulation tiles were attached to the sides
of the air intake trunks (and misidentified
as dielectric panels). The aircraft could also
carry test instrumentation pods on the wing
pylons.

Mikoyan MiG-25PU-SOTN TV tracker aircraft

In order to calculate trajectory guidance algorithms for the Buran space shuttle, LII converted a MiG-25PU *Foxbat-C* trainer ('22 Blue', c/n
N22040578) into a research aircraft designated MiG-25PU-SOTN (*samolyot optiko-tele-
vizionnovo nablyudeniya* – optical/TV surveillance aircraft). The aircraft's mission included:

• advanced research of trajectory guidance
algorithms for the space shuttle at altitudes
below 20,000 m (65,620 ft) as part of the
shuttle's total in-flight simulation complex
(which also included three Tu-154LL flight control system testbeds/approach trainers, see
Chapter 2);
• verifying the monitoring techniques for
the shuttle's automatic flight control system;
• training Buran pilots and navigators/sys-
tem operators;
• acting as a chase plane for the Buran during flight tests.

The aircraft was fitted with a KRL-78 radio
command link system (*komahndnaya rahdi-
oliniya*) integrated with the standard SAU-155
automatic control system, a B-218 data link
system, test instrumentation (data recorders)
and a TV tracking system for videotaping the
aircraft being shadowed. This equipment suite
was jointly developed for the MiG-25PU-SOTN
by NPO Molniya (the creator of the Buran) and
the Institute of TV systems.

The TV tracking system included a Sony
DXM-3P video camera, a 3800PS video tape
recorder, a DX-50 video monitor, a KL-108
transmitter and an MB-10 transmit antenna.
The ground control room was equipped with a
KL-123 receiver, short-range and long-range
receiving antennas and control and data recording gear. The front cockpit housed the video
camera and associated equipment. External
identification features were the extra aerials
under the nose and on the fuselage spine.

Lining up on the 'target aircraft', the pilot
of the MiG-25PU-SOTN extended the landing
gear and flaps and throttled back the engines
at about 18,000 m (59,055 ft) in order to follow the same steep glide path, get the 'target'
in his viewfinder and start shooting. Until the
real thing flew, the aircraft which doubled for
the space shuttle were the abovementioned
MiG-25RB '02 Blue', a modified MiG-31 ('97

The MiG-25PU-SOTN chase aircraft

Red') and the BTS-002, or GLI-Buran (CCCP-3501002) – a full-scale Buran fitted with four Lyul'ka AL-31 turbofans for taking off under its own power.

Stage A of the programme comprised 15 flights for checking the TV tracking system's function, range and sensitivity to interference. It turned out that signals from the aircraft reached the ground control room undistorted in about 85% of the cases and the worst interference was caused by the radio altimeters of other aircraft flying nearby.

Stage B was held in September-October 1986 and served to optimise the data link transmission parameters and to determine the effect of the 'target aircraft' on picture quality. The optimum distance to the target was judged at about 10 m (33 ft). Therefore, an additional TV monitor was installed in the rear cockpit; the front cockpit canopy was modified and a Betacam video camera used.

Finally, Stage C was held near Yevpatoria on the Crimea Peninsula on 11-16th October 1986 and included ten flights with the updated and complete equipment suite. Yet the heyday of the aircraft was yet to come: it was when the actual Buran lifted off on its one and only unmanned space mission. Test pilot Magomed Tolboyev flying the MiG-25PU-SOTN intercepted the shuttle on its subsequent re-entry and flew chase during its glide and automatic approach all the way to touchdown, videotaping the entire sequence.

Mikoyan MiG-31 approach technique testbed

The abovementioned MiG-31 coded '97 Red' (c/n unknown) was used by LII for verifying the landing approach technique developed for the Buran space shuttle. The technique involved a very steep approach followed by a flareout at extremely low level.

Sukhoi Su-7 space shuttle trainer

A Su-7 (version and tactical code unknown, c/n 0604) was used by LII in 1983-88 for training crews in piloting the Buran space shuttle at the re-entry and glide stage.

Sukhoi Su-17UM3 testbed/TV tracker aircraft

In 1983-84 a Su-17UM3 trainer coded '19 Blue' (c/n 17532361…19) was used by LII as a testbed during preliminary evaluation of the Buran's stability and control characteristics. Later the aircraft was used for optical/TV tracking and filming during the trials of the Buran's turbofan-powered analogue – the BTS-002.

Su-17UM3 '19 Blue' flies chase for the BTS-002 experimental aircraft during the Buran's development programme.

Miscellaneous testbeds

Antonov An-12 testbed

The test and research missions performed by the An-12 in the Soviet Union included investigating the integral and spectral properties of the heat signature of various aircraft types.

This MiG-29 coded '10 Blue' was used for investigating equipment reliability issues.

Antonov An-32 testbed

A photo published in the Western aeronautical press showed an example of the An-32 *Cline* tactical transport fitted with streamlined wingtip pods. This was allegedly a testbed of some sort. No further information is available.

Mikoyan MiG-25PD testbed

Upon completion of the trials MiG-25PD '305 Blue' (c/n N84042474) was converted under an unidentified test and development programme. The radar and IRST were removed (in so doing the dielectric radome was replaced by an identically shaped metal nosecone), and small antenna fairings appeared on the fuselage spine and ahead of the port mainwheel well.

The first Mi-8LL EMP measurement helicopter, a modified Mi-8TV, in LII's hangar, showing the forward and lateral antenna arrays.

Mikoyan MiG-29 equipment reliability research aircraft

After completing a special test programme with 'aircraft 919' in order to ascertain the causes of the all-too-frequent equipment malfunctions, LII began further research in order to deal with the MiG-29's teething troubles.

This time the 'dogship' was a standard *Fulcrum-A* ('10 Blue', c/n 2960516767) borrowed from the Air Force's 4th TsBP i PLS in Lipetsk. This aircraft later became Mikoyan's demonstrator workhorse (apparently because the OKB had no aircraft of its own free for this purpose), eventually crashing spectacularly at the 1989 Paris airshow; test pilot Anatoliy N. Kvochur ejected safely.

Mil' Mi-8LL (Mi-8TV) EMP measurement aircraft

A Mi-8TV *Hip-E* armed assault helicopter delivered to LII was converted into a research vehicle designated Mi-8LL for measuring electromagnetic fields and electromagnetic pulses (EMP). Its range of tasks included determining the patterns of emitter antennas, measuring the intensity of electromagnetic fields and determining dangerous radiation levels, and pinpointing the locations of active jammers – that is, an electronic counter-countermeasures (ECCM) function.

The helicopter had narrow triangular housings pointing upwards from the tips of the external stores outriggers (the BD3-57KrV pylons were removed) and a sensor array carried on a truss-type frame in front of the flight deck glazing. Additionally, the standard sliding side windows of the flight deck were replaced with makeshift angular faceted observation blisters.

Mil' Mi-8LL (Mi-8MT) EMP measurement aircraft

Later, an uncoded Soviet Air Force Mi-8MT *Hip-H* operated by LII was converted into a research vehicle likewise designated Mi-8LL for measuring electromagnetic fields and EMP emitted by ships. The helicopter featured a wide, shallow flat-bottomed antenna housing under the flight deck and carried a truss-type air data boom on the nose.

The other Mi-8LL EMP measurement aircraft – a converted Mi-8MT. The nose-mounted sensor array and air data boom are clearly visible.

Mil' Mi-8P (?) research/survey aircraft

The Aviation-90 trade fair held at the VDNKh fairground in Moscow in November 1990 featured an intriguing model depicting a Mi-8P passenger helicopter with a towed antenna system. The model was presented by the Moscow-based NPO Vzlyot avionics house and GONTI (presumably *Glavnyy otdel naoochno-tekhnicheskoy informahtsii* – Main Scientific & Technical Information Department, possibly part of TsAGI). In reality the model probably did not depict the actual aircraft (in other words, the helicopter used to carry the actual system – if it existed in hardware form – was not necessarily a Mi-8P).

The system looked like a large cigar-shaped container on wheels (for ground handling) with a stabilising parachute which ensured the required position of the container relative to the helicopter in flight. The parachute had a rather strange shape with fins, looking like the rear end of an airship. According to the accompanying data plate the heliborne flying laboratory was intended for carrying flexible and inflatable wave guides, and for carrying a flexible dipole antenna system around transmitter or receiver aerials on the ground with a view to determining their characteristics.

Sukhoi Su-7U testbed for the 'Kometa' towed target

A production Su-7U *Moujik* trainer coded '24' (c/n unknown) was used in the tests of the *Kometa* (Comet) towed target and associated TL-70 podded ram-air turbine driven winch (*toorbolebyodka*) which found use on the Su-25BM *Frogfoot-A* target tug.

Sukhoi Su-22UM3 aeromedical research aircraft

Between 1973 and 1975 a Su-22UM3 trainer (c/n 17532367220?) was used by LII for testing *Vydokh* (Exhalation) equipment to record the physiological parameters of the pilot.

Tupolev SB-2M-103 'trawler'

A production SB (ANT-40) fast bomber powered by Klimov M-103 engines – that is, the version designated SB-2M-103 – was equipped as a 'flying trawler' for pulling down enemy telephone lines. To this end it was fitted with a grapnel on a cable. However, this was obviously a bad idea, since the low-flying aircraft with no armour protection would be extremely vulnerable even to small arms fire.

Yakovlev Yak-28P testbed

A late Yak-28P coded 57 Red was modified by LII for some kind of test work, with two small projections (possibly cameras or aerials) on top of each wingtip fairing.

Su-7U '24' with a TL-70 winch and Kometa towed target on the centreline pylon.

414